# Astrophysics on the Threshold of the 21st Century

# Astrophysics on the Threshold of the 21st Century

edited by

**N.S. Kardashev**

*Astro Space Center, P.N. Lebedev Physical Institute*
*Russian Academy of Sciences, Moscow*

Translated from the Russian by
Dean F. Smith

GORDON AND BREACH SCIENCE PUBLISHERS
Philadelphia • Reading • Paris • Montreux • Tokyo • Melbourne

Copyright © 1992 by OPA (Amsterdam) B.V. All rights reserved. Published under license by Gordon and Breach Science Publishers S.A.

**Gordon and Breach Science Publishers**

5301 Tacony Street, Drawer 330
Philadelphia, Pennsylvania 19137
United States of America

Post Office Box 161
1820 Montreux 2
Switzerland

Post Office Box 90
Reading, Berkshire RG1 8JL
United Kingdom

3-14-9, Okubo
Shinjuku-ku, Tokyo 169
Japan

58, rue Lhomond
75005 Paris
France

Private Bag 8
Camberwell, Victoria 3124
Australia

"The Shklovsky Phenomenon" by N.S. Kardashev and L.S. Marochnik was originally published in Russian in Volume 6, Number 6 of the journal *Priroda* and in English in Volume 7, Number 4 of the journal *Astronomy Quarterly*.

**Cover**: Two images of the Crab Nebula are represented. The left is the radio image, constructed by T. Cornwell and J. Uson on Very Large Array of National Radio Astronomy Observatory at the wavelength 3.6 cm. The right is the X-ray image constructed by J. Trumper and colleagues from Max-Planck-Institut für Extraterrestishe Physik on the Rosat satellite in band 0.1...2.4 keV. The principal investigator of the Rosat satellite is Doctor Aschenbach.

I.S. Shklovsky was the first to explain all spectra of the Crab Nebula, from radio to X-ray, by the synchrotron emission. S.B. Pikel'ner constructed the first dynamical model of the Crab Nebula with acceleration by the internal magnetic field.

**Library of Congress Cataloging-in-Publication Data**

Astrophysics on the threshold of the 21st century / edited by N.S.
  Kardashev ; translated by Dean F. Smith.
    p.    cm.
  Includes bibliographical references and index.
  ISBN 2-88124-817-9
    1. Astrophysics.  2. Astronomy.  I. Kardashev, N. S.
QB461.A7755  1992
523.01--dc20                                                            91-38328

No part of this book may be reproduced or utilized in any form or by any means, electronic or mechanical, including photocopying and recording, or by any information storage or retrieval system, without permission in writing from the publisher. Printed in the United Kingdom by Bell and Bain Ltd.

This book is dedicated to
Joseph Samuilovich Shklovsky and Solomon Borisovich Pikel'ner

J.S. Shklovsky
(1916–1985)

S.B. Pikel'ner
(1921–1975)

# Contents

| | | |
|---|---|---|
| Preface | | ix |
| 1 | Two Great Astrophysicists: Some Personal Reflections<br>H.C. Van de Hulst | 1 |
| 2 | The Shklovsky Phenomenon<br>N.S. Kardashev and L.S. Marochnik | 7 |
| 3 | Words about Pikel'ner<br>L.S. Marochnik | 25 |
| 4 | Reflections on the Soviet-American VLBI Program<br>K.I. Kellermann | 37 |
| 5 | The Secrets of the Solar Atmosphere<br>H. Zirin | 53 |
| 6 | Escape Processes in Planetary Atmospheres<br>V.I. Moroz | 63 |
| 7 | Decametric Radioastronomy<br>S.Ya. Braude | 81 |
| 8 | New Developments in Cosmic Gas Dynamics: Galactic Shocks, Hot Protogalaxies and Galactic Superwinds<br>L.S. Marochnik and A.A. Suchkov | 103 |
| 9 | Observational Evidence for Magnetic Fields in the Galaxy and Galaxies<br>R. Wielebinski | 117 |
| 10 | Recombination Radio Lines<br>R.L. Sorochenko | 131 |
| 11 | Cosmic Masers: Yesterday, Today and Tomorrow<br>V.S. Strelnitskii | 151 |
| 12 | Gravitational Collapse of Massive Stars, Supernovae and SN 1987A in the Large Magellanic Cloud<br>V.S. Imshennik | 167 |
| 13 | Radio Emission from Supernovae<br>V.I. Slysh | 189 |

## CONTENTS

| | | |
|---|---|---|
| 14 | Stellar Winds and Supernovae in the Interstellar Medium<br>*T.A. Lozinskaya* | 223 |
| 15 | Joseph Shklovsky and X-ray Astronomy<br>*H. Friedman* | 245 |
| 16 | Quasars and Active Galactic Nuclei<br>*B.V. Komberg* | 253 |
| 17 | A Lucky Chance for Cosmology<br>*V.N. Lukash and I.D. Novikov* | 277 |
| 18 | The Last Love<br>*I.L. Rosenthal* | 295 |
| 19 | Searching for Planetary Systems<br>*B.F. Burke* | 303 |
| 20 | Space Radiointerferometry and Gravitational Waves<br>*V.B. Braginsky, N.S. Kardashev, I.D. Novikov and A.G. Palnarev* | 315 |
| 21 | Radio Astronomy of the Next Century<br>*Y.N. Pariiskii* | 331 |
| 22 | On Astronomy for the Twenty-first Century<br>*N.S. Kardashev* | 357 |
| Index | | 377 |

# Preface

This book is dedicated to Joseph Samuilovich Shklovsky (1916–1985) and Solomon Borisovich Pikel'ner (1921–1975), who were responsible for some remarkable developments in 20th-century astrophysics. The book, prepared by their colleagues, friends and students, covers a wide range of fundamental problems in modern astrophysics (solar and solar system physics, physics of different scale astronomical objects, evolution of the universe and the search for extraterrestrial intelligence), and prospects for the progress of astronomy in the 21st century are delineated. The book also contains some reminiscences about the history of science, particularly in the Soviet Union.

    J.S. Shklovsky was born in the small town of Glukhov in the Ukraine. In 1938 he graduated from Moscow State University; he then went on to the Sternberg Astronomical Institute, and in 1967, to the Space Research Institute in Moscow. His scientific interests included the creation of a general theory for solar corona, the theory of its emission in all ranges of the electromagnetic spectrum, the theory of emission of cosmic radio waves from neutral and ionized interstellar matter in continuum and in atomic and molecular lines, and the theory of synchrotron emission in radio, optical, X-ray and gamma-ray bands by relativistic electrons in the envelopes generated by supernova explosions and in active extragalactic objects – radiogalaxies and quasars. He is also a founder of the large scientific school of astronomers united in searching all electromagnetic ranges. His books include *Solar Corona* (1951), *Physics of Solar Corona* (in English, 1962), *Radioastronomy* (2nd ed., 1955), *Cosmic Radio Waves* (in English, 1960), *Supernovae and Their Remnants* (1976), *Stars, Their Birth and Death* (3rd ed., 1984), *Problems of Modern Astrophysics* (2nd ed., 1988), *Universe, Life and Intelligence* (1st ed., 1962; 6th ed., 1987) and *Intelligent Life in the Universe* (co-written with C. Sagan, 1966). In 1991, in what would have been his 75th year, a collection of Shklovsky's short stories (*Eshelon*, in Russian, and *Five Billion Bottles of Vodka up to the Moon*, in English) was published.

    S.B. Pikel'ner was born in Baku, Azerbaidzhan, and he graduated from Moscow State University in 1942. His collaboration with J.S. Shklovsky began in 1941, as a staff member of the Crimean Astrophysical Observatory in Simeiz until 1959, and then as a professor at Sternberg Astronomical Institute. S.B. Pikel'ner was a pioneer in the development of magnetohydrodynamical models of chromosphere and active solar phenomena, multicomponent interstellar matter, generation of stars, the theory of structure of gas nebulae and envelopes of supernova remnants on the basis of the physics of radiative shock wave crossing, and the theory of galactic halo formed by relativistic particles, hot gas and magnetic fields. His books include *Physics of Interstellar Matter* (1959), *Interstellar Matter* (co-written with S.A. Kaplan, 1963),

*The Basics of Cosmic Electrodynamics* (1966), *Physics of Solar Atmosphere Plasma* (co-written with S.A. Kaplan and V.N. Tsytovich, 1977), and *Physics of Interstellar Matter* (co-written with S.A. Kaplan, 1979).

J.S. Shklovsky and S.B. Pikel'ner were very sincere, kind and sociable people who influenced the lives of their colleagues and students. A look at this side of their lives is also presented in this book. But most of the book describes the achievements made in astronomy in the 20th century, the great, unresolved, fundamental problems that still exist, and the prospects for the development of astronomy in the 21st century.

Nikolai S. Kardashev

(From left to right) Van de Hulst, Shklovsky and Pikel'ner at the International Astronomical Congress at Moscow State University in 1957.

# 1

# Two Great Astrophysicists: Some Personal Reflections

## H. C. van de Hulst
*Leiden Observatory, Leiden*

When I think of Pikel'ner and Shklovsky, my thoughts do not rush forward to the next century but drift back to 30 or 40 years ago. It was a happy time of my life. I did not imagine the great advances still to come: the disclosure of the entirely new world of X-ray and gamma-ray astronomy, the discovery of pulsars and quasars, the fantastic advances in infrared astronomy, and the new opportunities granted to theorists by the simple existence of high-speed computers. And, although it was obvious that the discoveries by radio astronomy were still on a steep upgrade, I did not foresee that this new field would become so prominent that in a number of topics it would take over the leading role from optical astronomy.

About 1951 I completed a handbook article on the solar chromosphere and corona, and of course I read "everything" in the literature from east and west. I was fascinated by the detailed structures in the corona and by the attempts to register their minute changes from various sites during one period of total eclipse. However, the idea that some 40 years later a space probe (Ulysses, earlier named Solar-Polar Monitor, still earlier Out-Of-The-Ecliptic-Mission) would be ready to be launched for a swing by Jupiter, to look at the Solar corona from entirely different aspect angles, was too bold even to occur to me as a future perspective.

I think those post post-doc years are so happy because you can feel at ease: the student's view that all important things have been figured out by older and wiser persons is gone, and the professor's view that most of the important advances are made by bright youngsters has not yet come. It was in that level stage that I met Pikel'ner and Shklovsky, first by their published works, later in person.

The first papers by Pikel'ner which I studied were on what I remember as a clever model study of the evaporation of the corona. The term 'solar wind' had not yet become common, but the idea that the outer parts of the corona would gradually drift out was obvious and the eclipse photographs were witness of changes with time and place. A year later I was asked to referee a nice paper by Chapman covering this transition of the outer corona to the solar wind. I advised against immediate publication since Pikel'ner who had covered similar ground perhaps more thoroughly, had not been cited. Naively, I expected that this would lead the author to a closer study and a revision but, instead, the paper appeared unchanged in a different journal.

When I finally met Pikel'ner at the 1955 Dublin meeting of the International Astronomical Union he confirmed the mental image I had formed: a quiet but alert person with a wide interest, thinking twice before he asserted anything. We had lots to talk about: not only the corona but also interstellar space. For this was the same meeting where (as president of commission no. 34) I arranged a spontaneous session with Spitzer and others about the gaseous halo, which suddenly had become (even literally!) a hot subject. I suppose this was Pikel'ner's first journey to the west. In order to communicate, we found an empty class room and sat side by side during one and a half hours, each with a piece of paper scribbling diagrams, formulae and words. No precise memory is left but at one time Pikel'ner did not find the English word 'space', nor did I know the Russian word, but we did communicate through German 'Raum". However, improvement came rapidly and I do not remember any language problems during the Moscow 1958 IAU meeting. Pikel'ner then invited me to dinner at his home, an unusually brave act at the time.

Shklovsky was far more exuberant: a quick thinker, always ready for a joke, also at his own expense. I had, of course, been aware of his theoretical study on the 21-cm line of 1952 which went in some respects deeper than my prediction published in 1945, and later of his book on radio astronomy, the first edition of which appeared in 1952. In fact, having given the first university courses on radio astronomy (Leiden 1950, Harvard 1951) I made too grandiose plans to convert the course notes into a book, which never materialized. So I was happy to see that some of my pet didactic ideas had found their way into Shklovsky's book and thus reached a wider audience.

There must have been 6 or more International Symposia to which Shklovsky was asked to come for an invited talk. However, he never appeared, clearly not having obtained permission to travel. Burbidge (one of his symposium substitutes) once started his review with the words "My name is not Shklovsky". But when the world astronomers gathered in Moscow for the IAU in 1958, Shklovsky invited some 15 close colleagues in radio astronomy to his office. With some pride he showed his private "library" of radio astronomy books and reprints, stretching more than a meter on the corner chimney piece. When the buses after the IAU meeting were ready to leave, Shklovsky, shaking hands with Pawsey, loudly proclaimed: "my heart is in Australia." which clearly did not improve his chances to actually get there.

By 1968, when IAU Symposium No. 39 on Cosmical Gas Dynamics was held at the Crimea, both Pikel'ner and Shklovsky were there and participated in the lively discussion. By that time they were old friends among many new ones and I do not remember particular events worthy to record. Shklovsky liked to draw. A precious self-portrait from 1958, scribbled on a note-book sheet is reproduced in Bracewell's paper in Sullivan, "The early years of radio astronomy (Cambridge Univ. Press 1984)." In my collection I found a new-year's greeting with flying elephants (Fig. 1). The year cannot be retraced, but I suppose it must have been 1959, or so, soon after the first sputnik.

Memories fade much more rapidly than one would like and they may also become dead wrong. Therefore, I prefer to corroborate what I think did happen by referring to correspondence and archival notes. Advances in science in which one is intimately involved usually leave a written trace of conversation notes, letters asking clarification of particular points, or early versions of a paper with scribbled comments by a friend. The sad thing is that all of that is lacking. The apparent rule for the Soviet scientists during the cold war years was: send only published papers and standard new year's greetings.

By first withdrawing a little we can make a broad jump. In the same way, a quick glance at recent history may help us to focus on the future. So how might the preceding pages inspire us to face with creativity our duty to shape the astronomy of the decades to come? Let me try a few comments. They are like strategic comments to a chess game: to make the good moves is up to the players.

1. Predictions remain hazardous. I started this paper by stressing the misconceptions I once had. About 1948 I was on the pessimistic side because I saw signs of saturation in some fields of astrophysics and did not foresee the marvelous opportunities and splendid advances soon to come. Other colleagues were overly optimistic: I know an excellent scientist who about 1965 predicted that by 1985 batteries of X-ray telescopes would be operating on the moon.

2. The opportunities are plenty. Not only the new fields, which just begin to be explored, but also the more traditional astronomy profits from the new possibilities of instruments and computers. Think how the shaky methods of old photometry have been eased by interference filters and CCD's. We have a large harvest waiting, but only a limited capacity to reap. Given money for equipment and salaries, astronomy could be stepped up by a factor three: keep 3 IUE's and 3 infrared air planes busy with first-rate research and compress into 5 years the space science program now spread over the next 15 years. I do not actually plead that this should be done but it is good to be aware of this situation and to adjust our strategy to it.

3. Big science is here to stay. A growing fraction of astronomy requires huge amounts of money and manpower and a long lead time. It cannot thrive on personal inspiration alone but needs a well-structured process of decision making. This process involves besides financial and political experts a varied crowd of scientists, who are asked -in a committee or panel- to form an opinion on the scientific values of the expected results. An almost impossible task indeed, but it is definitely better to let the scientists themselves have a good try than to leave the decision to prestige-oriented politicians or to back room deals. So, if you are asked to serve, do not hesitate. By now the choice often is between 2–4 projects, all of which must be classed as excellent. This does not make the work easier or less important but it may ease our conscience about letting secondary factors enter into the decision.

4. The universe has become less orderly. This is another effect of the increased

accuracy of observation, which reveals differences where earlier we would be content to assume identity. Each object is an individual that does not have a precise par; this is true for interstellar clouds and for galaxies and is beginning to be true for stars. People in a nostalgic search for "grand" laws may be at loss, or may yield to the temptation of considering their models more important than reality. However, most astronomers now realize the advantage of this situation, namely that it shows the dynamism of the universe. Objects that have come to rest and to full equilibrium are exceptions rather than rule; stars still come relatively close to it. Just like a waterfall is less simple to describe but actually more revealing than a quiet pond, such moving, shrinking, or expanding objects invite a deeper scrutiny than an object conforming to the classical ideal sphere. Both the works of Pikel'ner and Shklovsky show fine examples of how to interpret such non-equilibrium situations with invention and caution. I wonder, however, whether education and popularizing do not still linger too much with the order ideal.

5. Good publication is to science what transportation is to agriculture. Without it the harvest may be largely lost. The public, who puts up a large part of the money spent on science, has the right to learn what happened with it. I feel a little jealous when I see those original Russian editions lying on my desk: Shklovsky, Radio Astronomy 1st ed. 1952 10,000 copies, 2nd ed. 1954 25,000 copies; Pikel'ner, The Interstellar Medium, 1st ed. 1958 13,000 copies. They are announced as popular books but actually by their thoroughness have been helpful to a generation of professional astronomers. Do we live up to this example? Sometimes it seems to me that modern science writing is spread on a long scale. At one end are popular books which dilute the science beyond recognition, fast food without nutrition. At the far other end are so-called reviews which actually are updates understandable only to a small group of specialists and not even to their professional colleagues in adjacent fields. Publications tend to gravitate toward either end of this scale, leaving a relative sparsity of old-fashioned thorough monographs in the middle.

Publication of current results poses formidable problems. The obvious need for other forms of information storage besides printed pages is being met but this will not fully prevent the tendency to further growth of journals. It is paradoxical that the astronomical harvest, which is made possible largely by public funds, is in this vital facet dependent at one side on the free-market whims of publishing houses and at the other side on the career policies, which in many countries stimulate the publication of many short (and sometimes flimsy) papers. I hope someone will at one time find the courage and insight to study this complex problem and come up with useful answers. These should certainly include an even stricter refereeing system. Quality counts.

6. Good education and wide open lines of communication are the keys to further progress. One delicate item in the education toward scientific research is that students should learn to see the difference between a chess game and a chess problem. In a chess game you may bluff: a wrong move may turn out right if the opponent does not see why it is wrong. In a chess problem the art of composing and the joy of solving lies in the

duty to win *also* against the very best moves of the opponent. Living science should be like problem solving, but students may sometimes get the impression that it is like a game.

Good communication does not have to be organized. Just remove the impediments. The influence of Pikel'ner and Shklovsky has been great in east and west. It might have been greater with fully open communication and both their joy and the quality of their work might have benefited.

In summary, my hope is that future educational systems will continue to give students the opportunity to develop and mature, and that the world political system will leave opportunity for an open and efficient communication, so that the main limiting factor will be as it was known in ancient times: *ars longa, vita brevis.*

# 2

# The Shklovsky Phenomenon[†]

## N. S. Kardashev
*Astro Space Center, Lebedev Physical Institute, Moscow*

## L. S. Marochnik[‡]
*Space Research Institute, Moscow*

Joseph Samuelovich Shklovsky [hereinafter "J.S.S." Ed.], a man who had an appreciable and often decisive influence on modern astrophysics, would have been seventy years old on July 1, 1986. This [article] is not an obituary. Obituaries are written in accordance with established canons that can not be applied to J.S.S.

He was born in the town of Glukhov in the Ukraine. In 1933 he entered the physics-math department of Vladivostok University and two years later transferred to the physics department of Moscow University (MGU). In 1938 this young physicist-optician was accepted as a graduate student by the astrophysics department of the P.K. Shternberg State Astronomical Institute (GAISh) at MGU, an institute with which he was to be associated his entire life. Then there followed the start of the war, evacuation to Ashkhabad (because of his poor eyesight he was not sent to the front), then return to Moscow and GAISh, and many years at the leading edge of the post-war revolution in astronomy.

The last forty years have been characterized by unprecedented development of observational methods and technology such as radio telescopes and interferometers and infrared, ultraviolet, X-ray, and gamma-ray astronomy. New classes of objects have been discovered: radio stars, radio galaxies, quasars, infrared sources, pulsars, cosmic ray bursters, sources of gamma bursts, background radiation at all wave-lengths, and, in particular, background radio radiation.[††] "The question 'What is this?' must be central."

---

[†] Translated from the original Russian-language article (which appeared in *Priroda*, 6, pp. 8495; 1986) by R. A. McCutcheon, Section Manager, Computer Sciences Corp., System Sciences Division (1100 West St., Laurel, MD 20707); the English version is printed here with the gracious permission of the authors of the original paper and the editors of *Priroda*.

[‡] Prof. N. S. Kardashev is a Corresponding Member of the USSR Academy of Sciences; Dr. L. S. Marochnik is located at the Space Research Institute of the USSR Academy of Sciences.

[††] The 2.7°K background radiation was discovered in 1965 by A. A. Penzias and R. W. Wilson [translator].

This is the main testament that Shklovsky left to younger researchers. He was one of the first astronomers to recognize the necessity of studying any astronomical object at all wavelengths in the electromagnetic spectrum. Only under such conditions can one hope to construct a proper model and understand the object's nature.

Shklovsky devoted his candidate's (1944) and doctoral (1949) dissertations to the physics of the solar corona. His first works concerning solar ultraviolet and radio-wave radiation appeared in 1946, at a time when rocket astronomy in the high energy part of the spectrum and radio astronomy in the long wavelength region were making only their first steps.

His study of galactic radio emissions appeared in 1947. In that same year he participated in an expedition to observe the total star eclipse in Brazil, the first such expedition to be equipped with a radio telescope. Beginning in 1950 Shklovsky took part in the first infrared observations using an image convertor. His pioneering work, *Galactic Infrared Radiation*, was published in 1953.

At MGU during the winter of 1952–53, J.S.S gave the world's first series of lectures entitled "Radio Astronomy." His audience included not only students, but also many scientists from various institutes. In his lectures Shklovsky revealed a new world; it turned out that radio waves are emitted not only by radio stations, but by practically all astronomical objects in the universe. The sun, moon, planets, stars, galaxies, and other as yet unknown sources are sending signals, the nature of which is strongly coupled with the unusual physical processes taking place within these objects. This radiation allows one to determine the most important parameters of those regions where the radio waves are being generated.

Shklovsky understood that new observational methods were needed if astrophysics was to master the new spectral bands. For this reason, in 1953 Shklovsky created a department of radio astronomy at GAISh which soon became an "all wavelength department."

From the very beginning, that is starting in 1957, J.S.S. took an active part in establishing and developing space research in the USSR. He understood that only through such research could astronomy truly extend all wavelengths. In the autumn of 1959, the Soviet Union's second lunar probe released a sodium cloud, thus creating an experimental "artificial comet." Solar rays acted on the sodium atoms and caused them to produce a resonance fluorescence that could easily be observed and studied from the earth. At first these experiments served as a satellite location indicator, but later they were used to study the characteristics of interplanetary medium and the upper layers of the earth's atmosphere. These experiments simulated a process already familiar to J.S.S. from the physics of the interstellar medium. In 1960 J. S. Shklovsky was awarded a Lenin prize for his "artificial comet" experiment.

In 1966 Shklovsky was elected a corresponding member of the USSR Academy of Sciences, and in 1967 he formed the department of astrophysics at the academy's newly created Space Research Institute. The department of astrophysics included laboratories studying the cosmos throughout the entire electromagnetic spectrum.

## 2. THE SHKLOVSKY PHENOMENON

Almost all of Shklovsky's works rely first and foremost on observational data and contain new, well-defined physical ideas. Typically these works would develop a theory to that level needed to calculate the theory's probable effects and evaluate the possibility of verifying these effects observationally. J.S.S. was one of the first astronomers to analyze the fundamental connection between physical processes that determine the properties of radiation which are characteristic at different regions of the spectrum. From this foundation he created the modern theory of the solar corona. Later he carried out a detailed investigation of the spectra of galactic and extra-galactic sources and thereby identified regions of ionized gas and regions with relativistic electrons. Shklovsky's work explaining the radiation in the radio and optical spectrum of the Crab Nebula (a supernova remnant in our galaxy) via a single mechanism and an analogous work concerning the radio galaxy Virgo A were particularly significant. J.S.S. carried out another work of fundamental significance when he studied the background radiation left over from the epoch when the universe was in a super-dense state. This work, which was performed less than a year after the discovery of the background radiation field, explained the anomalous high intensity of the optical interstellar CN spectral lines as a result of the background radio radiation acting on the CN molecules. Thus it was proved that the background radiation truly does extend beyond the boundaries of the solar system.

Shklovsky's development of a new method to determine the distance to planetary nebulae and his determination of the fundamental physical parameters of these nebulae were especially important. This was the work that evidently caused J.S.S. to think about the variability of radio sources formed during powerful explosions. In his last years Shklovsky returned to the problem of planetary nebulae. he proposed a hypothesis according to which the nuclei of these nebulae might be stars in a presupernova state. Cooling, they lose stability, which might lead to a supernova explosion of the first type. J.S.S. devoted an extensive series of works to the physics of explosions of supernovae, galactic nuclei, and quasars, and he studied the characteristics of the gas clouds and relativistic particles formed during such explosions. According to his hypothesis, the asymmetry of these explosions explains the high velocities of pulsars and the fact that old neutron stars must form an extended halo around galaxies. It is possible that as yet unidentified sources of gamma-ray impulses are associated with these objects. Such explosions in galactic nuclei could lead to the ejection of super-massive black holes.

Another important theme connected with Shklovsky is the development of spectral line studies in radio astronomy. In 1948, basing himself on the pioneering work by the Dutch astrophysicist H. van de Hulst, J.S.S. computed the intensity of the principal radio line (the 21-cm line) of atomic hydrogen—the main component of the interstellar medium—and showed that it could be detected. He foresaw the possibility of observing many molecules present in the interstellar medium. The four hydroxyl lines (OH) near 18-cm were detected ten years, and the CH line seventeen years, after his prediction. Shklovsky proposed a hypothesis by which molecular line radiation might be anomalously high in those regions of the Galaxy where young stars and planetary systems are being formed. These ideas also defined one of the themes of modern astrophysics.

Shklovsky played a great role in providing philosophical interpretations of the latest achievements in modern science (including astrophysics) and the related problems of extraterrestrial civilizations, man's role in the cosmos, and his role in transforming the universe. He was able to expound only a small part of his views on these problems in a series of articles and in the book *Intelligent Life in the Universe*. (M. V. Keldysh aided greatly in the publication of the latter).

Shklovsky's great talent as an orator and philosopher, the originality of his thoughts and the simplicity of their presentation, his temperament as an orator and his goodwill toward those thirsting for knowledge, his numerous appearances before both specialists and the general public—all of this gained him both great fame and recognition.

J.S.S.'s most characteristic traits as a scientist were his limitless interest in the facts, his quest for the essentials, his striving for simplicity in understanding the phenomena of nature, and his talent for always being on the leading edge of research.

He created a school of contemporary evolutionary astrophysics that encompasses the entire spectrum. Many astronomers both in the USSR and abroad have come under his strong influence.[†]

## THE HAMBURG ACCOUNT

He began in 1932 as a sixteen year old foreman on the Baikal-Amur Mainline (BAM; BAM was being built even then).[‡] He finished as the head of a scientific school that is known throughout the world, as an honorary member of many academies, and as a cavalier of the gold Bruce Medal, which among astronomers and astrophysicists is rated just as highly as is the Nobel Prize among physicists.

L. D. Landau once said bitterly: "I was born too late. I should have been born six or seven years earlier." Saying this, he had in mind the fact that when the "golden age of physics"[††] began (in 1925), he was only seventeen years old. In 1930, when he went to Copenhagen to work with Niels Bohr, the entire foundation of quantum mechanics had been laid by the only slightly "older" W. Heisenberg, P. A. M. Dirac, W. Pauli, and others. In this sense J. S. S. was lucky: he was born in time.

Over the course of its existence, the oldest of sciences, astronomy, has gone through

---

† For more details concerning J.S.S.'s fundamental works, see E. P. Aksenov *et al.* 1985, "In Memory of Joseph Samuilovich Shklovsky," *Uspekhi fiz. nauk (Advances in the Physical Sciences)*, **146**, pp. 719–720.

‡ BAM is the Soviet Union's second major east-west rail line in eastern Siberia; it was nominally completed in the mid 1980's. [translator]

†† This expression belongs to P. A. M. Dirac, who used these words when accepting the J. R. Oppenheimer Prize.

## 2. THE SHKLOVSKY PHENOMENON

two revolutions, each of which changed it in a fundamental way. The first occurred in December–January of 1609–10 when Galileo first turned his telescope toward the heavens (some historians maintain that the precise date of this event was 7 January 1610). Starting from this moment, man could study the universe with more than just the unaided eye.

Astronomy's second revolution began in the second half of our century. We are its contemporaries. The earth's atmosphere is opaque to almost all electromagnetic waves coming from the universe. There are only two windows of transparency: one in the optical and other in the radio band. Modern radio astronomy is obliged to this second window of transparency for its existence. However, all other radiation remains inaccessible to observation from the earth. The second revolution in astronomy began with the advent of space exploration, which has allowed man to take his instruments beyond the limits of the earths atmosphere and to measure signals that the universe is emitting at almost all bands of the electromagnetic spectrum.

Paraphrasing Dirac's words on the "golden age of physics," one can say the Shklovsky was not late for the "golden age of astrophysics." Rather, he stood at the sources of both Soviet and international radio astronomy and contemporary all-wave astrophysics. however, let J. S. S. tell how the second revolution in astronomy invaded Moscow and how he found himself at the revolution's cradle:

It all began for me when for some reason I needed to see my former graduate at GAISh, N. N. Pariiskii—a very dear man. That was early in the summer of 1946. After several unsuccessful attempts to meet him, I learned that he was at FIAN. At that time the Physics Institute of the Academy of Sciences (simply, FIAN) had not yet moved to its current location on Lenin Prospect. In the conference hall I remember being struck by the fluorescent lamps, which were among the first in Moscow. I went into the conference hall, looking in vain for Nikolai Nikolaievich. Alas, I could not find him in the hall, which was overfilled with people, and involuntarily I began to listen to the speaker, a middle-aged man with a colonel's shoulder straps. . . . He was speaking (evidently this talk had a review character) of how, during the recently ended war, officers of the British Royal Air Force had discovered that the sun emits radio waves in the meter band. This news literally staggered me. The speaker had already gone on to another, purely technical topic, whereas I, sitting at the back of the large conference hall, thought fixedly about what this unusual astronomical phenomenon might mean. At that time I had already been working for three years on problems of the solar corona and understood plasma physics to some extent (although I had always preferred spectroscopy, which seemed to me to be more concrete).

Evidently, what occurred to me in the FIAN conference hall in that distant, second post-war summer was a peculiar type of resonance. I was internally tuned to this information. Somehow or other by the end of the talk I already understood what sort of natural phenomenon these radio emissions must be. (You know, less than a half hour had passed since I had found myself in that hall.) In a lifetime one sometimes experiences (alas, too rarely!) such minutes of enlightenment.

In my subsequent scientific life I had such experiences only two or three times (Shklovsky 1988).†

That is how it all began. What happened then? There followed a series of discoveries, each of which was sufficient for the name of its author not to be forgotten by science.

*Priroda [Nature]* is read primarily by people working in science. They all know that amongst themselves professionals judge each other according to the Hamburg account‡

A typical question that one hears is: "Who? What did he do?" What, in other words, is attributed to him in science? Shklovsky understood this very well, having once said of G. A. Gamow:

I consider G. A. Gamow to be, perhaps, the greatest Russian physicist of the 20th century. In the final analysis, all that remains of a scientist are the concrete results of his work. Using a soccer analogy, it is not elegant feints and dribbling but rather goals scored that have real significance. This shows the severity of science. Three outstanding "goals" made Gamow's name immortal:

1) the theory of alpha decay or, more generally, the theory of "sub-barrier" processes (1928);

2) the theory of the "hot universe" and, as a consequence, the prediction of background radiation (1948), the discovery of which in 1965 marked the beginning of a new era in cosmology;

---

† We all knew J. S. S. to be an excellent story teller or, rather, even a novelist. he loved to tell stories, did so both willingly and skillfully, and always and everywhere attracted a large audience when doing so (without exerting even the slightest effort). Each of his stories was a small novella, "written" with the bright strokes of a master, always with an unexpected, paradoxical end. They are somewhat reminiscent of the stories by I. Babel and the novellas by O. Henry. Many of them are preserved in the family's archive. Thanks to this, we can still "hear" the live Shklovsky, from time to time giving him the floor. In the remaining text, these excerpts will be given without further special explanations.

‡ J. S. S. often used this term, which comes from the book by his namesake and distant relative, the writer V. B. Shklovsky, *The Hamburg Account*; J. S. S. recalled that

". . . there the story is told of how before the revolution, when there was neither television, nor hockey, nor many other 'achievements' of our restless 'go-go' century, people would go out of their minds over the 'world' championships' of French wrestling. Even Blok and Kuprin as well as school children were carried away by it. In Odessa, Ekaterinoslav, and Samara—literally, in one word, everywhere—the circuses organized 'world championships.' It would be arranged in advance that today Lurikh knocks out 'the terrible African wrestler Bambulu' in six minutes, whereas the day after tomorrow everything will be reverse. This was nothing but a gathered at a tavern in Hamburg where the proprietor himself was an old wrestler. There they would wrestle for real, without the public and press. Thus among themselves they had the 'Hamburg account' of victories and defeats."

3) the discovery of the genetic code phenomenon (1953), which is the foundation of modern biology.

Using the same soccer terminology, one can say that judging from the number of "goals" scored, Shklovsky was one of the most productive Soviet astrophysicists and, without a doubt, was in the first ranks of the symbolic "world all-star team."

Shklovsky predicted that the 21-cm line, which is emitted by all unexcited hydrogen atoms in the Galaxy, should be observable. The line was detected two years later. This literally made it possible to count all the hydrogen atoms in the interstellar medium and to study the medium's kinematics and dynamics. 21-cm radio observations are now the most effective and widespread method of studying the dynamics of the Galaxy and its nearest neighbors.

Shklovsky explained the luminescence of the Crab nebula (one of the most interesting objects in the sky) at all wavebands, from optical to radio, by one physical mechanism: synchrotron radiation of electrons in magnetic fields. the principle non-triviality of this explanation was that for the first time a "non-classical," fundamentally new mechanism for producing optical radiation was introduced into astronomy. Previously astronomers had dealt only with the "usual" thermal radiation of various cosmic objects. just like his article on "21-cm," this work ushered in a new phase in the development of astrophysics.

Shklovsky explained the nature of planetary nebulae by proving that they are a normal phase in the evolution of a particular type of star, the so-called red giants. Shklovsky also proved that the nuclei of planetary nebulae evolve quickly into white dwarfs and that, in fact, most white dwarfs originate in precisely this manner. This work was no less than ten tears ahead of its time and pointed to a new, previously unknown route for the evolution of matter in the universe.

His scientific legacy comprises more than two hundred articles, not mentioning books, almost all of which were written without co-authors! There is no need to explain to professional scientists what that means.

In spite of what would seem to be his "extreme individuality" in science, Shklovsky, as was mentioned previously, was the head of an outstanding scientific school. He nurtured many of his students when they were still literally in their "university diapers." However, more about this side of the Shklovsky phenomenon can be found in the previously mentioned article in "*Uspekhi fizicheskikh nauk*" [*English Translation: Soviet Physics-Uspekhi*].

## THE SUM OF EXCLUSIONS

J. S. S.'s relations with the cinema were complex. Being an artist by nature and a refined connoisseur of art, he had a very negative opinion of certain representatives of the film industry. He had his reasons for this. Several times during his long life in astronomy, he served as a consultant, reviewer, or in some similar role, and there were always misunderstandings and blunders resulting from the unobliging attitude or bad faith of people in the film industry. For example, there is the story of A. Tarkovskii's film, *Soliaris*. Shklovsky aided the production of this film to a great extent. his friends in the humanities asked him to gather representatives from the astronomical community and bring then to *Mosfilm* for discussions and to support the film:

It was fiercely cold, and there was a biting wind. It was not so simple to collect fifteen "shareholders" and take then by public transit to the entrance of Mosfilm. I convinced Iakov Borisovich Zeldovich, who was both an academician and three-times a Hero of Socialist Labor, to come as the "chief usher." Immediately there was a problem: the entrance passes had not been ordered. This occurred in spite of the fact the day before some personage close to Tarkovskii had called me at the institute, insistently invited us, and assured us that the organizational part of out visit had been taken care of. For fifteen minutes we stood around in the cold at the entrance. In vain I rushed about from window to window, trying to force some resolution. You can imagine how my colleagues looked at me! In despair I jokingly asked Iakov Borisovich (although it was no joking situation) to go to the window and show off his three gold stars. "Nothing will come from it. They will think that they're props!" answered Ia.B, thereby revealing a subtle understanding of the film business. We went away empty-handed. Several days later I learned that the discussion of Tarkovskii's film had been postponed and that no one had deigned to inform us. Tarkovskii did not even apologize.

Later the famous Italian director Antonioni, who directed "The Red Desert," "The Eclipse," and other no less outstanding films, came to Moscow and invited J. S. S. to visit and talk to him at his hotel. Not surprisingly, Shklovsky gave a not very polite refusal. As a result there was a telephone call from Mosfilm, and it was agreed that at 3:00 p.m., a time designated by Shklovsky, the maestro would come to GAISh.

The girls rushed into my cluttered office where there were three tables besides mu own. In addition there was a huge, broken easy chair. They feverishly began to clean things up, preparing to receive the celebrity.
"Stop it," I roared. "We will receive him in the style of Italian neo-realism—so to speak, under the Sicilian sky!"
At precisely 3:00 p.m. a motorcade of luxury automobiles entered the institute's courtyard. The maestro had come with his film group in the company of a very poorly qualified translator whose services I immediately declined. Antonioni turned out to be a very nice, somewhat sad, and not so young man dressed with exaggerated simplicity. he spoke English just as poorly as I do, and this of course helped our mutual understanding.
"How can I help you?" I asked.
"You see, I came up with an idea to film a fairy tale. In this film, some children are playing

in a city courtyard (a real brick *cul-de-sac*) and fly a kite that goes into space. Could that really happen?"

"You have thought up a charming tale, maestro, and in a fairy tale anything is possible." In any case, Antonioni knew his fairy tales no worse than I did. But no, he was interested in finding out whether, from the scientific point of view, such a thing could really happen!

"I must disillusion you. From the scientific point of view this could not happen."

"I understand," said Antonioni, "that according to the scientific point of view today this could not happen. But perhaps in 200-300 years science will not exclude such a possibility?"

"I'm afraid that even in 100 years the position of science will not change on this question. Could it be that these children have equipped their toy with some sort of annihilation-gravitation engine?"

No, the maestro had not created any sort of engine. It would have destroyed his idea. I began to explain to him that only primitive people and modern, civilized savages burdened with semi-knowledge believe (indeed, *believe*) in the unlimited possibilities of science. In fact, real science is the sum of exclusions. For example, all of physics consists of three exclusions:

a) it is impossible to build a perpetual motion machine of the first and second types;

b) it is impossible to transmit any signal at a speed faster than the speed of light in a vacuum;

c) it is impossible to measure simultaneously the coordinated and speed of an electron.

Antonioni darkened. These exclusions obviously were not to his liking.

It seems paradoxical to us, but this profound idea relates not only to real science. Perhaps when we talk of real order or real intelligence, we are really talking about a sum of exclusions (on an unconscious level, of course)? Shklovsky's sum of exclusions was absolute, just like physics.

## GOODNESS MUST BE CONCRETE

J. S. S. was not a saint. It is useless and unnecessary to canonize his image. He was too talented—talented in everything. Everyone who somehow or other came in contact with him felt the charm and scale of his personality. V. S. Berdichevskaia (GAISh) remembers:

". . . the scientific pedants did not forgive his boldness and risk taking. But our youth, it seems to me, should study such fearlessness. He was always surrounded by his friends and students—a large group of talented young astrophysicists. However, his pointed words, striking out like a Pushkin epigram, . . . earned him not a few enemies."

Very long ago (about forty years ago), Shklovsky had the good fortune to participate in a solar eclipse expedition. The expedition went to Brazil on the steamship *Griboedov*.

A. A. Mikhailov, the patriarch of Soviet astronomy and a very educated, intelligent man, directed the expedition.

It was especially difficult for me and the other young participants, who had no experience with fashionable receptions and who did not know the fine points of table etiquette. What sort of fine points could I have known when throughout the war I trained my character along the lines of stoicism: to bring home my scrap of bread ration intact...I was constantly making a fool of myself. My troubles began with ordering: the menu was in French. So as to simplify matters, I always sat next to Alexander Alexandrovich Mikhailov—the head of our expedition. Of course, it was not so easy to sit next to him all the time. I repeated his orders mechanically. Soon, however, I became convinced that this strategy was faulty, because it deprived me of the possibility of sampling the unbelievably tasty broiled meat dishes. Alas, A. A. and I had tastes that were polar opposites—he was on a strict diet. Thus I turned to dangerous self initiative, going to the chief for consultation at critical moments. I remember picking rather unsuccessfully at some exotic fish with a fork.

"What are you doing?" hissed A. A. quietly.

"I am trying to use my fork. After all, you can't eat fish with a knife," I babbled.

"Yes, you must use a knife, a special fish knife that is lying to your left!"

Just imagine! Another time, in response to one of my stupid questions, A. A. said quietly but distinctly:

"In general, J. S., you must use more initiative. Work according to the principle: at the table a man should be as far removed as possible from a dog. A dog eats in this way."

A. A. bent low over his plate and, to the amazement of everyone around us, began to wolf down his food with his hands. "But a man eats like this." He leaned back in his chair, holding his knife and fork in his almost completely extended arms. After such an explanation I no longer went to A. A. for consultation.

Several weeks later, when we were already sailing on to Argentina, I took revenge on A. A. In the passenger's lounge during after-dinner conversation, I somehow decided to demonstrate my erudition by reciting from memory a wonderful aphorism by Anatole France.("... *in some senses our civilization has retreated beyond the Paeleolithic: primitive people ate their old men, whereas we elect them academicians.*") A. A., who was present during this, did not even raise an eyebrow (he was, after all, of the old school), but afterward he always had a way, cold attitude toward me. . . .

On this subject Ya. B. Zeldovich once said with utmost precision: ". . . you can't throw out a song, not even if you are approaching a difficult part. Shklovsky's very personality polarized those surrounding him. Along with true friends, students, and followers, he also had enemies. He could deeply offend even those people who were favorably disposed toward him. The phrase 'he was a Man' contains a sub-text: nothing human was alien to him. Death sums up all that has passed. . ."

Yes, he could offend with a "Pushkin epigram," could offend an eminent colleague with an ambiguous accusation of ignorance, could fight for his scientific and ethical

## 2. THE SHKLOVSKY PHENOMENON

positions regardless of personalities. Nevertheless, even if it is paradoxical, goodness was the fundamental trait of his character.

His students include two corresponding members of the Academy of Sciences and ten doctors and thirty candidates of science.[†]

Indeed, Shklovsky helped almost all of them. Today no fewer than ten leading Soviet astrophysicists are obliged to him for their start in science. There were so many talented young people for whom, "not sparing his own stomach," he fought for (and obtained!) a Moscow residence permit or apartment. He always had time for young scientists, students, and even school children.

Concerning the "problem of goodness," we should listen to Shklovsky himself:

Whenever I go from home to the *Nauka* publishing house (more precisely to that publisher's astronomy editorial board), and the driver of trolley bus No. 33 announces (not always, of course) "Academician Petrovskii Street," the stop at which I must get out, without fail I become sad. I am deeply obliged to the man for whom the former Exhibition Lane has been renamed. He reinstated me at Moscow University when, in 1952, I along with several of my unfortunate colleagues was driven out of the GAISh.[‡] Two years later he used his authority to give me straight from the rector's fund an unbelievably luxurious three-room apartment in a fourteen-story building belonging to MGU on Lomonosovskii Prospect. Prior to that my family and I had been cooped up for nineteen years in one room in the Ostankino barracks. . . I was able to create a very viable department and fill it with talented young people exclusively thanks to Ivan Georgievich's self-sacrificing help. He provided housing to my hopeless young co-workers. Later, when the space age began, he helped us so many times! He had a perfect feeling (just as musicians have perfect pitch) for real science even if it was in an embryonic state.

For twenty-two years Ivan Georgievich directed the most outstanding university in the country. Nothing was closer to him than the university, which to him was both a home and family. For the university he even gave up his beloved mathematics. . . .

His fate was profoundly tragic. It's an old story: a good man at a hard place in difficult times! You have to understand how difficult it was for him. I witnesses dozens of good deeds performed by this remarkable man. From this, being well acquainted with statistics, I can affirm with complete confidence that the number of good deeds he performed over the whole period during which he was rector must be on the order of $10^4$! How many people can claim

---

[†] The Soviet candidate of science degrees is given upon completion of graduate studies and, on an American scale, is somewhere between the M.S. and Ph.D. The Soviet doctor of science degree is given to established scientists who have written a major treatise. It is usually given ten or more years after a scientist has received his candidate's degree. It has no equivalent in the American system. [translator]

[‡] The early 1950's (until the death of Joseph Stalin in 1953) represented the pinnacle of the anti-semitic campaign against so-called "cosmopolitans" in all spheres of Soviet society. [translator]

such a lifetime total? A once popular poet wrote the following "humanitarian" lines: "good must be done with the fists." That is a lie! Before all else must be concrete. There is nothing worse than spineless, abstract kindness. Our "radicals" should learn this simple truth. It would be just if the following simple inscription were placed on Ivan Georgievich's gravestone at Novodevichy Cemetery: "Here rests a man who performed 10,000 good acts."

## THE MECHANICS OF ARISTOTLE

Shklovsky was not only the father of new theoretical directions in science; he was also a man who organized global experimental studies in the new fields of astrophysics.

Bit by bit, beginning in 1953, I formed a department that was conditionally called the department of radio astronomy, although we did not limit ourselves just to radio astronomy (*in fact this was a department of "all-wavelength" astrophysics—N. K., L. M.*). It was formed out of our talented youth who deeply felt the revolution taking place in our science—astronomy. It was devilishly difficult to form this department. . . .

It is always difficult for the new and progressive to find its own road. Indeed, the cosmic era of mankind's history had begun—soon the first satellite was launched—but life became more and more difficult. In general, I went straight to the rector on all matters both big and small concerning my department. This, of course, can not be considered normal. But what else would you have me do? In 1968-69 the main, most creative and active members of the GAISh department of radio astronomy (including myself) abandoned the walls of their alma mater and moved to the Academy of Science's newly organized Space Research Institute, where they received more or less normal working conditions.

One should not think, however, that J. S. S., who deeply felt the problems of each of his students and co-workers as well as his own, walked a path strewn with roses from that time on. This was far from the case. However, as became apparent, a discovery awaited him on his path. The secret of motion is timeless. Thousands of years had to pass before Galileo grasped its essence. Later Newton formulated the first law of mechanics, according to which "everything moves with uniform, rectilinear motion until such time as outside forces remove it from this state." Nevertheless, Shklovsky thought that he also had managed to establish the **true significance** of the mystery of motion.

The problem's essence rests in the eternal mystery of motion. The great Aristotle believed that motion (uniform and rectilinear, of course) can take place only because some force is acting constantly on the moving body. If the force ceases to operate, the sooner or later the body will stop! The great Galileo, and after him the no less great Newton, came to the radical conclusion that uniform, rectilinear motion does not require any kind of force to sustain itself! This is the celebrated law of inertia that many millions of scholars on all continents learn by rote without due understanding.

## 2. THE SHKLOVSKY PHENOMENON

J. S. S. believed that the essence of his discovery to be that all our activity occurs, unfortunately, according to Aristotle's law of mechanics, not the law of Newton and Galileo. Just neglect the organizational aspects of an experiment and see what comes of it.

In fact, a good example of the formal action of Aristotle's mechanics is the motion of a body in a viscous medium. If the body (or a research paper, a job, or, for that matter, almost anything) is to be moved forward, one must constantly push on it. That is the essence of Aristotelian mechanics!

## HELLO, IS ANYBODY OUT THERE?

Many people throughout the world know Professor J. S. Shklovsky primarily as the author of the renowned book, *Intelligent Life in the Universe*. This book first appeared in Russian in 1962. Since then it has gone through five editions, and a sixth (posthumous) edition is now being prepared. It has been translated into many languages.[†] Interest in it has never waned, since for every new generation it is just as much a revelation as it was for its first readers.

I am especially proud that my book appeared in an edition for the blind—using Braille script! The four thick volumes printed on paper like cardboard make a strange impression.

*Intelligent Life in the Universe* is a book about the universe, the possible existence of life in the universe, and, most importantly, the possible existence not only of life, but of intelligent life. Shklovsky wrote this book "in one breath." Back then, in 1962, i the period of *Sturm und Drang*, it seemed that the gigantic successes of all-wavelength astronomy could not help but lead to the discovery of signals of intelligent origin.

This did not happen. The universe was silent, and Shklovsky's point of view changed. He came to the conclusion that our civilization is, more than likely, unique.

We do not wish to cite here all the *pros* and *cons* of one or another point of view. In the final analysis, this article is not about that problem, which is as old as civilization, but about J. S.S. In his opinion (Shklovsky 1977),

... at the present time, which is characterized by tremendous successes in astronomy, an affirmation of out practical cosmic singularity is much better founded in concrete

---

[†] The English language edition is an expansion that was co-authored with Carl Sagan: J. S. Shklovsky and C. Sagan, *Intelligent Life in the Universe* (San Francisco: Holden-Day, Inc., 1966); note that, for the most part, the Russian epigraphs were not included in this edition.

scientific facts then the widespread, common, traditional, and now dogmatic opinion about the multiplicity of inhabited worlds. . . . It seems to me that the possibility of our practical anthropocentrism, at least in the local system of galaxies, is immeasurably richer on the philosophical, ethical, and moral plane than the traditional "Hello, is anybody out there?"

We all know now that our world could be on the edge of catastrophe and that one of the main reasons for this is lack of mutual understanding. Well, if in addition to this we are alone in the universe, then...

In this connection we will relate the history of the epigraphs in Shklovsky's book, *Intelligent Life in the Universe*. The book consists of three parts (twenty-seven chapters). The first part is "The Astronomical Aspect of the Problem,;" the second is "Life in the Universe;" and the third if "Intelligent Life in the Universe." Each part is opens with an epigraph.

The literary critic Ben Sarnov gave me a good epigraph for the general astronomy chapter. ("And by a dread, dread change of direction/ To other, one or another/Unknown universes has the Milky Way been cast." This is from Pasternak.) The epigraph for the futurology chapter proved more difficult .... Not long before this I received a letter from my old, now deceased friend S. D. Solovyev. Among other things, this letter included the following lines: "Recently I re-read Aseev's new poems. He has started to write better in his old age. Read the following stanza that I have adjusted slightly:

> And isn't it curious, devil take it
> What will happen to mankind after us?
> What will happen then?
> What sort of dresses will they sew?
> For whom will they clap?
> To what planets will they sail?"

But surely that was just the epigraph I needed! It was only while correcting the galley proofs that I remembered Solovev's note about the "slightly adjusted stanza." Did that mean that these lines I had taken a liking to were not Aseev's originals? It c could lead to a scandal! ... With great difficulty I found Aseev's book where these lines had been published. My worst fears proved correct: after "For whom will they clap" Aseev had written the onomatopoeia "tim-tom, tim-tom, tim-tom!" But for me the entire meaning was contained in Solovev's line, "To what planets will they sail?" I had to throw out this ending and end the stanza with "For whom will they clap?" However, in the later editions, after Aseev's death, I resurrected Solovev's ending. . . . May those who protect the inviolability of poetic intention and an author's copyright forgive me.

Understandably, this epigraph was for the third part. For the second part Shklovsky chose a verse from the cycle *To a Blue Star* by N. S. Gumilev, a well-known poet from

## 2. THE SHKLOVSKY PHENOMENON

the beginning of this century who also was leader of the Russian acmeist movement and the husband of A. A. Akhmatova:

> On the distant star Venus
> The sun is gold and flaming,
> On Venus, oh, on Venus
> The trees have blue leaves...

In subsequent editions the epigraphs for the first and second parts did not change. In the final (fifth) edition J. S. S. changed the epigraph for the third part completely, replacing N. Aseyev with I. Ilf: "On such a planet it is pointless to waste time!"

Such a change is natural, the reader may think. It precisely reflects Shklovsky's changing attitude toward the problem of extraterrestrial civilizations. This is true of course, but that is not all. One can not exclude the possibility that the excerpt from Ilf's *Notebook* appeared not only in connection with the author's changing position regarding extraterrestrial civilizations, but also in connection with the history of the second (Venusian) epigraph.

The first Soviet rocket was launched to Venus in February 1961, and *Izvestiia* printed an article by J. S. S. dedicated to this very important event. The article began with [Gumilev's] long-forgotten verse.

I was very proud of my act and, bursting out with lofty feelings, sent Anna Andreevna Akhmatova the clipping from *Izvestiia*, accompanying it with a short, respectful letter.[†] Specially for this I learned the address of the Moscow friends with whom she always stayed when she was in the capital. I waited a long time for an answer. After all, she must have rejoiced at such an out of the ordinary event! Weeks, months passed. I learned without a doubt that Akhmatova had been in Moscow. Alas, I never did receive an answer from her, although I did find out for certain that she had received my letter.

Indeed, "on such a planet it is pointless to waste time" if even those people who have the same feelings about their era, share the same disposition, and belong to the same spiritual movement do not understand each other very well

But who knows?
Many years later I discovered the reason for Anna Andreevna's silence. It turns out the Gumilev

---

† N. S. Gumilev was executed as a counter-revolutionary in 1921 and was considered a "non-person" until the post-Stalin era; Shklovsky's pride is due to the fact that he was one of the first to resurrect Gumilev's memory in print. [translator]

dedicated the cycle *To a Blue Star* to another woman! This is simply striking. To the end of her days she remained a lady and never became an old woman!

## IS GOD ON THE SIDE OF LARGE BATTALIONS?

How could I have known that the spring and first half of the summer of 1947 would be the brightest and, perhaps, happiest time of my complicated life, which is now approaching its finish. In that third post-war spring, when I was filled to the limit with good health, youth, and an unshakable faith in an infinite and joyous future, I thought that the forthcoming expedition to the Tropic of Capricorn, to the fantastically beautiful Brazil, was only the beginning. I was certain that there would be much more that would be good, that would excite the soul, but that was as yet unknown. After my poverty-stricken, pre-war youth, after the difficult deprivations of the war years, finally the world was suddenly opening before me. It was the world as I had imagined it in childhood in my native Glukhov when I used to wait with baited breath to receive the next issue of the magical journal *The World-Wide Pathfinder* with its numerous appendices. Later there were journals *Around the World, The World-Wide Tourist*, and the collected works of Jack London with their striped brown paper covers. Avidly reading the *Marakatov Abyss* by Conan Doyle or, let us say *The Cruise of the Snark* by London, I was thousands of miles from my native Chernigov life. The salty waves of the sea, the whistle of the wind in the ship's rigging, the brave people burned by the tropical sun—that is what I lived for then. Overall, the feeling that has remained with me from childhood is that of a parade of astonishingly bright and lush colors. There is one summer morning that has engraved itself it my memory for my entire life. Waking up, I looked for a long time through the window where the lush, green leaves of an old pear tree were projected against the bright blue sky. In my artistic endeavors at that time I had no green paint (poverty!) and had to mix blue and yellow. What was I doing? After all, blue and green are the colors of the sea and plain. . . . But then came my bleak, poverty-stricken youth. The muse of the distant travels went away somewhere to the sphere of the unconscious. Living in distant Vladivostok, whenever I glanced at a map of our country I would always shudder: "Where has this all brought me!" During the war the maps of the fronts called forth completely different emotions, at first dreadful and later inspiring hope.

The war ended. . . . I was greedily carried away by science. I was very lucky that the start of my scientific career coincided exactly with the advent of the epoch of *Sturm und Drang* in the science of the heavens. The "second revolution" in astronomy has arrived, and I understood this with my entire being.

This is where my childhood dreams about distant countries helped me! Often I felt myself to be like Pigafeta and Oreliana, laying the path to a mysterious, secret, beautiful country. I am deeply convinced that without those childhood day-dreams while reading *The World-Wide Pathfinder*, London and Stevenson, I never would have accomplished what I did in science. In science I was a strange mixture of an artist and a conquistador. Such phenomena appear only during epochs when the habitual, fixed conceptions are broken and replaced by new ones. Such a style of work has already become impossible. In our days the Napoleonic rule that "God is on the side of large battalions" is rigorously enforced.

## THE PAULI PRINCIPLE

Among the fundamental exclusions of physics that Shklovsky did not mention during his conversation with Antonioni (see "The Sum of Exclusions") is the Pauli principle. The essence of this principle is that in an atom of hydrogen, for example, the spin of an electron can be in only one or another direction.†

"Either-or," they joke in Odessa. The Shklovsky phenomenon can not be fully understood without the Pauli principle.

"Drop it, J. S. S. You are destroying your nerves over trivia! Protect your health!" said one colleague—not jokingly—when he visited an agitated Shklovsky in the hospital.

"I can only be in one of two states," answered J. S. S. "Either living, as you all know me, or not living. I know of no other states."

He died on 3 March 1985, from a stroke.

## REFERENCES

Shklovsky, J. S. (1977) "An answer to Lem (On critical Comments in the Article by Lem Concerning My Article 'The Possible Uniqueness of Intelligent Life in the Universe')." *Znanie-Sila*, No. 7, pp. 41–42.

Shklovsky, J. S. (1982) "History of the Development of Radio Astronomy in the USSR", Series *Cosmonautics, Astronomy*, No. 11 (Moscow: *Znanie*), pp. 7–8.

---

† Of course, in its general form the Pauli principle is formulated somewhat differently.

Поздравляю Вас с новым годом

Н. Шкловски

A New Year card drawn by J.S. Shklovsky, a talented caricaturist. It symbolizes the Russian love for dreaming which is sometimes realized in terrible reality.

# 3

# Words about Pikel'ner

## L. S. Marochnik

*Space Research Institute of the USSR Academy of Sciences, Moscow*

### AD INFINITUM

Solomon Pikel'ner, the Soviet Union's foremost astrophysicist, died in 1975 in a shabby Moscow hospital from . . . appendicitis.

Of course, Soviet medicine is a most inspiring theme for Mr. Hitchcock. That is a different story; I would like to talk about other things here.

The death of S. B. Pikel'ner was a natural, logical continuation of his life, of the biology of his character or soul, if you will.

The deepest intellectuality and, as a consequence, tact in his relations with colleagues and pupils, and, indeed, anyone was in his makeup. It was this characteristic that prevented him from turning to his neighbors at a critical moment; he couldn't allow himself to bother them with his complaints of pain; "it can't be too serious," he said to his wife before going to the hospital.

He died at the age of 54. In 1991 when this book is published he would have been 70 years old.

He was my teacher, although I am now two years older than Pikel'ner was when he died, and now I, as an astrophysicist, am able to judge his contribution to the most diverse fields of science (he was one of the last universalists) which his life illuminated. I shall discuss this below, although, in fact, these remarks are of a completely different nature.

During a life that was too short Pikel'ner wrote more than 140 scientific papers and eight books. His interest included all astrophysics ranging from solar physics to cosmology. He was attracted to classical studies and had ideas that were only experimentally verified many years later.

I do not want to repeat the many articles (see, e.g., references /1,2, and 3/) written about Pikel'ner's scientific interests; nevertheless, Pikel'ner's most important achievements are as follows:

1. Predicted the presence of a Galactic gaseous corona (1953).

2. Explained the cause of the approximate equality of the energy densities of the magnetic field and cosmic rays in the galactic disk by the "valve" mechanism permitting an excess of the latter in the halo (1961).
3. Explained the causes of the decay of the interstellar medium into cold and hot phases by the nonmonotonic dependence of its pressure and density of the Van der Waals type (1967), a study changing the character of the science of the interstellar medium.
4. Proved (together with P. V. Shcheglov) the existence of a stellar wind according to observations of velocities in HII regions (ionized hydrogen) (1968) and constructed a theory of two-layer shock waves arising from the interaction of the stellar wind with the interstellar gas.
5. Gave a theory of the formation of the radiation spectrum behind a shock wave with illumination in a partially ionized gas (1954); this study introduced into astrophysics the new idea of a shock mechanism of interstellar gas emission.
6. Gave the first correct theory of dissipation of the solar corona (1950), a forerunner to the theory of the solar wind. Taking into account small corrections both theories give coincident results.
7. Gave a theory of nonthermal radio emission of the solar corona (with J. S. Shklovsky) (1951).
8. Explained (simultaneously and independently of E. Parker) the origin of the solar chromospheric network (1962).
9. Explained the whole spectrum of solar flare emission (which now also relates to other stars) as a single process of the interaction in the chromosphere of particles accelerated in the corona (1974).

This list is not complete. It would be easy to continue.

However, I would like to note the main thing. Pikel'ner's contribution to the science which illuminated his life cannot be covered by a simple list of his scientific results although, as J. S. Shklovsky wrote about G. Gamov, only goals are important, if one uses football terminology. However, especially in the case of Pikel'ner, this is not true (perhaps, as is often said, this is the exception that proves the rule). No list of achievements can possibly express the value of the influence on astrophysics *ad infinitum* of Pikel'ner's books on cosmic electrodynamics, the interstellar medium, and solar physics; his ideas were widely and freely propagated by him among young and older colleagues, alike.

## THE DELPHIC METHOD

At some stage I was told the history about the glow of Soviet Science of Science (science sociology) with a translation from an English publication which used the figurative term "Delphic method." Obviously the first of the translators did not understand the sense of this term and thought that this was the method of some kind of Doctor Delphic relating to the solution of any kind of social problem. The Delphic method then wandered from article to article and from book to book of Soviet sociologists until after several years a hair-splitter returned to the first story and explained that the meaning of the Delphic method at that time amongst Soviet sociologists was simply the method of the Delphic oracle.

From the end of the 1950s, the method of the Delphic oracle was also affirmed with the appearance at Moscow University of Solomon Pikel'ner in Soviet astrophysics. Professor Pikel'ner was this oracle.

In the sea of lies and Pharisaism flooding the country, the self-realization of young people could take place only through science. Of course, things were not so primitive. In fact a genuine enthusiasm for science existed and a true faith in the "internal voice" determined the choice of science as means of existence or, more precisely, a way of life. However, I believe that it is not wrong to say that many honest and intelligent people of our generation entered science, preferring this path to any other and subconsciously trying, at the Freudian subcortex level, to avoid or, to reduce to a minimum any interaction with the Communist system.

In science (astrophysics in this case) Moscow University Professor Solomon Pikel'ner existed as an oasis amidst an Orwellian nightmare. Both Pikel'ner and Shklovsky were undoubtedly the leading astrophysicists of the country. At the same time, they were very different people. One can understand what Shklovsky was like from the preceding articles. Pikel'ner was just short of being the complete opposite. He was always fair with people, intelligent, sensitive, and punctilious to the limit in relations with colleagues and students, and often attended even to unimportant people coming to him in crowds for scientific consultations with questions, and, indeed, simply with their often nonsensical ideas. It always remained a mystery to me how he managed to have time for everything. Later I understood: a very talented person simply thinks, understands, and solves scientific and other problems much faster than an ordinary person.

One could say that Moscow University Professor S. B. Pikel'ner was a figure of such sharp contrast with the Soviet reality surrounding us (the contemporary youth) and such an outstanding personality (in a professional sense) among his colleagues that it is easy to understand his great popularity. He, if you will, was a unique oracle in the Soviet astrophysics only beginning to appear from behind the Iron Curtain. This is a typical situation: there is a scientific argument whose cause is irrelevant among two, three, or more participants and all are confused. So what? "Give it to Pikel'ner," someone says. All turn to him and after several minutes studying the problem it is completely clear or, in any case, the reasons that it is not possible to obtain a simple answer are formulated.

During the time he worked at the Shternberg Astronomical Institute at Moscow University as well as generally in Russian astrophysics, Pikel'ner worked smoothly and became the usual Delphic oracle. Everyone from graduate students to professors went to

Pikel'ner, telephoned him, rating him as a higher authority in any unclear scientific problems, not necessarily astrophysical ones.

## "VIEWS FROM LONDON"

To a Western reader the heading above means nothing and, in any case, he does not understand why it is put in quotation marks. However, to a Soviet colleague it means a lot. In dark times of developing socialism BBC transmissions broadcast to Soviet listeners were especially popular when they were presented by Anatolii Maksimovich Goldberg. His rubric was "Views from London." The program discussed eventx in the USSR as they were viewed from far-off London.

At the time of my first meeting with Solomon Pikel'ner (1958) the phrase "Views from London" was the current joke of intellectuals (whether from science, art, or literature). In those years I lived in a remote province in Dushanbe. I was rarely in Moscow where the "great" science was done. A typical dialogue of those times was:

Hello, old man! Have you been in the capital long?

What is your notion about collisionless shock waves?

I don't have the slightest idea. I am viewing this from London where there is neither literature nor information.

My first impressions of Pikel'ner were, as it were, impressions "from London," from the far-off Tajik province to where fate led me after finishing university.

These lines were written about Solomon Pikel'ner, but initially I was forced to write something about myself at that time, beginning to enter science as a young person, because initially I want to present to you Pikel'ner through the eyes of a young provincial, as it were, viewing from Tajik "London" in 1958.

I finished university as an experimental expert in ferroelectrics and not in astronomy, let alone astrophysics which I had never heard of because even a course in general astronomy was not taught. Thus, I did not, and could not, know any astronomical names.

Arriving at the Pedagogic Institute in Kulyaba (in the Tajik interior), the department manager gave me the job of lecturing on such theoretical physics courses as quantum mechanics and statistical physics, and a course on general astronomy. All these courses were to be taught according to the standard official program. The students hardly understood the Russian language with which I gave the lectures, but they diligently and silently wrote my concepts. For this reason and also my youth and lack of teaching experience, I did not immediately recognize their dreadful level of preparation.

Only in the second or third lecture on quantum mechanics, coming to the theory of operators, did I understand with horror that not one of the students was able to divide a fraction by a fraction (quantum mechanics was taught as the fourth course of the Department of Physics and Mathematics).

## 3. WORDS ABOUT PIKEL'NER

From this time in lectures carrying the name "quantum mechanics," I began to teach them basic arithmetic and then elementary school algebra. The successes of my students were no less conspicuous in the region of astronomy which I learned, you understand, in parallel with them. I will never forget the single question which one of the students (apparently the brightest of them) gave me in a consultation before an exam.

"Mualim,"† he said, "when you throw a stone upwards, doesn't it fall to the Earth?"

"Of course."

"Then why doesn't the Sun fall to the Earth?"

I cannot resist the temptation and will be distracted to tell the unexpected continuation of this old story.

Several years ago a professor from sunny Tajikistan published the sensational book *Phenomena Disclaiming Terrestrial Gravity*. No one would ever have known about this "work" if the author had not sent this book to Roald Sagdeev‡ with the touching note "To a dear teacher . . .". Remembering that I had once lived in Dushanbe, Sagdeev asked me to examine this "work" by such a respectful but unknown author.

You can imagine my surprise when I realized that the book was written by the same student (although, of course, this is only my hypothesis), who somehow became a professor and wrote an answer to the question that had troubled him in his youth: "Why the Sun does not fall to the Earth, but a rock does?"

The answer turned out to be simpler than blinking an eye: a rock falls under the influence of atmospheric pressure, but on the sun the atmospheic pressure is absent.

Let's go back to 1957–1958. I was enlightened when, studying astrophysics in the process of giving lectures, I was attracted to it, made my first contributions during the year in Kulyaba, and went to Moscow to show them to someone.

When I appeared, as they say from the street at the Shternberg Astronomical Institute, a young fellow similar to me immediately told me that I should go to Pikel'ner since there was no one else to turn to. Before me there is a completely gray-headed person with a suntanned face and an unbelievably young, very young face. He sat next to me and very attentively listened for nearly two hours, asking questions, extremely tactfully and respectfully, referring to me (and I felt quite correctly as a greenhorn) by name and patronymic, as an equal which embarrassed me. I want to say immediately that this was not an affected manner. It was the natural manner (which I knew for many years) of Solomon Pikel'ner, already number one astrophysicist in those years.

My second encounter with Pikel'ner occurred several days later at his home where he

---

†Mualim means teacher in the Tajik language. In the East they do not use instructor (professor) with a name or title, but only what is important in the sense of old Eastern traditions. Teacher!

‡ Professor R. Z. Sagdeev was the director of the USSR Academy of Sciences Space Research Institute and he was also the head of the Soviet Space program.

invited me to help me improve one of my articles, which he immediately accepted for the *Soviet Astronomical Journal.*†

It was a snowy Moscow winter. I had shoes with treads that continuously collected wet snow. Remembering this on entering the professor's apartment, I carefully stamped my feet. During a heated scientific discussion the forgotten snow melted and I saw under myself, more precisely under the stool on which I sat with my feet on the crossbeam, a large and absolutely unbecoming puddle. To say the least, I was embarrassed, but I said nothing. Remember that I was a youth from Kulyaba and very shy.

I was dumbstruck by the overwhelming horror and ashamed, but I could not move from the stool because of nervous tension. Pikel'ner immediately understood everything. He went away, returned with a rag, and muttering something like "please don't worry, Leonid Samoilovich," and quickly wiped up the cursed puddle. I soon asked to leave because after what had happened, I was in no state to work.

After a year I became a postgraduate student of Pikel'ner, but I continued to live in Tajikistan (now in Dushanbe) because to transfer to Moscow, as a Soviet understands, but probably not a Western reader, was impossible due to the visa problem. Briefly speaking, not completely understanding everything that I did (because of my naivete and youth), I wrote Pikel'ner personal letters with endless questions, ideas, etc., because I could only be in Moscow once or twice a year due to the extremely limited financial resources of the Institute where I was working.

He answered all my letters with long letters typed by himself with additional discussions concerning all my problems. Remember that there was no one busier than him either in the Shternberg Astronomical Institute nor in Soviet astrophysics in general. Unfortunately, almost all this archive is lost. It so happened that of all these remarkable letters I miraculously saved only three. I will give only one part of one of the letters.

In one of the letters I informed my scientific advisor that a son had been born to me. He immediately proposed that I should receive his money for scientific advising because now it was more useful to me than to him. Naturally, I refused. At the end of his reply after two pages of scientific discussion he wrote me:

Concerning money. You accepted it this way in vain. I am actually more guilty than you and should incur the expenses according to law.† However, if you consider that you cannot accept it as a gift, you should take the money on loan until you begin to receive pay with a doctor's

---

†From that time until his death S. B. Pikelner was the permanent scientific secretary of the main Soviet astronomical journal—*Astronomical Journal* (Russian) thus, determining its character. With all his unparalleled activities he considered it his duty to run a journal, apparently understanding full well that he could entrust this to no one else, in spite of a large, multi-titled titled editorial board.

†He had in mind a delay with my thesis defense due to the objections of one of my examiners.

degree. You can return it then since the money is simply not necessary to me; since the time of my confirmation as a professor (several months ago) I have received 450 rubles a month and this is completely adequate for our family. . .

I have loved this person all my creative life (in science) starting with the "View from London" and then completely unwaveringly since.

## PLATO IS DEAR TO ME

After the death of Solomon Borisovich I learned of a startling story about how a student asked Pikel'ner to give him an excellent mark without an examination (for some reason he was unable to prepare for it) which was necessary to obtaining a higher salary. Probably, it was an able student of Moscow University. The student explained to the professor that he would pass this course and, naturally, with an excellent mark, but later in a specified period of time. In other words, it was a gentlemen's agreement.

The student never passed the examination, relying on the well-known gentleness and sensitivity of the professor (everyone knew that he wouldn't harm a flea). With surprise I found out that Pikel'ner showed unexpected harshness in relation to this person and convinced the Scientific Committee to stop his progress in science. This incident startled me. In contrast, I remember passing my doctoral exam with Solomon Pikel'ner.

He gave me several days to study the proofs of the first edition of his well-known book *Cosmic Electrodynamics*, saying that this material would be sufficient. You understand that this was a course I never attended and I had no special astronomical education, only what I had been exposed to in the process of working.

I read seriously and three to four days after I obtained proofs I decided that, in general, I had mastered this material. The day of the examination arrived. I was nervous, but everything was straightforward and simple. Solomon Pikel'ner arrived with a blank examination paper and said that it was not worth bothering the remaining members of the commission for their time and they would agree with his mark. After this he proceeded to fill the blank paper.

"Well, what should we write for the first question? What do you think about shock waves? Do you know this, Leonid Samoilovich?"

"I know," I said, being completely certain that I knew.

"Good," Solomon Pikel'ner said, writing on the chart the mark "excellent."

"For the second question, let's write about 'causes of field line freezing in magnetohydrodynamics.' You know this?"

"Of course."

Pikel'ner again wrote on the next chart, "excellent."

"The third question, let's write about 'Solar Radio Emission.' How about this?"

"I don't understand this very well."

"Never mind," Solomon Pikel'ner said, writing a mark of "good."

After about a month it was necessary in my work to consider shock waves in cometary atmospheres and I understood that the answer "I know" to the question about shock waves in the examination was, mildly speaking, incorrect. When I told this to Solomon Pikel'ner on our next meeting, he smiled and answered, "I knew that you would study it when it was necessary."

I later used this method with my students and doctoral candidates. It worked without a hitch.

Returning to the story about the student with the destroyed gentlemen's agreement, I should say that it seemed amazing at first glance only from naivety. I will explain in more detail.

In agreement with his style of relations with colleagues and people in general, Pikel'ner carried out much of his work ahead of time. Thirty years ago when venerable scientists demonstrated a haughty, unhurried style of education with colleagues and students, Professor Pikel'ner, in particular, was quick in speech and actions, outwardly unemotional, composed, and efficient. He always tightly economized others' time, recognizing that he didn't have enough of his own. Essentially, this was the contemporary Western style. This was unusual. You involuntarily began to imitate him.

The gentleness, tact, and intelligence of this collocutor contrasted sharply with his business-like style. One only spoke about business. You simply felt that you did not waste your collocutor's time. However, in all these serious business-like conversations you always felt the gigantic iceberg of his intellect which he did not demonstrate to you in 99 cases out of 100.

In a final analysis, in spite of the business-like style, the feeling remained that you were dealing with an extremely gentle person and it conformed to reality. As everyone knew, he was the only person who had no enemies because his tact was unlimited.

However, this person became inflexible concerning things of pure science and those things clearly related to it. It was not only his students who did not always understand this.

The problem of the spiral structure of the galaxy is a classical one. The names of Jeans, Heisenberg, Weizsäcker, Chandrasekhar, and Fermi are connected with it. However, it is well known that all of the attempts of these famous researchers to interpret the phenomenon of spiral structure of galaxies ended unsuccessfully. It became clear at the beginning of the 1960s that there was a significant magnetic field in the gaseous spiral arms directed, primarily, along the arms.

In a series of articles published in those years in the *Astronomical Journal* (Russian), Pikel'ner succeeded in creating a consistent magnetic theory of this phenomenon.

Several months afterwards, we discussed with him the gas dynamic effects arising under the influence of the stellar density waves discovered by Lin and Shu. I asked: "How about your magnetic theory?"

He answered that "I attempted to show that they [Lin and Shu] were wrong and I understood that they were right."

... But truth is dearer still.

## JESUS CHRIST SUPERSTAR

As you will remember, the action of this great rock opera of Andrew Lloyd Webber begins in the desert. You hear this improbable music, view the screen (or stage if you were luckier than I) and gradually begin to recognize a sympathetic fellow, not simply a sympathetic fellow, but Jesus Christ and that he (Jesus Christ) is actually a superstar.

I know many people who, when the speech turns to Pikel'ner, talk about him not simply "with aspiration," but as a saint, always ready to share his talent with those around him regardless of their beliefs. We, the contemporary youth, considered him to be nothing less than the superstar of our beloved science, astrophysics.

A funny coincidence is that the action here also began in the desert, a different one, the Kara-Kum. This point in time was preserved for us in the memories of Joseph Shkolovskii about the life of the young Pikel'ner at the time of the evacuation of Moscow University to Ashkhabad in October 1941.

> It began as a completely fantastic, happy, and hungry one, our life in Ashkhabad which was in no way similar to any other one. There were many things in the ten months of this life. There were turtles which I loved in Kara-Kum going 20 km in one case in the desert and there was also the death of Deli Gel'fand in this same desert. There was our school used as a dormitory at 19 Engels Street close to the Russian bazaar. And there were a lot of other things. For example, giving lectures in the office of Part-Pros[†] to a single student of the fourth course Mone Pikel'ner, subsequently being adorned with our astronomical science. My heart contracts with pain when I recognize that Solomon Borisovich Pikel'ner, the best of the people that I knew, died almost ten years ago. It is both happy and sad that to the end of his days he invariably related to me as a student to a teacher. Then in the unforgettable year of 1942 the student and teacher, differing little in age and terribly ragged (Mone more so and barefooted) in the empty although luxurious building of the Part-Pros (destroyed by a strong earthquake in 1948) we developed fine points of the Schwartzshield-Shuster model of the formation of spectral absorption lines in the solar atmosphere.

In naming this section "Jesus Christ Superstar," I am not being blasphemous. You can convince yourself of this by reading the pronouncement of J. Shklovsky at the burial. He would have called it:

---

[†] Part-Pros is a big building of the local Communist party.

## "Words About Pikel'ner"

"We lost the best astronomer of the country. The sharp pain has not disappeared and to the end of my days I will live with 'a clear recognition of the pain' of this irretrievable loss. He was an amazing person. At his funeral I remembered a book ny a distant relative, Victor Shklovsky, *The Hamburg Account*, written 50 years ago. ". . . there the story is told of how before the revolution, when there was neither television, nor hockey, nor many other 'achievements' of our restless 'go-go' century, people would go out of their minds over the 'world' championships' of French wrestling. Even Blok and Kuprin as well as school children were carried away by it. In Odessa, Ekaterinoslav, and Samara—literally, in one word, everywhere—the circuses organized 'world championships.' It would be arranged in advance that today Lurikh knocks out 'the terrible African wrestler Bambulu' in six minutes, whereas the day after tomorrow everything will be reverse. This was nothing but a gathered at a tavern in Hamburg where the proprietor himself was an old wrestler. There they would wrestle for real, without the public and press. Thus among themselves they had the 'Hamburg account' of victories and defeats."

"So it is, as a professional astrophysicist, I can verify to the young generation of astronomers that Moscow University Professor Solomon Borisovich Pikel'ner according to the Hamburg account was the best astronomer of the country. No one else saw the essence of cosmic processes in this way and no one else felt simplicity in the complex. No one else had such a super-light reaction at the perception of something new. No one so exactingly, thoughtfully, and mainly respectfully regarded the business with which his life was connected. He understood the complex fundamental importance of a problem to the uttermost and all the same returned to it many times to see if he could find out something more from a new point of view. I have not known another person who had so *spatial* an imagination. He always thought in three dimensions and hardly like the overwhelming majority of theoreticians who in the best case see the world projected on the piece of paper on which they carry out their calculations.

"A picture of all the grandiose multiple phenomena connected with solar activity that he created was always in his memory. The interstellar medium changed from a boring, static construction far from reality to a *living*, inhomogeneous, and continuously changing one, as it were, *breathing*.

"He was a great worker. This was in combination with such talent and fast reactive comprehension! The idea of inactivity was meaningless to him. He was continuously active like a machine. He divided his invaluable time into two unequal parts; the larger part was for limitless and constant help to people, far and near; the smaller part (which was large in absolute measure) went to science and creativity.

Almost everyone (but there were important exceptions) loved and exploited him. However, he did not imagine another life. I know that he did not relate to all people identically. However, no one noticed this because he had a huge reserve. For all the gentleness and faultless willingness to help everyone, he was a person with a purpose and, when the matter concerned the basics of our science and questions of ethics, was of the highest principles.

"He was anecdotally modest. His modesty was organic like that of Professor Shain who had a strong influence on forming the character of young Pikil'ner. In this regard they both resembled Chekhov,[†] known as people who perform nothing loud, pompous, or showy. This person, who was the pinnacle of our science, was turned down *five times* in elections in the USSR Academy of Sciences. Of course, this was not the first case in the history of this honorable establishment. The meaning 'Hamburg account' is not known to the majority of Soviet academicians. However, what is important is 'for all, for this one . . .'

"We will never see again his high inaudibly floating figure, his bashful smile, and his youthfully settled hair. No one asks anymore what you don't understand yourself. There is no more confidence that Solomon Pikel'ner has already examined it! We thought that it would be eternal. However, his ruthless and blind death was point less. Our sorrow is unlimited."

## References

1. Aksynov, E. P., et al. (1976) *Usp. Fiz. Nauk* **119**, 2, 377.
2. Shklovsky, J. S. (1976) *Quart. Journ. Roy. Astron. Soc.* **17**, 342.
3. Shklovsky, J. S., and Kaplan, S. A. (1976) *Phys. Today* **29**, 65.

---

[†] Anton Chekhov was a famous Russian writer.

# 4

# Reflections on the Soviet-American VLBI Program

## K. I. Kellermann
*National Radio Astronomy Observatory, Charlottesville, Virginia*

### INTRODUCTION

I met Joseph Shklovsky for the first time during my initial trip to the Soviet Union in 1965. I still have warm memories of this visit which began scientific collaborations and friendships that have continued for 25 years, including periods of great stress between the governments of our two countries. Although at the time, there was little formal interaction between Soviet and American radio astronomers, many of us were pursuing similar interests, but from different directions. Ultimately our paths would continue to cross in strange and unexpected ways.

The 1960's were exciting times for radio astronomy. At laboratories around the world, great strides in instrumentation were taking place. The use of large electronic computers, which was just beginning to play a major role in the development of radio interferometry, would soon lead to radio images of unprecedented angular resolution and image quality. The ability to detect remarkably faint radio sources promised, incorrectly as it turned out, to resolve the long standing debate on the interpretation of the radio source counts. Quasars has just been discovered and were being found at increasingly great red shifts, far surpassing those of any previously observed galaxy. After decades of searching, the first radio stars were detected in the 1960's, while radio and radar observations of the solar system turned up unexpected results on almost every planet. New techniques for radio spectroscopy were being developed that would verify Kardashev's (1959) prediction of recombination lines and Shklovsky's (1949) prediction of radio lines from interstellar molecules. The detection of the 3 K microwave background radiation by Penzias and Wilson (1965) would change, in a fundamental way, observational and theoretical cosmology. Finally, for the first time, serious scientists were contemplating techniques for interstellar communication with extraterrestrials.

By the mid 1960's radio astronomy was being increasingly concentrated toward centimeter and millimeter wavelengths with emphasis on large and expensive filled aperture steerable antennas and arrays or steerable antennas to obtain high angular resolution. The discovery of pulsars in 1968 revitalized low frequency radio astronomy. It led to the development of sophisticated instrumentation for the detection and analysis of periodic dispersed signals, and opened up a whole new area of research in astrophysics and relativistic physics.

The discovery of pulsars, quasars, interstellar molecular masers, and the CBR in the 1960's, as well as many of the earlier discoveries, such as Jovian and solar radio bursts and the pioneering work of Jansky, were serendipitous or accidental. This is in contrast to the popular image of science, often encouraged and nurtured by professional scientists and teachers of science, where progress develops logically from theory to experimental confirmation. As a new science based on a rapidly developing technological base, the history of radio astronomy has been very different. Neutron stars and the cosmic background were discussed long before their actual serendipitous discovery. However, theoretical considerations played little role in their discovery which were often the result of the right observations, but made for the wrong reasons. Personal and laboratory rivalries have played an important role in getting the important observations made in a timely way, and there are also examples of missed or greatly delayed discoveries because theoretical ideas pointed in the wrong direction.

Joseph Shklovsky was one of the first astrophysicists to turn his attention to the implications of the discoveries made by radio telescopes. His simple, but always elegant, papers have served as a framework for the interpretation of new observational discoveries and have suggested fruitful directions for new research. His successful prediction of the decrease in flux density of the Cas A supernova remnant was a milestone in radio astronomy which established synchrotron radiation as the basic radiation mechanism for non-thermal emission. Later he made the revolutionary suggestion that compact extragalactic sources such as PKS 1934-63 should vary on time scales of the order of a year. His theoretical formulation of the time variations from an expanding cloud of relativistic electrons has been the basis for all other work on radio variability (Shklovsky 1965). His classical paper on M 87 was the first to speculate that one-sided jets might be due to bulk relativistic motion (Shklovsky 1963). These pioneering papers provided much of the early inspiration for the development of Very Long Baseline Interferometry (VLBI) in the United States. It is fitting that his former students are now leading the effort to extend VLBI techniques to extraterrestrial baselines, and in a spirit characteristic of Shklovsky, they are already discussing further extensions to baselines of interplanetary dimensions.

In this paper I will trace the background of some of the events which have led to the long standing collaboration in VLBI between NRAO and Soviet scientists and discuss some of the technical, scientific, and social implications for extending these techniques in the next century to even higher resolution.

## RESOLUTION OF RADIO TELESCOPES

Perhaps the most important unanticipated development in radio astronomy has been the advancements made in high resolution imaging. Due to the long wavelengths involved, for many years radio telescopes appeared to be limited to poor angular resolution. In fact, the reverse is true; the long wavelength radio waves pass relatively unaffected through the terrestrial atmosphere while optical telescopes are limited by *seeing*. Moreover, because the precision needed to build diffraction limited instruments at radio wavelenghs is not as demanding as at optical wavelengths, radio telescopes may have essentially unlimited resolution. Since Karl Jansky's first observations in 1932, improvements in technology have dramatically increases the resolution as well as the sensitivity of radio telescopes. Sophisticated new techniques for analyzing radio interferometer data effectively eliminate any effects of image distortion from the atmosphere to give radio images with extraordinary quality and angular resolution which is several orders of magnitude better than available by any other technique on the ground or in space. The radio telescope is now the instrument of choice for high resolution and high fidelity images of many types of celestial objects.

In the mid 1950's Jodrell Bank (Morris, Palmer, and Thompson 1957) and CSIRO (Mills 1953) experimented with radio linked interferometers to obtain interferometer baselines of tens of kilometers. For the first time, the resolution of a radio telescope was better than that of an unaided human eye. The radio results were limited by the lack of any phase data and the inadequate number of baselines to form an image, but they showed that most radio sources were resolved and had dimensions greater than ten seconds of arc. Previously, it had been speculated that the observed cosmic radio emission might originate in unusually strong "radio stars" which were expected to be of such small angular size that interferometer baselines of thousands of miles would be needed to resolve them. Due to the lack of any stable independent local oscillator reference, Hanbury-Brown and Twiss (1954) developed the tape recording intensity interferometer which they had anticipated using on transatlantic baselines. But, the Jodrell Bank radio linked interferometers were successful in resolving nearly all sources with baselines of well under a hundred kilometers. At the same time improved position measurements were allowing radio sources to be identified with galaxies. the "radio star" hypothesis was being abandoned and with it the incentive for pushing tape recording interferometry techniques to very long baselines.

At that time, technical restraints limited the long baseline radio linked interferometers to meter wavelengths. At such long wavelengths, the most compact sources are opaque, and the observed radio emission comes primarily from larger scale structure. The success of the radio linked interferometers in resolving most sources and the growing evidence that they would all turn out to be galaxy-size radio galaxies discouraged the further extension of interferometer baselines beyond what could be achieved with cable or simple radio links. Further work on tape recording radio interferometers at Jodrell Bank was not pursued while further development of intensity interferometry was concentrated to optical wavelengths where high resolution was clearly needed to resolve stars.

Roger Jennison (1958) realized that some phase information could be extracted from the so called *phase closure* relations, even for interferometers with no phase stability. However, two decades were to pass before the computational techniques and computing power were available to produce milliarcsecond images from tape recording interferometer data (Readhead and Wilkinson 1978, Cotton 1979) and to exploit the full potential of connected element interferometers such as the VLA (Schwab 1980) and MERLIN (Cornwell and Wilkinson 1981).

## THE NEED FOR HIGHER ANGULAR RESOLUTION

Shortly before my 1965 visit to Moscow, Gennadi Sholomitskii (1965) had just announced the detection of rapid time variability of the radio sources CTA-21 and CTA-102. This very important discovery was not initially accepted in the West for several reasons:

1. The observations were made using an instrument designed for spacecraft tracking which was largely unknown to western scientists. Due to security restrictions, few experimental details were given in Sholomitskii's paper.
2. Observations of these sources at nearby wavelengths by others (including this author) did not show any evidence for flux density variations (e.g., Maltby and Moffet 1965).
3. It was recognized by many, including Sholomitskii and Shklovsky, that variability at surprisingly short time scales and at such a long wavelength (30 cm) and great distances was "theoretically impossible." This was, and still is, used an argument against cosmological redshifts. But in 1965, it was used as an argument against the observation (e.g., Maltby and Moffet, 1965).

About this time, Nicolai Kardashev was thinking about the technical capability of advanced extraterrestrial civilizations. He had speculated, on the basis of their peculiar spectra, that CTA-21 and CTA-102 might be artificial sources. Indeed the reported "102 day period" for CTA-102 implied a remarkable knowledge of the rotation period of the earth on the part of the extraterrestrials. These innovative speculations, which were widely reported in Pravda as well as in newspapers around the world did not contribute to the credibility of Sholomitskii's observations.

We now understand that the lack of confirmation by western observers was just bad luck. Bad luck for Sholomitskii because he did not receive proper credit for his important discovery, and bad luck for us because we did not pay sufficient attention to the

## 4. REFLECTIONS ON THE SOVIET-AMERICAN VLBI PROGRAM

important clues implied by Sholomitskii's data. It took about 15 years before it was recognized that CTA-102 as well as other sources do indeed do vary at decimeter wavelengths on time scales reported by Sholomitsky and by roughly the amount he claimed (e.g., Fanti et al. 1979). Unfortunately, it did not happen to vary when we were observing it at Caltech in the early 1960's.

The variability of extragalactic radio sources was soon confirmed by Bill Dent (1965) at the University of Michigan. He measured variations in the quasars 3C 273, 3C 279, and 3C 345 over a several year period. Dent's observations were at a much shorter wavelength and suggested somewhat longer time scales than Sholomitskii's CTA-102 observations, but they still challenged the cosmological interpretation of quasar redshifts or the acceptance of synchrotron radiation as the source of radio emission. It was clear that we needed to better determine the radio spectrum of the variations and if possible measure directly their angular size.

Together with Ivan Pauliny-Toth, I had planned to use the new 140 foot radio telescope to study radio source variability and to extend my earlier studies of radio source spectra to shorter wavelengths where the most compact sources might be relatively strong. After years of delays and frustrations the 140 foot radio telescope had just been completed at the then enormous cost of about 15 million dollars. Although the completion of the 140 foot antenna was to ultimately establish the NRAO as a "user facility," at this time there was relatively little interest among the potential users. We were fortunate in being able to obtain large amounts of observing time to determine the centimeter wavelength spectra of many sources and to follow their variability over a wide range of wavelengths. The only competition for observing time, came from Peter Mezger and his colleagues who had just confirmed the detection of Kardashev's recombination lines.

We were able to measure variability in a number of sources at centimeter wavelengths (Pauliny-Toth and Kellermann 1966). Although we did not see any variations in CTA-102, we did observe comparable rapid variations in 3C 120 and NRAO 140 at an even longer wavelength of 40 cm. We were able to interpret our observations in terms of a generalized form of Shklovsky's expanding source model. However, the apparent energy requirements were unrealistically large and it was difficult to reconcile conventional synchrotron theory with the expected effect of inverse Compton cooling of the relativistic electrons.

Arguments based on light travel times suggested angular sizes far to small to be measured by connected element interferometers. In addition, the discovery that some sources have remarkably high self absorption cutoff frequencies and showed interplanetary scintillation even at short centimeter wavelengths gave added support to the need for interferometer baselines of thousands of kilometers. About this time, there was a growing realization that the so called "mysterium" lines were due to cosmic OH masers, and that they also were likely to be so small that interferometer baselines would need to be extended to thousands of kilometers.

## THE BEGINNINGS OF SOVIET-AMERICAN VLBI

The first published discussion of an independent oscillator-tape recording interferometer were the papers by Matveyenko, Kardashev, and Sholomitskii (1965) and Slysh (1965). Their interest in Very Long Baseline Interferometry appears to have been aroused by the 1963 visit of Sir Bernard Lovell to the Evpatoria tracking station where he met Leonid Matveyenko and Gennadi Sholomitskii. They discussed the possibility of UK to USSR interferometry using tape recording interferometers. At that time however, high speed tape recorders and stable oscillators were not available in the USSR and there was little interest among senior Soviet scientists in pursuing these ideas. In the UK, Lovell's interests turned to the search for flare stars. A VLBI system was later developed at Jodrell Bank and used between Jodrell Bank, Sweden, and Puerto Rico, but the system was not reliable and was ultimately abandoned. The pioneering work begun by Hanbury-Brown, Twiss, Jennison, Palmer, Thompson, and Morris with the radio linked interferometers was not exploited until the development of MERLIN (Multi-Element-Radio-Linked-Interferometer) in the late 1970's.

In his book, *Out of the Zenith*, Lovell (1973) has suggested that the development of VLBI in the United States may have originated from his 1963 discussions in the USSR and were passed on by Shklovsky during his visit to the United States in 1968. But, by 1968, we had already completed a series of transcontinental as well as intercontinental VLBI observations between the United States and Europe and between the United States and Australia which confirmed the small size of AGN's and quasars. The presence of components that were still unresolved at milliarcsecond resolution, and the even smaller dimensions which were suggested by the rapid time variability and high frequency spectral cutoffs implied that observations at short centimeter wavelengths would be necessary to achieve sufficient angular resolution from the surface of the Earth. At that time the only radio telescope known to us outside the United States which could be operated at short centimeters wavelengths was the 22 meter reflector at Puschino.

In February 1968, Marshall Cohen and I wrote to Victor Vitkevitch at the Lebedev Physical Institute to inquire about the possibility of using the Puschino antenna for VLBI. When we did not receive any response we naturally assumed that our proposal was considered to be unrealistic so we were surprised when we received an encouraging reply nearly six months later. We later learned that Vitkevitch had passed our letter along to Leonid Matveyenko, was able to talk Soviet military and government authorities into giving permission for the proposed observations. Matveyenko informed us that the Soviet Academy had given permission to use the 22 meter antenna near Seimis rather than the one in Puschino. The reasons given at that time were related to the better weather in the Crimea and a more favorable baseline with respect to Green Bank. But, we suspected, correctly as we later learned, that the real reason for the proposed change was due to the sensitive location of the Puschino telescope not too far from Moscow.

Prior to receiving the reply from Moscow, we had not done any preparation in the

## 4. REFLECTIONS ON THE SOVIET-AMERICAN VLBI PROGRAM 43

United States to do VLBI with the USSR. We needed permission from the national Science Foundation to send the necessary equipment to the USSR as well as an export license from the Department of Commerce which required the approval of the various military and intelligence agencies as well as the State Department. At that time relations between the USSR and the US were not good, partially as a result of Soviet activities in Czechoslovakia, and the National Science Foundation initially expresses concern about the proposed collaboration. Even more concern was raised by US military and intelligence authorities who were worried about the military implications of accurately knowing the distance between Crimea and Green Bank. Fortunately about that time accurate geodetic data began to become available from satellite observations and so the VLBI observations were no longer critical.

When Shklovsky visited NRAO in October 1968 to present the second Jansky Lecture we traded stories of the negotiations which were taking place in both countries between the radio astronomy and military/intelligence communities. I think we both began to appreciate that the differences between Soviet and American scientists or between Soviet and American bureaucrats was less than the difference between the scientists and the bureaucrats in either of our countries.

Leonid Matveyenko and Ivan Moiseyev visited us in January 1969 to discuss plans for the joint observations. By that time the political situation in Europe had stabilized, and we were able to go ahead with preparing the observations which we tentatively scheduled for October. We received our export license in August. I arrived in Moscow on September 10, five days before the scheduled arrival of three thousand pounds of equipment and magnetic tape from Green Bank. It was an exciting period in history; only a few months earlier Apollo 11 had landed on the moon. I was not surprised to find that the American lunar landing was a topic of great interest to Soviet astronomers, but I was not prepared for the enthusiastic interest among all the Soviet people whom I encountered—taxi drivers, hotel and restaurant personnel, shopkeepers, etc. Everyone took pride in the "giant step for mankind." Souvenirs of the Apollo mission which I was able to obtain from the American Embassy and distribute to friends and colleagues were greatly appreciated.

## REFLECTIONS

It was with mixed emotions of frustration and accomplishment that we still recall the numerous logistical, bureaucratic, and communication problems that we repeatedly faced during those first experiments in 1969. One of our first tasks in preparing for the observations was to synchronize our atomic time standard to US Naval Observatory time. We had planned to carry our Rubidium clock to Leningrad where we expected to receive the Loran C transmissions from Germany. But this proved unsuccessful due to interference from a powerful secret "Soviet Loran" station as well as continual difficulties in keeping our clock running on several attempted airplane flights between Leningrad and Crimea. At one point, following repeated unsuccessful tries at keeping our batteries charged during the long trip between Crimea and Leningrad, we even speculated that our problems might be due to a different definition of "positive" and "negative" electricity in our two countries.

In the end we needed to be rescued by our radio astronomy colleagues in Sweden who braved a major storm to synchronize a running atomic clock at a Swedish defense laboratory. There was no time for the niceties of export licenses and customs declarations. A passenger ticket was purchased for the clock which travelled from Stockholm to Leningrad in a first class seat. I remember the look on the face of the Soviet customs agent at the Leningrad airport when he peered into the open wooden box to see a ticking clock and flashing light. He could not know that this indicated that our battery was about discharged and not that an *atomic* clock was about to be detonated. I panicked because I realized that we had to get to a source of commercial power immediately or else lose the time. But, the look of terror on my face was clearly misinterpreted by the customs agent who ordered us to take our equipment and leave immediately.

Communications between the US and USSR twenty years ago was difficult at best. Telephone calls took many hours or even days to complete and were often abruptly terminated. Our speculations that this was due to the intervention of the KGB or CIA were supported when one of our data tapes was "lost" in transit. I still have visions of teams of CIA and KGB intelligence agents trying to decipher our tape which contained a hundred million random *ones* and *zeros*.

My strongest memories of our 1969 experiments, however, are of the cooperation and support of many friends and government authorities who helped us through the complex bureaucracies to transport our equipment around the Soviet Union, to arrange telephone and telex connections, and to obtain last minute airplane tickets between Leningrad and Crimea. I also remember the moments of despair, when nothing seemed to be working, that were relieved by the early morning trips to the Yalta sherry bar with John Payne and Ivan Moiseyev. These early VLBI experiments between the United States and the Soviet Union were probably the first large scientific collaborations between our two countries that required the exchange of advanced instrumentation. They set a new record for angular resolution (Broderick et al. 1970), but, even more important prepared the foundation for our continued collaboration over the past two decades.

For about the past five years our joint program has been absorbed into the US and European VLBI network observations. No longer do we travel to distant observatories with clocks and tape recorders. No longer do we need to be experts on international telephone and telex communications to arrange for last minute changes in the observing schedule. No longer do we need to worry about customs and export licenses to ship our equipment and tapes among different countries. No longer do the individual scientists spend endless hours at the correlator trying to synchronize tapes or to find evasive fringes. Instead the observations are scheduled and organized by the VLBI consortia and the observations are made by the local staff who also arrange for the distribution of magnetic tapes. The data are correlated at the one of the VLBI processing centers in Germany or the US and analyzed using one of the several sophisticated software packages. But, sadly some of the problems which plagued us two decades ago such as crossed polarization or poorly recorded data are still common.

## RADIO SOURCE VARIABILITY AND SUPERLUMINAL MOTION

Much of the motivation for the development of radio interferometer with milliarcsecond resolution came from the desire to understand the apparent theoretical conflicts implied by the rapid flux density variations observed in distant quasars. The detailed images of the morphology and kinematics of quasars and AGN's which are now being made with the global VLBI networks are far beyond our simple expectations of a few decades ago. However, the theoretical problem remain, albeit in a more subtle form. Relativistic beaming models have been very popular, but still leave many unanswered questions. Indeed, the recent detection of unusually rapid flux density variation in some quasars such as 0917+624 have exacerbated the problems which concerned us following Sholomitskii's first observations of CTA-102 (Quirrenbach et al. 1989).

The VLBI observations have shown that superluminal component motion is very common in quasars and AGN's (e.g., Zensus and Pearson 1987). Components appear to move with apparent transverse velocities typically from five to ten times the speed of light, always in the direction pointing toward a large scale jet extending up to hundreds of kiloarcsecs away. This is generally interpreted to be the result of the finite signal propagation time from a suitably oriented relativistically moving source. When the relativistic motion is close to the line of sight, the source nearly catches up with its own radiation. This causes an apparent compression of the time scale and a corresponding increase in the apparent transverse velocity. Due to relativistic beaming, or Doppler boosting, of the radiation, the luminosity is enhanced by orders of magnitude for an observer oriented close to the line of sight. It is attenuated for an observer located in the opposite direction.

The idea that bulk relativistic effects might be important in radio galaxies and quasars appears to have been introduced independently by several workers. Shklovsky (1963) was the first to suggest that Doppler boosting from an intrinsically double jet, could cause the appearance of the apparent single sided jet seen in M 87. Later, Ozernoi and Sazinov (1969) interpreted the multi-peaked spectra observed in compact radio sources as the result of Doppler shifted "multiple components flying apart with relativistic speed". Both Rees (1966) and Woltjer (1966) pointed out that bulk relativistic motion would resolve the apparent inconsistency of the rapid time scales of the observed intensity variations and the minimum dimensions set by Compton cooling. In 1969, Ginzburg and Syrovatskii (1969) published their definitive discussion of the effect of bulk relativistic motion.

An early difficulty with the relativistic beaming interpretation of superluminal motion was the surprisingly large fraction of compact sources which show superluminal motion, whereas simple geometric arguments suggested that only a few percent of the relativistic beams should be fortuitously oriented near the line of sight. The presence of symmetric extended radio components suggested that they were supplied by energy from the *central engine* by two symmetric beams. But it was difficult to understand the comparable luminosity of the approaching and receding, or even stationary, components. Differential Doppler boosting should cause the luminosity of the approaching component to be greatly enhanced. This apparent discrepency is usually discussed in the context of the *twin exhaust model* (Blandford and Konigl 1979) where the emission from the core is seen at the stationary point where the approaching relativistic flow becomes opaque so the radiation is boosted by the same amount as that of the moving component. Superluminal motion is observed between this stationary point in the *nozzle* and moving shock fronts or other inhomogeneities in the relativistic outflow.

The so-called *unified models*, which interpret the variety of observed properties as simple geometric effects, have been only partially successful. In its simplest form, relativistic beaming models make straightforward predictions about observational relations between the apparent velocity and the Doppler boosting of the luminosity. Discussion has centered about the nature of the unbeamed or *parent* population. Scheuer and Readhead (1979) speculated that radio loud QSO's are the Doppler boosted subset of the much larger number of optically selected quasars, while Orr and Browne (1982) have considered that compact sources are the Doppler boosted components of extended radio sources. However, detailed observations of radio cores and jets are not completely compatible with the effects expected from simple relativistic beaming models (e.g., Kellermann *et al* 1989, Schilizzi and de Bruyn 1983, Saika 1981, 1984).

Compact superluminal jets are always aligned in the same direction as the more extended jets, including in some cases (e.g., 3C 273 and M 87) optical jets. Thus, the interpretation of the one sided appearance of compact jets as the result of differential Doppler boosting of an intrinsically two-sided relativistic flow would appear to require that the large scale jets are also moving with bulk relativistic motion. This would be surprising since it is difficult to see how the relativistic flow can continue without modification for up to hundreds of kiloparsecs away from the *central engine*. However, measurements of Faraday depolarization in front of extended radio sources indicate that the least depolarization occurs on the side with the jet, as expected if the jet is seen only on the nearest side due to differential Doppler boosting (Laing 1988). There is also more direct observational evidence in the radio galaxy 3C 120 that the relativistic flow continues out at least several kiloparsecs away from the AGN (Walker et al. 1988). It will be important to confirm this result.

A further complication is raised by the apparent continuity of properties from the radio through the IR, optical, and high energy parts of the spectrum. If the radio luminosity and morphology is primarily the result of bulk relativistic motion and orientation and are not intrinsic properties, then the observed characteristics at other wavelengths must also be interpreted in this way. But, quasars with misdirected beams should still have bright emission lines in the presence of weak continuum emission, and such *naked* quasars are not seen. Moreover, it is not clear how the subliminal sources or the sources which contain both stationary and superluminal sources (e.g., Pauliny Toth et al. 1987) fit into this simple scheme. Ironically, the compact radio jet in M 87 appears stationary and is certainly subliminal, although it was on this basis of this one-sided optical jet that Shklovsky first postulated the importance of relativistic beaming effects.

The relativistic beaming interpretation is also challenged by the AGN's extreme properties. Arp (1987) has emphasized the unlikely probability that the unique object 3C 120 should just happen to be properly orientated to show superluminal motion. Similarly, the quasar 3C 273 is unique; it is the brightest quasar in the sky at all wavelengths. The a priori that this unique object is properly oriented along the line of sight to show superluminal is small, unless of course the optical, IR, X-Ray, and $\gamma$-ray luminosity is also Doppler boosted. But, 3C 273 is unique even in the strength of its emission lines and it is difficult to imagine scenarios which allow the line emission to be enhanced by bulk relativistic motion.

The relativistic beaming model has provided an elegant mechanism to interpret the flux density variations, superluminal motion, and the absence of inverse Compton scattered x-rays. Although, the simple ballistic and continuous flow models make straightforward predictions that are not consistent with the observations, more complex models can be devised which can be brought into quantitative agreement with the growing wealth of observational material. This includes intrinsic differences in luminosity or relativistic flow velocity, dispersion in the direction of particle flow which may be greater than the observed opening angle of the jet, and differences in the velocity of moving shock fronts and the velocity of the particle flow so that apparent discrepancies between the observed superluminal motion and differential Doppler boosting may be relaxed (Lind and Blandford 1985). But, additional complexity means that the models lose much of their predictive ability and are thus less attractive or useful.

It is encouraging to note the AGN's and quasars with the greatest redshifts have the smallest measured rates of angular separation, consistent with a small spread in intrinsic velocity and a cosmological interpretation of quasar redshifts, (Cohen *et al.* 1988). More refined measurements may provide an independent measure of the Hubble constant.

## EXPECTATIONS AND SPECULATIONS

Within the next few years several dedicated radio telescopes and networks of radio telescopes will come into routine operation for VLBI. These include the ten element VLBA, three Italian VLBI antennas, three antennas in China, the Soviet QUASAR array, the Cambridge link between Merlin and the EVN, the Australia Telescope element at Mopra, and a dedicated VLBI antenna in Poland. These antennas will form powerful global VLBI systems with greatly improved (u, v) coverage having correspondingly improved dynamic range and image quality. The recent advances made in millimeter VLBI will be extended to include more antennas such as the VLBA and IRAM antennas to give greatly improved sensitivity with resolutions of the order of a hundred microarcseconds.

By the end of the decade, VLBI baselines will extend beyond the limits imposed by the size of the Earth to increase the available resolution of radio interferometers by an order of magnitude. Space VLBI will be very expensive and much less flexible compared with ground based VLBI. To an extent ground based millimeter VLBI may achieve the same resolutions as the planned centimeter wavelength space VLBI missions for a very much lower cost.

However, some critical observations can only be made from space. The brightness temperature limit of a radio interferometer depends only on the length of the baseline, independent of wavelength. For terrestrial baselines it may not be possible to measure values significantly in excess of the limit of $10^{12}$ K characteristic of incoherent synchrotron sources. Experimental space VLBI observations using the TDRSS satellite already suggests that some sources may exceed this limit (Linfield et al. 1989). For continuum radio sources, it is only from space that we can explore brightness temperatures in excess of the inverse Compton limit to make the measurement critical to understanding the importance of relativistic beaming. For OH and $H_2O$ masers, where the frequency is fixed, the resolution can be improved only be extending the baselines to space.

Ultimately, phase fluctuations due to the propagation through the interstellar medium may limit the maximum useful resolution that can be obtained. Even on terrestrial baselines, the effect of the interstellar medium can be a problem at low galactic latitudes. But at high latitudes and at millimeter wavelengths it seems baselines of the order of an AU might be used to obtain resolutions better than one microarcsecond. Even at longer wavelengths, it may be possible to use self calibration or adaptive optics techniques which have been used so effectively to reduce the effect of atmospheric phase fluctuations on terrestrial instruments.

Are there any microarcsecond sources in the sky? The inverse Compton limit for incoherent synchrotron sources of $10^{12}$ K for a one microarcsecond source corresponds to a flux density of about one microjansky. Observations with the VLA have detected about one source per arcmin$^{-2}$ above a limit of 10 microjanskys (Fomalont et al. 1988). Little is known of the properties of microjansky sources; essentially all of the powerful radio galaxies and quasars have much larger flux density.

The construction of space interferometers with microjansky sensitivity will be a formidable task, but does not appear to be beyond a reasonable extrapolation of current space engineering technology. The 21st century will no doubt see a space array of 100 meter or larger antennas, system temperatures close to the limit set by the cosmic background radiation and bandwidths of the order of a gigahertz, to give sub microjansky sensitivity with microarcsecond resolution. Some microarcsecond sources, such as pulsars, coherent emission from stars, Doppler boosted quasars, as well as some maser sources may have brightness temperatures well above the inverse Compton limit and could be sufficiently strong to observe with small space antennas working with large ground-based radio telescopes. Of special interest will be the quasars whose rapid variability indicate brightness temperatures as much as a million times greater than the inverse Compton limit (e.g., Quirrenbach et al. 1989). RADIOASTRON will extend the physical baselines of radio interferometry by an order of magnitude over what has been achieved from the surface of the earth. The small size of the antenna and the highly elliptical orbit will limit the sensitivity and imaging power of RADIOASTRON, but, it will be the first step to later much more elaborate space radio telescope systems of the 21st century.

As we stand on the threshold of the 21st century the prospects for further progress are exciting, but also sobering. On the positive side the VLBA will bring true multi-frequency imaging capabilities, including full polarization ability. Frequent monitoring of source kinematics will be possible on a scale which was inconceivable just a quarter of a century ago. The non-specialist will be able to exploit radio interferometry for a wide variety of scientific investigations including both imaging and the direct measure of distances throughout the Galaxy and to distant galaxies. Before the end of this century, RADIOASTRON will increase VLBI baselines by about an order of magnitude. Future space VLBI missions are being discussed in the USSR, Europe and the USA, could give orders of magnitude further improvement in both resolution and sensitivity.

However, with the completion of the VLBA and the anticipation of expanded space VLBI activity, it will be increasingly difficult to maintain the hands-on traditions that characterized the early growth of VLBI. The astronomers may become even more isolated from their instruments and their data. Space VLBI, in particular will be big science with the associated bureaucratic controls and administrative constraints. The space elements will need support from the traditional ground based VLBI systems as well as from the global complex of space tracking networks. The operation and use of the ground VLBI networks, which have been so very successful, may become increasingly under control of the space agencies with their tradition of strong program management and may become subordinate to the more costly and more visible space programs. A particular problem will be the processing of data recorded with different recording systems such as the one being built for the VLBA which is compatible with MK III system now in widespread use, the VCR based system being developed by Canada for RADIOASTRON, and the Japanese system being built to support VSOP.

## ACKNOWLEDGMENTS

It is a pleasure to recognize many discussions with Leonid Matveyenko, Nicolai Kardashev, and V. Slysh on the background of VLBI development in the USSR. The discussion given in this paper about the early developments of VLBI in the U.S. are based, in part, on a paper presented on the occasion of Grote Reber's 75th birthday (Kellermann and Cohen 1988) and I am grateful to Marshall Cohen for permission to use this material and for many discussions of the topics covered in this paper.

## REFERENCES

Arp, H. (1987) *Astrophys. and Astron.* **8**, 231.
Blandford, R. D. and Konigl, A. (1979) *Astrophys. J.* **232**, 34.
Broderick, J. J. *et al.* (1970) Astron. Zh. **47**, 784; *Soviet Ast.-AJ* **14**, 627.
Cohen, M. H., *et al.* (1988) *Astrophys.* J. **329**, 1.
Cornwell, T. and Wilkinson, P. (1981) *Mon. Not. Roy. Ast. Soc.* **196**, 1067.
Cotten, W. D. (1979) *Astron. J.* **84**, 1122.
Dent, W. A. (1965) *Science* **148**, 1458.
Fanti, R., *et al.* (1979) *Astron. and Astrophys Supl.* **36**, 359.
Fomalont, E. B. *et al.* (1988) *Astron J.* **96**, 1187.
Ginzburg, V. L. and Syrovatsii, S. I. (1969) *Ann. Rev. Astron. and Astrophys.* **7**, 375.
Hanbury-Brown, R. and Twiss, R. Q. (1954) *Philo. Mag.* **45**, 663.
Jennison, R. C. (1958) *Mon. Roy. Ast. Soc.*, **118**, 276.
Kardashev, N. S. (1959) *Astron. Zh.* **36**, 838; (1960) *Sov. Astron. A-J.* **3**, 813.
Kellermann, K. I. and Cohen, M. H. (1988). *J. Roy. Ast. Soc.* **82**, 5.
Kellermann, K. I. *et al.* (1989) *Astronom. J.* **98**, 1195.
Laing, R. (1988) *Nature* **331**, 149.
Lind, K. R. and Blandford, R. D. (1985) *Astrophys. J.* **295**, 538.
Lovell, Sir. A. C. B. (1973) *Out of the Zenith,* Harper and Row, New York.
Maltby, P. and Moffet, A. T. (1965) *Astrophys. J.* **142**, 409.
Matveyenko, L. I. Kardashev, N. S., and Sholomitskii, G. V. (1965) *Radiophysics* **8**, 651.
Mills, B. Y. (1953) *Aust. J. Physics. Res.* **6**, 452.
Morris, D., Palmer, H. P. and Thompson, A. R. (1957) *Observatory* **77**, 103.
Orr, M. J. and Browne, I. W. A. (1982) *Mon. Not. Roy. Ast. Soc.* **200**, 1067.
Ozeroni, L. M. and Sazinov, V. N. (1969) *Astrophys. and Space Sci.* **3**, 395.
Pauliny-Toch, I. I. K. and Kellermann, K. I. (1966) *Astrophys. J.* **146**, 634.
Pauliny-Toch, I. I. K., *et al.* (1987) *Nature* **328**, 778.
Penzias, A. A. and Wilson, R. W. (1965) *Astrophys. J.* **142**, 419
Quirrenbach, A. *et al.* (1989) *Astron. and Astrophys.* **226**, L1.
Readhead, A. C. S. and Wilkinson, P. N. (1978) *Astroph. J.* **223**, 25.
Rees, M. J. (1967) *Mon. Not. Roy. Ast. Soc.* **135**, 345.
Saika, D. J. (1981) *Mon. Not. Roy. Ast. Soc.* **197**, 1097.
Saika, D. J. (1984) *Mon. Not. Roy. Ast. Soc.* **208**, 231.

Scheuer, P. A. G., and Readhead, A. C. S. (1979) *Nature* **277**, 182.
Schilizzi, R. T., and de Bruyn, A. G. (1983) *Nature* **303**, 26.
Schwab, F. R. (1980) *Proc. 1980 Optical Computing Conf., SPIE Proc.* **231**, 18.
Shklovsky, J. S. (1949) *Astron. Zh.* **26**, 10.
Shklovsky, J. S. (1963) *Astron. Zh.* **40**, 972; (1964) *Soviet Astron. A-J* **7**, 972.
Shklovsky, J. S. (1965) *Astron. Zh.* **42**, 30; (1965) *Soviet Astron. A-J* **9**, 22.
Sholomitskii, G. (1965) *Astron. Zh.* **42**, 673; (1965) *Soviet Astron. A-J* **9**, 516.
Slish, V. I. (1965) *Usp. Fiz. Nauk* **87**, 471.
Walker, R. C., et al. (1988) *Astrophys. J.* **335**, 668.
Woltjer, L. (1966) *Astrophys. J.* **146**, 597.
Zensus, J. A., and Pearson, T. J. (1987) *Superluminal Radio Sources*, Cambridge Univ. Press, Cambridge.

# 5

# The Secrets of the Solar Atmosphere

## H. Zirin
*Big Bear Solar Observatory, California Institute of Technology*

## 1. A LITTLE HISTORY

While the solar photosphere is well-observed and reasonably well-understood, the atmosphere above is transparent and not easily observed. This is of course true of all stars, but we know so little about the more distant objects that stellar astronomers are not so concerned about the the atmospheric details, while solar astronomers spend much of their time trying to grasp what goes on in the transparent part of the solar atmosphere. When I first met Shklovsky and Pikel'ner in the back of the Hall of Columns in Moscow at the IAU grand opening in 1958, we immediately started an argument on the temperature of the corona, preventing bystanders from hearing the ponderous welcome speeches taking place in the Hall. For this I feel little guilt; the ritual of pompous welcoming speeches is surely a relic of medieval times. I thought at that time that the chromosphere was hot, mostly because the generally accepted view was that it was cold. Shklovsky explained to me that the chromosphere could not be hot because the radio emission was too low. I, hardly understanding the significance of radio emission, ignored this fact. So have other solar astronomers, although Athay and Thomas made it clear in their text, but without realizing the upper limit it placed on the emission measure of the chromosphere.

The tale begins with Lockyer's discovery of the lines of the unknown element now known as helium in the chromospheric flash spectrum, and Grotrian's identification of the coronal lines many years later. Until then I guess people felt the temperature of stellar atmospheres simply dropped off to absolute zero. These spectral properties told us that the corona was at the remarkable temperature of a million degrees (until then the highest temperature actually recorded in the cosmos). So if the corona was so hot, it was reasonable that a chromosphere should lie somewhere between. The solar element is very hard to excite, and a hot chromosphere provided a natural explanation for the extraordinary limb-brightening of helium; the D3 line is invisible against the disk and one of the strongest lines in the chromosphere. So we found a room at MGU (Moscow Univ.)

and argued and argued. I think everything I said was wrong, and I'm not so sure that Pikel'ner and Shklovsky were correct, either. But we learned by arguing, and the critical role of the radio observations stayed with me. Of course my two Russian colleagues understood the radio much better; as Shklovsky relates in his memoirs, he and Ginzburg had made a memorable trip to Brazil in 1946 to understand the solar radio emission.

In the following years a series of discoveries shed some light on the structure of the atmosphere. Leighton and his students showed that there was a network associated with a supergranular photospheric velocity pattern, with magnetic fields clustered at the edges. Cragg, Howard and Zirin (1963) showed that the spicules, high-velocity jets into the atmosphere, were clustered in these places. Parker (1958) showed that the only stable solution for the solar corona was a steady outward expansion; thus the Sun's atmosphere was connected to the interstellar medium. In 1962 I tried to fit the newly measured XUV spectrum of the sun into the picture (Zirin and Dietz, 1963) and found to my astonishment that Shklovsky was right; the high temperature component responsible for the beautiful lines of CIV, OV, etc., was a trivial fraction of the solar atmosphere. The heating of the chromosphere was much less of a problem than we thought, although the temperature still reversed and the corona was indeed quite hot. The back of an envelope should have told us that; these great resonance lines are no stronger than the coronal resonance lines formed at much lower densities. Since a single scale height of a barometric atmosphere contains as much material as all the overlying layers, the transition zone must be (allowing for the $N^2$ factor in emissivity, less than one chromospheric scale height). Of course, we had the spicules, and it was generally felt that these were the source of the high-excitation emission.

Despite this evidence model after model that completely violated the upper bounds set by radio and UV poured out of the modelling computers. This has been primarily a result of NASA policy; UV data came from spacecraft, and its analysis was liberally funded; no one worried much about solar radio emission, especially from the quiet Sun. In a science where complex structure plays such a great role, the divorce between observation and theory is particularly unfortunate, especially when coupled with a narrow outlook. Several quite competent solar radio astronomers admitted to me that they knew nothing of the XUV, and the reverse is obviously true of the XUV modellers.

But what of helium? If the weak UV lines signal a tiny transition zone, how do we explain the enormous strength of the helium lines? Both Goldberg and the Nikolskys had suggested that UV fluorescence was the source for helium, but a quantitative estimate was lacking. In 1974 I got a look at the chromosphere in D3 through our new universal filter. A picture is worth a thousand words and many calculations; I saw a *uniform* bright band elevated about a thousand km above the photosphere and disappearing in coronal holes. The helium was much lower that the spicules (which extend to 1000 km and show only a weak height gradient). The dark band below the He layer of course marked the levels to which the UV could not penetrate.

But how could the weak UV flux produce the huge D3 flux (and that of the other He lines)? Because the D3 emission is really resonance scattering of photospheric photons. All the UV has to do is get the He atoms up to the triplet levels and the photosphere does the rest. This picture is confirmed by the absence of He emission under coronal holes and the sharp limb brightening of the He lines.

## 2. WHITHER THE CHROMOSPHERE?

While the foregoing means that our picture of the chromosphere has improved considerably since the days of Joseph Samuilovich and Solomon Borisovitch, the truth is that our understanding leaves much to be desired. And we must face the fact that if we don't understand the solar atmosphere, we don't understand the atmosphere of any star. Most astronomers, including Shklovsky, solved this problem by walking away from problems of the Sun and turning to galaxies; but the fact remains that the galaxies are made of stars, which we obviously understand less than the Sun.

Foremost is the question of the heating, a subject that occupied Pikel'ner for much of his life. After the significance of the network became known, it was easy to guess that the network boundaries were the probable locus of heating. Of course the first "overlappograms" from Tousey's instrument, as well as the Harvard results confirmed this. In fact, it is fairly clear that heating takes place wherever the magnetic field is vertical and relatively strong, a fact that led Pikel'ner to propose his models of Alfven wave heating of the chromosphere. But the Skylab X-ray pictures revealed that trapping is as important as heating; the solar wind carries away material in open field lines, so much of the corona we see is concentrated in closed field geometry not necessarily because it is heated there, but because it escapes otherwise.

Shklovsky always sought to utilize new wavelength ranges to probe the universe. Usually they are first available for the Sun because it is the easiest first target. For various reasons this new data takes years to enter the mainstream. For example, in recent years an entire new wave-length range, with a series of emission lines in the 12 m region were discovered by infrared astronomers. These were explored by Noyes and Brault (1983), Deming et al. (1988) and Zirin and Popp (1989). In addition, important balloon spectra have been obtained by Clark and co-workers from balloons in the hidden region between 12 and 800 $\mu$.

The emission lines are due to recombination to n = 7 of MgI, SiI, H, He and other elements. These lines show strong secant-type limb brightening which produce severe constraints on the atmospheric model:

1. The temperature must increase upwards at the level where lines are formed.
2. Because the line is 1.4 times brighter than the continuum at the limb, it must be formed at a temperature no less than 1.4 times hotter. If the temperature commonly quoted for 12 $\mu$ is 5000°, the temperature for the 12 $\mu$ line is 7000°. More likely the

quoted temperature is wrong (like almost every other sacred source), and the temperature is only 5600° (Zirin and Popp).
3. The density of formation must be fairly high, or else no emission line will be formed. This requires a density $N \approx 10^{12}$. This would require a temperature of 7–8000 deg at a height of 300–400 km above the surface.

The secant distribution generally requires that this rise occur everywhere, and observations show the lines to be present in emission everywhere. However, the resolution of the Fourier Transform Spectrometer at Kitt Peak, where these measurements are made, is severely limited, and it is possible that an ensemble of hot spots might fit the data. In this case the temperature in the hot spots would be proportionately greater.

The behavior of the lines further in the IR shows the expected rising temperature pattern. While the MgII lines are much stronger than the H lines at 12 μ, the H lines rapidly dominate in the lower wavelengths; the H lines up to n = 14 have now been observed. That means that the temperature is rising from the point where the continuum around 15 μ is formed, because the lines at 12 μ show both absorption and emission.

Early radio maps of the Sun were made by turning off the RA drive of the telescope and letting the Sun drift across the beam. This almost always produced a scan with "ears", spikes near both limbs. This effect was thought to be due to the upward increase of temperature, but in fact was due to the increase of projected active region distribution toward the limb. In 1968 Mike Simon pointed out to me that the improved radio telescopes, which could scan N-S or separate out the active regions, did not show the ears. Pursuing the matter, we soon found that there was almost no evidence for limb brightening (Simon and Zirin 1969).

Well, there must be *some* limb brightening; the presence of emission lines can only mean that the temperature increases upwards, and we must see higher at the limb. With the development of the yet more sensitive radio telescopes, this proved to be the case; limb brightening was observed at some wavelengths, however it was orders of magnitude less than earlier expectations.

At present most solar physicists place their faith in the so-called "VAL model", a valiant and massive attempt to model various regions of the atmosphere on the basis of different elements of the UV data. This model would produce radio temperatures twice those observed, absorption rather than emission on the 12 μ lines and other disagreements with observation. Nonetheless, a young astronomer recently told me he had full faith in VAL "because it was such an extensive and thorough calculation." Maybe so, but it does not fit the chromospheric data.

Recently we (Zirin, Hurford and Baumert 1990) used the Owens Valley Solar Interferometer to measure the quiet Sun radio temperature for 1–18 GHz. The temperature was measured many times, and the data calibrated to the Moon and point sources. The

## 5. THE SECRETS OF THE SOLAR ATMOSPHERE

resulting data fits well the formula

$$T_b = 141635 \, v^{-2.1} + 11000. \tag{1}$$

Below 10 GHz the spectrum is that of the optically thin million degrees corona, corresponding to a base density of $3.2 \times 10^8$ cm$^{-3}$ and a temperature $10^6$ deg. At higher frequencies the spectrum is flat at about 11000°, reflecting the true temperature of the upper chromosphere. Actually the brightness decreases from 11500° at 10.6 GHz to 10460 at 18 GHz, and one may assume it continues down to the temperature minimum. There is no evidence for a transition region. This brightness temperature is about a third of that which would result from the common chromospheric models. The UV calculations are not really wrong, but the form factors are different. As first suggested by Grebinskii (1987), the radio brightness is a linear function of the temperature, and the UV intensity, an exponential one. A limited fraction of hot elements will produce most of the UV with little effect on the radio.

Why have we not worked out, from such a great mass of data, the general parameter of such a close star? Partly because as recently as a few year ago, the fine structure of the atmosphere was neither recognized or resolved, and partly because it is hard to make models that take this structure into account. Further, the interface of surface and atmosphere is a sea of spicules whose temperature is still really unknown. And finally, most important, hardly anyone works at this field anymore; those that do crouch over the keyboards of their computers, hoping that the silicon chips will emit some truth about the star that the chips know little about.

## 3. THE CORONA AND THE SOLAR WIND

Parker (1958) showed that conductive heating of the outer corona was so strong that the atmosphere had to flow outward in the solar wind. That wind has now been found in many stars, and a peculiar problem arises. As Parker showed, in the absence of restraint the corona should rapidly flow outward, and, unless we have a source, the corona should disappear. This is in fact seen; when the magnetic field structure is open, a "coronal hole" forms, marked by high interplanetary wind speeds, and little or no corona near the Sun. The corona is observed only where we find closed magnetic field loops, which prevent it from flowing away. Shklovsky didn't believe this result, and thought Parker had made a mistake. I didn't believe it because the material flux proposed by Parker was far too high; but measurements showed that the interplanetary density near the Earth was much lower than formerly thought, and Parker's wind was really there, with lower flux than he had initially expected.

The existence of coronal holes shows that the sun cannot maintain the normal density in the free outward flow, and probably overproduces in the closed-field regions, whence there probably is some escape. Thus we cannot distinguish heating from containment by simply looking for concentrations of corona.

An astonishing aspect of coronal holes was revealed by Arthur Babin (1974) with the Crimean millimeter telescope, an instrument in which Shklovsky had a great interest. Coronal holes are less dense and probably cooler than the normal corona, and appear darker in most wavelengths, including long-wave radio. But at 3 mm, coronal holes are *brighter* than the average Sun. When this result first appeared I didn't believe it; when confirmation was obtained elsewhere, I still didn't believe it, until the big Nobeyama dish confirmed the result with beautiful images. Radio waves at this wavelength arise in the chromosphere. I know of no reason why they should be more intense in coronal holes.

Another surprise in the atmosphere is the fractionation of elements according to their first ionization potential (FIP). In 1986 Ed Stone explained to me that measurements of abundances in solar energetic particles (SEP) showed an underabundance of about an order of magnitude of high-FIP elements (such as C, N, O, Ne, A) in SEP from flares (Breneman and Stone 1985). Ionization and other data suggests these nuclei are swept up in the corona by the flare shocks. Meyer (1985) immediately made the reasonable proposal that the outer solar atmosphere is populated by dynamic processes that require ionized particles.

How Solomon Borissovitch would have enjoyed this news! A basis for a whole new model of the chromosphere-corona, based on the definite separation of charged from neutral particles! And now the material returned to the interstellar medium is not a "photospheric abundance", the "bonne a touts faire" of astrophysics, but a stream rich in Si, Mg, Ca. Nothing is said about the ultimate high-FIP element, He, which has its own wild variations.

Well, we shouldn't need SEP measurements to detect abundance anomalies in the solar atmosphere, all the studies of the past 50 yrs must have produced *some* evidence of height variations in the abundance. So I looked. Back in the old days there were many expeditions to measure the "flash spectrum" at eclipses; at HAO, where I worked, it was an endless preoccupation, permanently tying up shops and staff. Despite an immense effort, the data were not very consistent because data had to be obtained by differentiating photographic material. Also the flash spectrum is dominated by low-FIP lines; the only suitable high-FIP lines are the infra-red oxygen triplet. Somebody should measure then with a CCD at an eclipse, compared to the CaII IR triplet or something like that.

Although the high-FIP lines are rare in the flash spectrum, the solar UV is rich in them. And the calculations by Withbroe and others of abundances immediately reveal the underabundance of C, N, O, and Ne. Some of this has always been attributed to the sharpness of the transition zone, but it is now clear that much is due to the filtering out of high-FIP elements. For example, O is much more abundant than Mg, and the OVI line is produced at a much higher density than MgX. One would expect the OVI line to be 10 times stronger, yet the intensities are similar.

But UV lines are poorly calibrated. Can we find any coronal lines representing the high-FIP elements? It happens that of all the high-FIP elements only argon is represented in the spectrum. The only high FIP element for which was forbidden lines are observed is Argon, for which two lines (ArX λ5539 and ArXIV (686 eV; λ4412) are observed in the visible and others in the UV. λ4412 was observed by Zirin (1964) in a limb flare and gave an abundance similar to calcium. Jefferies et al. (1971) got a similar result at an eclipse. This data strongly suggests no differentiation in a flare.

In the quiet corona we have only the ArX line at λ5536 which has been copied from one table of coronal lines to another for years and in fact does not exists. Edlen (1983) predicts a wavelength of 5534 Å and lists (1942) a line at 5536 Å. Modern work moves the line to 5539 Å (Jefferies et al. 1971; Aly 1955) but in both cases it is marked as too weak to measure. There is good reason to suppose that the ArX line has never been observed, and certainly it is weaker than it would be with "photospheric abundances" (which are unknown in any case for Ar and Ne). In the UV and XUV one finds a weakening of the A lines except in flares.

By comparison, intense flares producing nuclear gamma rays are reported not to exhibit the underabundance of high-FIP elements, presumably because they dredge up unfiltered surface material and deliver a fair sample of photosphere. This conclusion is based mainly on SXR lines produced in the thermal phases of the flare and implies that the post-flare thermal cloud is the result of "evaporation" of material from the surface by non-charge-dependent processes.

What picture can we now draw of the continuous creation of the solar atmosphere by these processes? It confirms the role of the spicules, jets which occur only in regions of vertical magnetic field concentrated by the supergranulation flow. It means the atmospheric material is not the result of hydrostatic equilibrium processes, but of the dynamic injection of material. Possibly if we could distinguish the intranetwork and network regions in the chromosphere we might see some difference in relative abundances. Do the active regions play a role? The variation of the high-FIP abundances in the solar wind is not well-determined so we cannot say. But the chromosphere was long recognized to have a hydrostatic and ballistic part, the latter identified with spicules, which probably are the source of the corona as well. None of this is new, but the fractionation by charge clearly confirms this remarkable process. Where and how the heating enters this process remains to be understood.

Since solar winds represent the basic interchange of stellar and interstellar material, it is possible that this process is important in understanding abundances in the interstellar medium.

## 4. THE PROBLEM OF CORONAL HEATING

When I first met Pikel'ner and Shklovsky we always argued about the temperature of the corona. Shklovsky, who had independently developed the ionization theory, believed its result that the coronal temperature was around 700,000°. I, who had measured the temperature by line widths, believed the latter, which gave temperatures of 2 million degrees. I think Pikel'ner did, too. The problem was solved when Unsold suggested, and Burgess showed, that the obscure process of dielectronic recombination had to be included in the ionization calculation, and then the problem went away. Today this process is included in *all* ionization calculations.

Unfortunately little progress has been made in the theory of coronal heating since Pikel'ner's last attempts. This is for several reasons: theorists, even Pikel'ner, went off into more cosmic problems; most of the most important models have been modeled, and, most important, little progress was made in formulating the observational tests that would determine the suitability of one or another model of coronal heating.

So the typical paper on the subject contains a series of impenetrable equations always ended by the conclusion that this heating mechanism will work (If the theoretician had not convinced himself it worked he would not have submitted it) but no way the innocent reader could determine whether it should be substituted for the half-dozen other models. So I will here endeavor to outline some of the observational facts of coronal heating that may be relevant to understanding it.

First, like the cold virus, there seem to be several kinds of coronal heating. The corona is hot in active regions and hotter still in regions of flux emergence, which must be associated with magnetic field reconnection, since a new magnetic field replaces an old. Of course, in flares we get particularly great heating, to temperatures about $3 \times 10^7$ deg. Yet the quiet sun corona, though cooler than that in active regions, is still hot, in the 1,000,000° range. Thus under all kinds of conditions it is possible to heat this remarkable medium.

Is it possible that the outer regions of any stellar envelope radiate so poorly that there always are plasma phenomena that will heat it? Do we know of any star where the atmospheric temperature actually drops to a low level? If, as has been suggested by Ostriker, the interstellar material is at coronal temperatures, is there a boundary condition that establishes a high temperature?

Second, there is a remarkable relation between density and temperature in the corona. While one would expect constant pressure to be observed (and popular, incredibly wrong models of the transition region require pressure balance), one finds just the opposite; density is roughly linearly proportional to temperature. In the quiet corona we have $N = 5 \times 10^8$, $T = 800,000$. In post flare condensations, $N = 2 \times 10^{10}$. Of course this is not a natural law. It must flow from the fact that the heating and containment are both tied up with the presence of strong fields, and also with the source of dense, cool material below that boils off when high temperature is available. For this reason it is most important that X-ray images be obtained in several wavelengths so that temperatures can be deduced.

Considerable data on spatial distribution is now available. There are X-ray images giving the distribution of emission measure, radio images giving the integral of density times optical depth, and the famous Skylab slitless spectrograms showing loops with FeXV emitting at the top and NeVII at the base. Similarly, the Skylab X-ray images allow us to relate the emission measure of the quiet sun elements to that in active regions. The radio data makes this possible with a different dependence on N and T, so definite conclusions are possible. There is no evidence for any network structure in either case; it is believed, but not proven, that field lines spread out from sources in the network. The coronal brightness is sufficiently uniform that there is no possibility of diffusion from the active regions.

One important question in understanding the quiet sun coronal heating is the possibility of an "infrared catastrophe" of tiny active regions heating the corona with a mechanism similar to that in active regions. The X-ray images, which exaggerate contrast, show a range of bright points which may indeed be part of a continuum. Unfortunately the counts of bright points have been subjective, with no use of quantitative limits. Thus one cannot determine whether the bright points extrapolate to a general background. It is certain that the more intense points are bipolar, and radio images show somewhat weaker sources associated with the AR dipoles, but nothing with the random bipolar structures in the chromosphere. But if we follow a region of quiet Sun for a day there are always some small eruptions, which may be capable of producing the heating. Of course the main heating is in active regions, where flares do the job nicely.

For a long time theories of shock-wave heating of the corona were popular, even years after it became evident that the chromospheric and coronal heating took place only in the regions of strong magnetic fields. Then models of magnetic heating began to appear. Again, few are persuasive. The dominant factor is the inability of the thin corona to radiate the input energy.

## 5. THE STARS

Long ago Sir William Huggins wrote, referring to the news of Kirchhoff's interpretation of the solar spectrum "Here at last presented itself the very order of work for which in an indefinite way I was looking—namely, to extend his novel methods of research upon the sun to the other heavenly bodies."

What has solar research in recent years taught us about the stars? While there has been considerable development in the synoptic observation of late-type stars, in general this work has been oriented toward fairly simple measurements, such as the Ca K line. So far I know there has been no special effort to compare the K profiles with those of solar regions of different magnetic intensity. The important comparisons of activity with rotation rate have served to give a picture of the evolutionary state of the sun. Observations of the neutral helium lines in late stars have shown them to be well-correlated with X-ray flux, confirming the idea that the helium levels are excited by coronal back-radiation, and giving a wide sample of coronal activity in the stars.

There is an interesting contradiction in our concept of the effects of stellar activity on the stellar input to the interstellar medium. It has been found that the strongest winds are associated with open magnetic fields, coronal holes, which permit the free escape of coronal material. But the observation of strong X-ray emission or He absorption in a star means not only enhanced activity, but closed field lines, hence no stellar wind. It is also known that, coronal holes are most prominent on the sun when there is some activity: not too much, not too little. While the data surely exists, there is no published conclusion on the overall dependence of solar-wind flux on the level of activity.

Recent research shows that solar mass ejections, spectacular coronal eruptions associated with flares and prominence eruptions, are at least as great an input to the solar wind as the steady flow. This question becomes particularly interesting when one considers the vast number of M dwarfs in the Galaxy, of which at least half are flare stars far more active than the Sun.

Finally, a small number of stars display big outflows: huge radiation-driven winds are found in O stars, which are relatively rare, and low-temperature K giants. In studying the He 10830 line in stars, I found a number of cases of cool outflow, dense winds with P-Cygni type profiles flowing with velocities of a few hundred km/sec. If element separation occurs in any or all of these outflows, and if this source is significant compared to supernovae and other cataclysmic sources, it will have to enter our picture of the abundance balance between stars and the interstellar medium out of which new stars form.

So the stars indeed have personalities, but for the time being we will have to understand them through our own star.

## REFERENCES

Babin, et al. (1974) *Ivz. Krym. Astrofiz. Obs.* **55**, 3.
Brault, J. W. and R. Noyes (1983) *Ap. J.* **269**, L61.
Breneman, H. H. and E. C. Stone (1985) *Ap. J.* **299**, L57.
Cragg, T. et al. (1963) *Ap. J.* **138**, 303.
D. Deming et al. (1988) *Ap. J.* **333**, 978.
Grebinskii, A. S. (1987) *Astr. Zh. Lett.* **13**, 299.
Jefferies, J. T. et al. (1971) *Solar Phys.* **16**, 103.
Lindsey C. et al. (1986) *Ap. J.* **308**,448.
Meyer, J.-P. (1985a, b) *Ap. J. Supp* **57**, 151, 173.
Parker, E. N. (1958) *Ap. J.* **128**, 669.
Simon, M. and H. Zirin (1969) *Solar Phys.* **9**, 317.
Zirin, H. (1964) *Ap. J.* **140**, 1216.
Zirin, H. and R. D. Dietz (1963) *Ap. J.* **138**, 664.
Zirin, H. Baumert, B. M. and Hurford, G. J. (1990) BBSO#0314, submitted to Ap. J.
Zirin, H. and Popp, B. (1989) *Ap. J.* **340**, 571.

# 6

# Escape Processes in Planetary Atmospheres

## V. I. Moroz
*Space Research Institute, Moscow*

## INTRODUCTION

One of the most important factors in the evolution of planets is the escape of their gaseous atmospheres, that is, the loss of their atmospheres by removal into interplanetary space. We present a review of escape mechanisms and the basic data that relate to this problem. A significant contribution to the development of escape theory was made by S. B. Pikel'ner and I. S. Shklovskii who are the inspiration for this article.

There are a number of other reviews of this subject. For example, no sooner had the excellent lecture of Hunten (1990) been published than it was awarded the Kuiper Prize. We present here another viewpoint and introduce new facts in the hopes that this review will also be useful.

## 1. FIRST STEPS; UP TO JEANS

For a long time, only one form of atmospheric escape was known, namely thermal "evaporation". The basic idea is very simple. The molecules of an atmospheric gas follow Maxwell's distribution and some fraction of them (possibly small and possibly very large) have velocities larger than the ballistic one. These molecules are able to escape permanently into space.

The first scientific work on the escape of planetary atmospheres was most likely that of Waterton in 1846. He presented his paper to the Royal Society, but his work was dismissed, and it was not published until 1892.[†]

Soon after its publication, Stoney (1898,1900) and Cook (1900) correctly explained, by thermal escape, the lack of a lunar atmosphere and the deficit of hydrogen and helium in the Earth's atmosphere. However, they were unable to make quantitative calculations for the flow of escaping atoms.

---

[†]For details on this subject, see Chamberlain (1978), the bibliography to § 7.1.

For definitiveness, we need to mention that the problems we shall discuss here are primarily concerned with the atomic components of atmospheres and the term 'molecule' does not mean only multi-atom or diatomic molecules. The term "molecule" is simply taken in gas kinetic theory for all cases and we shall do the same where theory is concerned.

## 2. THERMAL ESCAPE ACCORDING TO JEANS

The main difficulty lay in the fact that nothing was known about the composition of the Earth's upper atmosphere, let alone that of other planets. Jeans (1916) saw a way that almost did not require this knowledge and which allowed one to obtain correct estimates (though only to the accuracy of several orders of magnitude) of the rate of loss of atmospheric gasses. We will reproduce his reasoning here.

For a Maxwellian distribution, the number of molecules that have a velocity magnitude in the range u to u + du is

$$dn = f(u)du = \frac{4n}{\sqrt{\pi}} e^{-y^2} y^2 du, \qquad (1)$$

where n is the density, $y = u/u_o$,

$$u_o = \sqrt{\frac{2kT}{m}} \qquad (2)$$

is the most probable velocity of the molecules, m is the mass, and T is the temperature.

We assume that the atmosphere is isothermal and spherically symmetric. In this atmosphere, we construct a boundary surface with radius $R_c$ such that it satisfies the following two conditions. Firstly, for the molecules that cross this boundary moving outward, the probability of collision with another molecule is far less than one. Secondly, the Maxwellian nature of the velocity distribution is preserved on this surface. We call this altitude the critical level, and will not concern ourselves at this point with the question of how to relate it to a specific height. We note parenthetically that these two requirements are mutually contradictory, but that this contradiction does not lead to large errors.

The escape velocity at a distance R from the center of a planet is given by

$$u_e = \sqrt{\frac{2\gamma M}{R}}, \qquad (3)$$

where M is the mass of the planet and $\gamma$ is the gravitational constant.

The flux of molecules moving outward through a unit surface at the critical level with

## 6. ESCAPE PROCESSES IN PLANETARY ATMOSPHERES

a velocity that is higher than the escape one and that leave the planet is equal to

$$F = \frac{n_c}{4\pi} \int_{u_e}^{\infty} \int_{1}^{2\pi} \int_{0}^{\pi/2} uf(u) \sin\varphi \cos\varphi \, du\,d\psi\,d\varphi, \qquad (4)$$

where $n_c$ is the density of molecules of a given type at the height $h_c = R_c - R_o$, and $R_o$ is the planetary radius.

Integrating Eq. (4) gives

$$F = \frac{n_i u_o}{2\sqrt{\pi}} f(X), \qquad (5)$$

where

$$f(X) = e^{-X}(1 + X), \qquad (6)$$

$$X = \left(\frac{u_e}{u_o}\right)^2 = \frac{\gamma M m}{kTR}. \qquad (7)$$

In the theory of evaporation of fluids and solids, the equation that applies has the same form as Eq. (5) but uses $u_e = \sqrt{2L/m}$ where L is the heat of evaporation. The analogy between these two cases is deep and obviously physical. In fact the escape process is sometimes called the evaporation of planetary atmospheres.

The density at height $h_c$ can be written in terms of the density at the surface of radius $R_o$

$$n_c = n_o \exp\left\{-\frac{\gamma m M}{kT}\frac{h_c}{RR_o}\right\} = n_o \, e^{-X_o + X}. \qquad (8)$$

This barometric equation includes the dependence of the accelerating force of gravity on height in the gravitational field of a spherical body. Substituting Eq. (8) into (5) gives

$$F = \frac{n_o u_o}{2\sqrt{\pi}} e^{-X_o} (1 + X) \qquad (9)$$

The quantity $R_o$ in Eq. (8) can refer to either the solid surface of a planet or to any atmospheric height where the density is known. The quantity $X_o$ in Eq. (9) is obtained from Eq. (7) with $R = R_o$. Jeans serendipity lay in his omitting from his analyses the unknown molecular densities at the critical level and then showing that the knowledge of this height was not very important. In fact, it affects only the linear multiplier $1+X$ in Eq. (9).

The total number of molecules in a spherical, isothermal atmosphere is

$$N = 4\pi R_o^2 n_o H, \qquad (10)$$

where

$$H = \frac{kT}{mg} = \frac{u_o^2}{2g} \qquad (11)$$

is the scale height and g is the acceleration of the force of gravity.

The number of molecules that are lost from the atmosphere per unit time is

$$\frac{dN}{dt} = 4\pi R_c^2 F. \qquad (12)$$

Combining Eqs. (9), (10), (11), and (12) and substituting $1 + X_o$ for $1 + X$ gives

$$\frac{dN}{dt} = \frac{N}{t_o},$$

where

$$t_o = \frac{u_o \sqrt{\pi}}{g} \frac{e^{X_o}}{1+X_o} \qquad (13)$$

which is the time taken to reduce the atmospheric mass by a factor of e.

In fact, the atmosphere is not isothermal and its composition varies with height. Thus we need to know the height of the critical level $h_c$ and also the density of the gas there. The height $h_c$ can be defined most naturally if we consider this height as the altitude at which the mean free path is equal to the scale height:

$$l = H. \qquad (14)$$

Consequently, the total molecular density at the critical level is equal to

$$n_s = \frac{1}{\sigma H \sqrt{2}}, \qquad (15)$$

where $\sigma$ is the collisional cross section.

## 6. ESCAPE PROCESSES IN PLANETARY ATMOSPHERES

Let us now consider the escape of some small component with density $n_c$ such that

$$n_s \gg n_c ,$$

The quantity H in Eq. (15) is the average for the gaseous components of the atmosphere at the critical level. Therefore it is the same for different components (the small dispersion will be due to the difference in the gas kinetic cross sections $\sigma$).

Note that to put the scale height in the right hand side of Eq. (15), is correct and the use of any other scale length did not occur immediately. Shklovskii (1951) used the radius of the critical level in place of H and Mitra (1952) used its height. Eq. (14) is not rigorous because the mean free path varies with height. The corresponding correction coefficient (Johnes 1922/23; Shklovskii 1951; Mitra 1952) can be computed easily, but it usually does not deviate much from unity. The modern choice for the characteristic scale (H) was first made, apparently, by Spitzer (1947).

Another small correction, made by Johnes (1922/23) and Mitra (1952) relates to how the probability of escape depends on the angle between the velocity vector and the zenith. In other words, there exists a "cone of escape".

It is generally not necessary to use all of these corrections since much more important ambiguities can always be found in calculations of this sort. The important fact to keep in mind is that, for a number of planets, we now know sufficiently well the structure of their atmospheres, the location of their critical levels, and the temperatures and densities at these levels. Therefore, Eq. (12) can be used directly to compute the rate of loss due to thermal escape.

In order to estimate the "lifetime" for non-isothermal escape of atmospheres, Spitzer (1947) added to the right hand side of Eq. (13) an additional correction factor

$$B = \frac{n_0 T_0}{n_0' T_c} , \qquad (16)$$

where $n_0$, $T_0$ are the density and temperature on the surface, $T_c$ is the temperature at the critical level and $n_0'$ is the surface density obtained by extrapolation from the critical level downwards along the isotherm with temperature $T_c$. An alternative approach, used by Moroz (1967), is to immediately use the empirical relationship between the density at the surface and at the critical level then

$$t_0 = H_0 \beta \sqrt{\frac{2\pi}{g_c H_c}} e^{X_c} \frac{1}{1+X_c} \left(\frac{R_0}{R_c}\right)^2 , \qquad (17)$$

where $H_0$ is the scale height at the surface, $\beta$ is the ratio of densities at the surface and at the critical level, and $g_c$, $H_c$, and $X_c$ all refer to that level.

Nevertheless, the Jeans equation in its original form (Eq. (13)) remains useful for making rough estimates for heavenly bodies for which the properties and compositions of the atmospheres are unknown, or not well known.

Table 6.1 gives the values of $t_o$ for the planets and the Moon that have been calculated from Eq. (13). It shows that the Earth and Mars do not retain hydrogen and helium. The presence of both gases in the Earth's atmosphere indicate that there are sources which compensate for the losses. As is well known, these are the dissociation of water molecules for H, and the radioactive processes in the core for He. The situation on Mars is analogous. Nitrogen, the lightest gas, is stable with regard to escape on both planets. Venus differs in that it can retain helium. Note, however, that nonthermal escape processes exist (see section 5) and these can drive off helium faster than thermal processes for conditions at Venus. The Moon, judging from this table, could also retain gases beginning with krypton, and Mercury could begin with argon. Neither of these bodies have atmospheres of heavy gases. Evidently, they are both swept away by the solar wind. The question of argon at Pluto is unclear. It could be stable to thermal escape, and the solar wind pressure is much smaller at this large distance than in the inner solar system. Jupiter and the other giant planets retain any gases securely.

Table 6.1. Values $t_o$ in Years for the Moon and Planets[a]

|  | Earth | Moon | Mercury | Mars | Venus | Jupiter | Pluto |
|---|---|---|---|---|---|---|---|
| $M/M_o =$ | 1 | 0.0123 | 0.0554 | 0.1075 | 0.815 | 317.8 | 0.0022 |
| $R_o$, km = | $6.38 \times 10^3$ | $1.74 \times 10^3$ | $2.42 \times 10^3$ | $3.39 \times 10^3$ | $6.052 \times 10^3$ | $7.13 \times 10^4$ | $1.15 \times 10^3$ |
| $T_c$, K = | 1800[b] | 400[c] | 750[c] | 400[d] | 300[e] | 1100[f] | 53[g] |
| H (1) | $4 \times 10^{-4}$ | $1 \times 10^{-4}$ | $1 \times 10^{-4}$ | $4 \times 10^{-4}$ | $1.5 \times 10^3$ | $10^{85}$ | $2 \times 10^{-4}$ |
| He (4) | 20 | $3 \times 10^{-4}$ | $1.4 \times 10^{-3}$ | 5 | $2 \times 10^{30}$ |  | $2 \times 10^{-3}$ |
| N (14) | $4 \times 10^{18}$ | 0.3 | $6 \times 10^2$ | $3 \times 10^{16}$ |  |  | $2 \times 10^2$ |
| O (16) |  | 1.3 | $9 \times 10^3$ |  |  |  | $2 \times 10^3$ |
| Ne (20) |  | $2.7 \times 10^1$ | $2 \times 10^6$ |  |  |  | $4 \times 10^5$ |
| Ar (40) |  | $2 \times 10^8$ | $4 \times 10^{18}$ |  |  |  | $6 \times 10^{16}$ |
| Kr (84) |  | $10^{24}$ |  |  |  |  |  |
| Xe (131) |  |  |  |  |  |  |  |

[a]The limit noted for each planet is for the lightest gas stable to thermal escape.
[b]The temperature at the critical level during the day close to solar activity maximum (Johnson 1966).
[c]The temperature was taken as the surface temperature during the day at the subsolar point.
[d]The temperature at the critical level during the day at high solar activity.
[e]The same in the case of Venus (Keating et al. 1985).
[f]The same in the case of Jupiter (Atreya et al. 1981).
[g]Surface temperature at the subsolar point calculated for an albedo of 0.5 (Tholen and Baie (1990); radius and mass taken from this source).

## 3. WHEN JEANS THEORY FAILS TO WORK: THE HISTORY OF TWO ERRORS AND A FEW EXPLANATORY REMARKS

The theory of atmospheric escape hardly progressed after Jeans up until the 1940s. For a long time it remained unnoticed and neither its limitations nor the possibilities of further developments were considered. In particular, too great hopes had been put on its cosmogenic applications. The works of S. B. Pikel'ner and I. S. Shklovskii are related to a clarification of the real role of atmospheric escape in cosmogony..

The work of S. B Pikel'ner (1950) was dedicated to the solar corona. For solar protons ($\bar{m} = 0.5$) at a distance of 1.2 $R_o$ from the center of the Sun,

$$X \cong 7$$

and Eq. (5) gives an estimate of the strong flow which would result in the escape of a significant part of solar material within the lifetime of our star. We know the magnitude of the proton flow of the solar wind and can thus immediately state that there has to be some mistake in the calculations. The actual flow is much smaller than that predicted by the Jeans formula. However, at the end of the 1940s the effect of thermal escape of protons on the evolution of the sun and stars was taken very seriously (Krat 1947). S. B. Pikel'ner clearly explained why the Jeans equation does not apply to protons in solar and stellar coronae: "they move in a plasma with speeds greater than the thermal speed. The protons gradually loose their energy in distant collisions and fairly rarely experience direct scattering that results in an abrupt change in velocity. We can not assume, as follows from the gas dynamic representation, that protons are strongly deviated by a braking influence of the medium. In order for a proton to leave the corona, its energy has to be sufficient to overcome the resistance of the medium and gravity." S. B. Pikel'ner made estimates taking into account the main physical facts of the calculation. It turned out that, in the case of the coronal plasma escape comes from a very broad layer from 1.2 to 3.0 $R_\odot$ and the flux of escaped protons is much smaller than had previously been thought.

The work of Shklovskii (1951) was also "produced" because of a common mistake. For many years, it was supposed that thermal escape could be explained by the difference in chemical composition of the atmospheres of Earth-like planets and the giant planets (see, e.g., Fesenkov (1951)). The following scenario seemed attractive. All planets were pictured as large gaseous spheres with small hard cores. Those that were closer to the Sun lost their primary gaseous envelopes by thermal escape and became Earth-like planets. The planets further from the Sun retained more of their gaseous envelopes and were thus giant planets. I. S. Shklovskii showed, however, that thermal escape could not remove a significant part of the planetary mass. In fact

$$F = \frac{1}{m}\frac{dM(t)}{dt} = -\frac{n_c u_0}{2\sqrt{\pi}} e^{-X} (1+X) 4\pi R^2 \ . \tag{18}$$

If we set $n_c = 1/\sigma R$, as was done by Shklovskii, and use $X = \gamma Mm/kTR$, we have

$$\frac{dM(t)}{dt} = 2\sqrt{2\pi}\, m^{3/2} (kT)^{-1/2}\, \gamma M e^{-X}(1+X^{-1}) \tag{19}$$

Let us now assume that throughout all time the compression $X = 2$. This is a very strong determining assumption. Then

$$\frac{dM(t)}{dt} = \frac{M(0)}{t_o} ,$$

where

$$t_o = \frac{3e^2}{2\sqrt{\pi}} \frac{(kT)^{1/2}}{\gamma m^{3/2}} = 4\times 10^{20} T^{1/2}\, s \cong 10^{13} T^{1/2}\, yr \tag{20}$$

for the case of hydrogen.

It is obvious that whatever its characteristics a "gaseous planet" will not lose any significant part of its mass. We note that this result does not annihilate at all the hypothesis of massive gaseous proto-planets. There exists, in principle, other possible mechanisms for the escape of the outer layers of such proto-planets, for example, a tidal instability. The hypothesis of very massive gaseous proto-planets similar to Jupiter is rarely taken seriously nowadays. However, the original atmospheres of Earth-like planets could have been much larger in extent than at present (Sasaku and Nakazawa (1990); see also the references therein).

Shklovskii (1951) presented a very important idea that allowed for further developments of the theory. This was that for $X < 2$, the kinetic approach looses its meaning and, in place of evaporation, hydrodynamic flow would take place. This hydrodynamic regime of escape has very different, very interesting properties and most likely played a large role in the early stages of atmosphere evolution (see § 7).

The escape of hydrogen atoms can go so rapidly that their density at the critical level decreases because the diffusion of these atoms from below is too slow (Hunten 1973, 1990); a detailed bibliography is given in the latter paper).This results in a hydrogen atom density profile that is significantly steeper than it would be without escape.

In summary, the basic conditions under which the classical theory of thermal escape is applicable are the following:

1. The planetary mass does not change during the escape process.
2. The atmospheric temperature profile does not change in a time of order $t_o$.
3. The parameter $X > 2$.
4. There are no restrictions on diffusion.
5. Distant interactions are negligible (this is essential for coronal plasmas).

## 4. THE EXOSPHERE AND THE HYDROGEN PLANETARY CORONA. NONTHERMAL ESCAPE

Above the critical level there exists a layer in which the atmosphere is very tenuous and which has the distinctive characteristic that there are almost no collisions. This region is called the exosphere. Note that if escape is neglected here, then the barometric Eq. (8) applies despite the absence of collisions and, consequently, despite the fact that it is impossible to apply the concepts of hydrostatic equilibrium in a physically sound manner. The problem lies in the fact that, in its most general form, this law is derived from the Boltzmann equation and not from hydrostatic equilibrium.

If escape is significant, then the barometric equation and the Maxwell distribution are violated. The fastest atoms escape and the others show up in satellite orbits. The velocity distribution of hydrogen atoms in the exosphere is the subject of several works by Chamberlain (see his book (1978)).

Of course the velocity distribution also deviates from a Maxwellian at the critical level. However, the differences are not very large and they result in errors in the computations of the escape fluxes that are not larger than several tens of percents (Byutner 1959).

Since hydrogen is the lightest component of the exosphere, its density decreases with height significantly slower than that of the heavier atoms. As a result, the exospheric hydrogen forms the hydrogen corona of planets which extends to many thousands of kilometers in the case of the Earth, Venus and Mars. The hydrogen corona scatters solar radiation in the line $L\alpha$ (by resonant scattering). Measurements of the luminescence excited provide data on the corona, on its structure, its density, etc.. The hydrogen corona of the Earth is the one that has been studied in the greatest detail. Comparison of the observed and theoretical distributions for hydrogen atoms here provides an indirect experimental check on the thermal escape calculations. Such indirect verification is very important as the flux of dissipated neutral ions can not be measured directly. This observation does not extend, at least occasionally, to the escape of ions (see § 6). In the case of the Earth, comparison of data with theory showed that only about 30% of the escaping hydrogen atoms can be explained by thermal escape. The remaining 70% requires an additional mechanism, a nonthermal one (Chamberlain 1978).

The hydrogen corona of Venus was discovered by a group of researchers under the direction of one of I. S. Shklovskii's former students, V. G. Kurt. The measurements were made by the Venera-4 spacecraft in 1967 and were first published in 1968. Venera-4

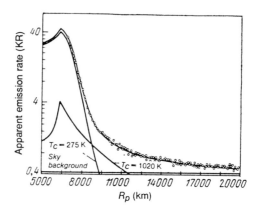

Figure 6.1. Intensity of Lα scattered by the hydrogen corona of Venus from Mariner 5 measurements and the theoretical distribution corresponding to the 3 components: T = 275 K, T = 1020 K, background (Anderson 1976). (After Chamberlain 1978).

was the first spacecraft to land on another planet. It was followed within a few days by the U.S. satellite Mariner 5 that flew by Venus. It also carried a photometer for measuring the Lα luminosity. These measurements in general confirmed the Soviet results. They showed, however, an additional important characteristic: namely, that the Venusian hydrogen corona has two components that differ in scale height. This is possible either because of a difference in atomic masses (hydrogen and deuterium) or because of a difference in temperatures.

It later became clear that the second suggestion is the correct explanation (see the review by Hunten (1990)). The Venusian hydrogen corona consists of two components (Fig. 6.1). One has a distribution in height which corresponds to the temperature at the critical level (300–350 K) and the other is much flatter (T ≅ 1000 K) which is called the nonthermal one. The flux of nonthermally escaping hydrogen atoms is several orders of magnitude higher than that of the thermally escaping ones.

For the case of Io, the loss of matter is very complex. Here, in addition to the exosphere or corona, there exists a torus which fills the region of the orbit of this satellite. The radius of the Hill sphere around Io is about 10,000 km. Atoms of oxygen, sulphur, and sodium that filter out from its range are located in the gravitational field of Jupiter and they are distributed along the orbit of Io. The material in the exosphere of Io is renewed very rapidly on a timescale of order $10^4$ s (Hunten 1990).

## 5. OTHER EXAMPLES OF NONTHERMAL ESCAPE AND THEIR MECHANISMS

A verification that the theory of thermal escape applies to helium on Earth may be made very simply: the magnitude of the flux from the core is well known and we also have data on the concentration of helium at the critical level ($h_c \cong 400$ km) and of the temperatures at this height.

Table 1 gives estimates of $t_o$ using the maximum values of the temperatures measured during the day and at maximum solar activity. For a more accurate estimate, the escape flux must be averaged over a 24-hour period and over a solar cycle. When such a calculation is made for the flow of helium that escapes the Earth's atmosphere, the value obtained, $2\times10^4$ cm$^{-2}$ s$^{-1}$, is almost 2 orders of magnitude smaller than the flux from the core from radioactive decay. This contradiction was first noticed by Spitzer (1947), but he thought that a short duration increase in temperature at the critical level would assure a balance. Nicolet (1957) and Bates and McDowell (1957) showed that this explanation would not work because it gives an excessively large escape flux of He$^3$. In another paper, Nicolet (1961) noted that the production of He$^+$ ions is approximately equal to their loss due to escape, but he did not find a specific mechanism. One of the possibilities is the formation of the unstable molecule He O$^+$ which decays into a fast He isotope (Bates and Patterson 1962). However, another explanation is now the generally accepted one, that the helium ions are carried away by the "planetary wind", a flux of plasma along open field lines in the polar regions (Axford 1968). In addition to the neutral exosphere, planets are surrounded by a plasmasphere which consists of "hot" ions. For Earth, these are basically hydrogen ions with densities up to $10^3$ cm$^{-3}$. The Earth's plasmasphere extends up to 10,000 km. It was discovered by instruments on the first Soviet lunar rockets (Gringauz et al. 1960). These are blocked from the low- and mid-latitudes by the magnetic field, and escape from high latitudes in the form of the polar wind. Helium is also carried off with the hydrogen.

It subsequently became clear that nonthermal escape plays a very important role in modern atmospheres and its mechanisms are very diverse. These are listed as follows:

1. Planetary wind, which was discussed above.
2. Charge exchange. This is very efficient between fast hydrogen ions in the plasmasphere and thermal neutral atoms. As a result, the plasmasphere generates a flux of disappearing atoms even at low latitudes.
3. During the recombination of molecular oxygen ions, superthermal neutral oxygen atoms are formed.

$$O_2^+ + e \rightarrow O^* + O^* \qquad (21)$$

This is a very important process, as it enables the escape of oxygen (see § 6). The velocities of the superthermal oxygen atoms reach values of 5.6 km/s, and the corresponding Jeans parameter is about 2 for the Earth and Venus, and 1 for Mars. Amongst the fast oxygen atoms in the exosphere, a fraction do not have sufficient energy for escape and these form the oxygen corona.

4. Collisions of fast oxygen atoms with thermal hydrogen atoms results in a dispersal of the latter, i.e. they may serve as a possible source of non-thermal hydrogen.
5. Ion-molecule reactions, e.g.,

$$O^+ + H_2 \rightarrow OH^+ + H^*, \tag{22}$$

$$OH^+ + e \rightarrow O + H^* . \tag{23}$$

6. Ionization with the subsequent "sweeping away" of ions by the solar wind.

It is assumed that in the case of hydrogen, process (2) is the most important process at Earth while for Venus it is processes (2) and (4) (Chamberlain and Hunten 1987). Process (3) can also play an important role at Venus (McElroy et al. 1982). Non-thermal escape of helium takes place not only at Earth, but also on Venus. It is assumed that in this case the dominant process is (6).

## 6. MARS

The flux of hydrogen atoms that are dissipated from the atmosphere of Mars consist of about $3 \times 10^7$ cm$^{-2}$ s$^{-1}$. This is equivalent to removing a layer of ice or water at the rate of about 200 g cm$^{-2}$ in 4 billion years assuming the flux remains unchanged. According to estimates, the present supply of ice on Mars consists of about $10^3$ g cm$^{-2}$ (Moroz 1978). Recently, the HDO line was identified from infrared spectra of Mars taken with a high resolution in the range of 3.5–3.7 μ (Owen et al. 1988). The ratio of D/H on Mars is 6 times greater than on Earth. This is naturally explained by different rates of escape of deuterium and hydrogen and, it would appear, enables us to evaluate the original water content of Mars more reliably. There is a difficulty, however, related to the fact that the permafrost may be so isolated by the low temperatures of the ground that it practically makes no contribution to the water vapor in the atmosphere. In this case the ratio of D/H can not be used to make inferences about the primeval supply of water on Mars.

In one way or another, hydrogen frees up a fair amount of oxygen even with the present rate of escape. Over a period of a billion years, much more oxygen would have accumulated than exists in the present Martian atmosphere. One of the possible mechanisms for the drain of oxygen is the oxidation of the solid surface material by

## 6. ESCAPE PROCESSES IN PLANETARY ATMOSPHERES

chemical reactions. Another stronger mechanism is nonthermal escape.

Calculations of the density of nonthermal oxygen atoms, O, that result from the reaction given in Eq. (21) for the Martian atmosphere show that the flux $\phi_O$ of these atoms is about twice as large as the flux of dissipated hydrogen $\phi_H$. McElroy (1972) advanced the idea which explained what seemed to be a strange coincidence. It is based on the fact that it is possible to automatically keep the equality

$$2\phi_H = \phi_O \qquad (24)$$

because of the following mechanism: if

$$\phi_O < 2\phi_H ,$$

then O and $O_2$ are accumulated which accelerates the recombination of $H_2O$, and the flux $\phi_H$ decreases. If, however,

$$\phi_O > 2\phi_H ,$$

then CO accumulates in the atmosphere, which "removes" O formed from $H_2O$, and the flux $\phi_H$ increases. The flux $\phi_H = 6\times10^7$ cm$^{-2}$ s$^{-1}$ evaluated from measurements of L$\alpha$ is equal to approximately one half of the ionization rate on the dayside half of the planet. The accumulation of CO does not arise from escape of nonthermal O atoms, but from their replacing the thermal atoms formed from $H_2O$. Even if $CO_2$ were present in an atmosphere that had an initial reservoir of oxygen, nonthermal escape would not change the quantity of $CO_2$ in the process of evolution; i.e., in the final analysis, the source of escaped oxygen is not carbon dioxide, but water. In later studies (McElroy and Donahue 1972; Liu and Donahue 1976), this idea was considered in greater detail. It became clear that difficulties exist, not the least of which is that the efficiency of the autocompensation mechanism of hydrogen and oxygen losses is open to question.

If the compensating flux $\phi_O$ is integrated over the entire planet, the value obtained is about $10^{26}$ s$^{-1}$. These are neutral atoms. Somewhat smaller though also significant values for the integrated fluxes for escaped ion streams were measured by the two Phobos missions; namely $2\times10^{25}$ s$^{-1}$ (Lundin et al. 1990) and $5\times10^{24}$ s$^{-1}$ (Verigin et al. 1991). These values multiplied by the age of the solar system are sometimes comparable to the quantity of oxygen in atmospheric carbon dioxide gas and this leads to the conclusion that the characteristic lifetime in the Martian atmosphere is short (Verigin et al. 1991). This is a clear misunderstanding insofar as we had discussed original reserves of oxygen bound up as ice

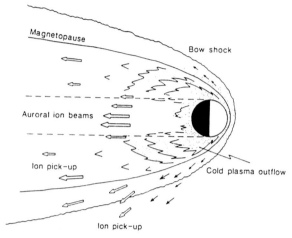

Figure 6.2. Schematic drawing showing the geometry of the escaping ion flows in the vicinity of Mars (Lundin et al. 1989).

$H_2O$. The escape of oxygen ions forms an additional part of the neutral ion flux, but it is unlikely to upset the balance.

We shall now say a few words about O and $O_2$ ion escape mechanisms. Evidently, it is analogous to the planetary wind mechanism (see § 6), but differs in that the flow is in the direction of the magnetospheric tail (see Fig. 6.2). This results from the nature of its structure (Fig. 6.2). The magnetic moment of Mars is either very small or nonexistent. In addition, a significant role may be played by the process of "sweeping away of the atmosphere" by the solar wind.

In addition to the D/H ratio in the Martian atmosphere, there is another isotropic anomaly which is evidently related to escape and this is the $N^{15}/N^{14}$ ratio, which is larger than that at Earth by approximately a factor of 1.7 (Near et al. 1976). The atmosphere of Mars is stable with respect to thermal escape of nitrogen (see Table 1) and therefore superthermal mechanisms have to be considered. These could be dissociative recombination of $N_2^+$ ions (analogous to Eq. (21)) or the dissociation of neutral $N_2$ molecules in collisions with electrons (McElroy et al. 1976; Near et al. 1976). From estimates we have from geological history, from 5 to 75 g cm$^{-2}$ of nitrogen should have escaped. If we use the ratio of $CO_2/N_2$ measured at Earth, then the density of $CO_2$ we obtain corresponds to a total pressure of 0.1 to 1.5 bar. This fact is one of the source points for the "martian paradise" hypothesis that billions of years ago Mars had a much denser atmosphere and a much milder climate.

## 7. HYDRODYNAMIC ESCAPE. THE FATE OF WATER ON VENUS

The concluding remarks made by I. S. Shklovskii in his 1951 paper turned out to be extremely important for our modern understanding of the early evolution of planetary atmospheres. If the Jeans parameter X < 2 in the original composition, then the escape regime applies in sharp contrast to the classical case. The essence of this process lies in the fact that a flow of mass leaves the planet; the whole atmospheres takes part in the flow and not just the atoms that find themselves in the tail of the Maxwellian distribution. This is the picture for atmospheres that are rich in hydrogen and are sufficiently hot. The dissipative flux can not have just any value because escape energy is needed. The energy source is solar ultraviolet radiation which is absorbed in the thermosphere. How much of it is absorbed, and where, depends on the vertical structure of the thermosphere, that is, on its temperature. A temperature profile is established such that the energy influx balances the outflow due to escape. The dynamics of the thermosphere under conditions of hydrodynamic escape was studied by Watson et al. (1981). The first distinctive characteristic of the hydrodynamic regime they found was the limit on energy. The second was that hydrogen in the hydrodynamic regime carries off all gases including the heaviest such as Xenon (although of course there is some fractionation according to mass as will be discussed below).

The hypothesis of primordial hydrogen atmospheres was originally put forth on the assumption that they were captured from the protosolar cloud. However, another scenario seems more plausible. It is believed that planets are formed of meteoric material. This parent material contained a lot of volatiles, the most abundant of which was water. Consequently, the processes at the time of formation of the planets (accretion) could themselves have been the main source of the gasses from which the primordial atmospheres were formed. A review of the relevant work was made by Muchin and Gerasimov (1989). These primordial atmospheres (protoatmospheres) contained a lot of water vapor which creates a strong opacity in the infrared. Estimates of the influence of water vapor on the thermal balance in the Earth's protoatmosphere showed that its surface temperature could have been very high, up to 1500 K (Matsui and Abe 1986a,b). Under such high temperatures, the interaction of water and metallic iron would result in the liberation into the atmosphere of a large amount of free hydrogen with an accompanying heating of the thermosphere which creates the typical conditions needed for hydrodynamic escape. Note however, that during the accretion stage the main source of energy for maintaining a high surface temperature has to be an internal source, or more precisely, the heat is generated in collisional processes. The greenhouse effect was either a secondary effect or it did not play a significant role. What could have happened after the turnoff of the "collisional" source? There are two possibilities:

1. The surface temperature decreased, the atmosphere became cooler, water condensed out creating an ocean, and hydrodynamic escape ended.
2. The surface and atmospheric temperatures remained high maintained by the greenhouse effect, water vapor remained in the atmosphere and hydrodynamic escape continued until almost all the water was used up.

Evidently, the former scenario applies to the Earth and the latter to Venus. The term "escape as a consequence of the greenhouse effect" is never used in connection with the prime role played by the greenhouse effect in enabling hydrodynamic escape during the later stages after accretion. In the papers of Kasting and his coauthors (see the review by Kasting (1989)) this mechanism of the early evolution of atmospheres was studied in detail. They confirmed the old idea that if the Earth were moved closer to the Sun it would become the same as Venus. The significant new idea they introduced was that of using the hydrodynamic escape mechanism to answer the question of how oxygen that was released after the hydrogen escaped was bound. Kasting took into account in his calculations the fact that in the presence of large quantities of water, $H_2O$, in a primordial atmosphere the wet adiabatic gradient should be used and not the dry one that applies now.

After the final stages of hydrodynamic escape, part of the remaining water was removed by nonthermal mechanisms. This is evidenced by of the large D/H ratio on Venus. It is higher than the value at Earth by approximately a factor of 100 (McElroy et al. 1982; Donahue et al. 1982). There were several attempts to estimate the mass of primeval water on Venus, taking into account all processes (McElroy 1982; Kumar et al. 1983; Kasting and Pollack 1983; Krasnopolskii 1985). The general result is that either the original mass of water was smaller than that on Earth or there was an additional source of water in the process of evolution, either internal or external (for example, comets). A review of the nonthermal escape processes on Venus is given in the book by Krasnopolskii (1986).

The increase in the fraction of heavy isotopes is distinctly weaker for hydrodynamic escape than that for Jeans escape, but is nevertheless a significant effect. For example, it provides a satisfactory explanation for the isotope ratio on Mars and this suggests that in the history of that planet there was a stage when the protoatmosphere was rich in hydrogen and hydrodynamic escape took place (Hunten et al. 1987). This mechanism also satisfactorily explains the isotope ratio of xenon on Earth.

## 8. CONCLUSIONS

The evolution of planetary atmospheres poses one of the most complex problems of solar system physics. It is also one of the most important problems because its understanding will help us understand the future development of our planet. There are many parts to this question. In fact we examined only one of these and one can see from the above what a tortuous path developments have followed, using unexpected, or indirect findings of delicate, yet important, effects. However, studies of atmospheric escape have advanced much further than other questions, such as the role of the interaction of the atmosphere and the surface, and the role of external sources (comets, the interplanetary medium). The reason is clear: the basic concept, the theory of escape, is a well defined basic physical problem and its examination has attracted many bright investigators amongst whom were S. B. Pikel'ner and I. S. Shklovskii.

## References

Anderson, D. E. (1976) *J. Geophys. Res.* **81**, 1213.
Atreya, S. K., Donahue, T. M., and Festou, M. C. (1981) *Astrophys. J.* **247**, L43.
Axford, W. I. (1968) *J. Geophys. Res.* **73**, 6855.
Bates, D. R., and McDowell (1957) *J. Atm. Terr. Phys.* **11**, 200.
Bates, D. R., and Patterson, T. N. L. (1961) *Planet and Space Sci.* **5**, 257.
Byutner, E. K. (1958) *Astron. Zh.* **35**, 572.
Chamberlain, J. W. (1978) Theory of Planetary Atmospheres, Academic Press, New York.
Chamberlain, J. W., and Hunten, D. M. (1979) Theory of Planetary Atmospheres, Academic Press, New York.
Cook, S. R. (1900) *Astrophys. J.* **11**, 36.
Donahue, T. M., Hoffman, J. H., Hodges, R. R., and Watson, A. J. (1982) *Science* **216**, 630.
Fesenkov, V. G. (1951) *Astron. Zh.* **28**, 221.
Gringauz, K. I., Moroz, V. I., Kurt, V. G., and Shklovskii, I. S. (1960) *Astron. Zh.* **87**, 64.
Hunten, D. M. (1973) *J. Atmos. Sci.* **30**, 1481.
Hunten, D. M. (1990) *Icarus* **85**, 1.
Hunten, D. M., Pepin, R. O., Walker, J. C. G. (1987) *Icarus* **69**, 532.
Jeans, J. H. (1924) The Dynamical Theory of Gasses, Cambridge University Press, London.
Johnson, F. S. (1966) Structure of the Upper Atmosphere, Near-Earth Space, F. S. Johnson (ed.), Mir, Moscow.
Johnes, J. E. (1922/23) *Trans. Camb. Phil. Soc.* **22**, 535.
Kasting, J. F. (1989) Atmospheres Lost due to the Parnikov Effect, Science of Planets, R. Z. Sagdeev, L. M. Mukhin, and T. M. Donahue (eds.), Space Research Institute, Moscow.
Kasting, J. F., and Pollack, J. B. (1983) *Icarus* **53**, 479.
Keating, G. M., et al. (1985) *Adv. in Space Res.* **5**, 11.

Krasnopolskii, V. A. (1985) *Icarus* **62**, 221.
Krasnopolskii, V. A. (1986) Photochemistry of the Atmospheres of Mars and Venus, Springer Verlag, Berlin.
Krat, V. A. (1947) *Dokl. AN SSSR*, **40**, 247.
Kumar, S., Hunten, D. M., and Pollack, J. B. (1983) *Icarus* **55**, 369.
Kurt, V. G., Dostavalov, S. V., and Sheffer, E. K. (1986) *J. Atmos. Sci.* **28**, 668.
Lundin, R., et al. (1989) *Nature* **341**, 609.
Lundin, R., et al. (1990) *Geophys. Res. L.* **17**, 873.
Liu, S. C., and T. M. Donahue (1976) *Icarus* **28**, 231.
McElroy, M. B. (1972) *Science* **175**, 443.
McElroy, M. B., and T. M. Donahue (1972) *Science* **177**, 986.
McElroy, M. B., Yung, Y. L., and Nier, A. O. (1976) *Science* **194**, 70.
McElroy, M. B., Prather, M. J., and Rodrigues, J. M. (1982) *Science* **215**, 1614.
Matsui, T., and Abe, Y. (1986a) *Nature* **319**, 303.
Matsui, T., and Abe, Y. (1986b) *Nature* **322**, 526.
Mitra, S. K. (1952) The Upper Atmosphere, Calcutta.
Moroz, V. I. (1967) The Physics of Planets, Nauka, Moscow.
Moroz, V. I. (1978) The Physics of the Planet Mars, Nauka, Moscow.
Muchin, L. M., and Gerasimov, M. V. (1989) Science of the Planets, Space Research Institute, Moscow.
Nicolet, M. (1957) *Ann. Geophys.* **13**, 1.
Nicolet, M. (1961) *J. Geophys. Res.* **66**, 2263.
Nier, A. O., McElroy, M. B., and Yung, Y. L. (1976) *Science* **194**, 63.
Owen, T., J. P. Maillard, C. de Bergh, and B. L. Lutz (1988) *Science* **240**, 1767.
Pikel'ner, S. B. (1950) *Dokl. AN SSSR* **72**, 255.
Pikel'ner, S. B. (1950) *Crimean Astrophys. Iz.* **5**, 34.
Prather, M. J., and M. B. McElroy (1983) *Science* **220**, 410.
Sasaku, Sh., and K. Nakazawa (1990) *Icarus* **85**, 21.
Shklovskii, I. S. (1951) *Astron. Zh.* **28**, 234.
Shklovskii, I. S. (1951) *Dokl. AN SSSR* **76**, 193.
Spitzer, L. (1947) The Atmospheres of the Earth and Planets, U. Chicago Press, Chicago.
Stoney, G. J. (1898) *Astrophys. J.* **7**, 25.
Stoney, G. J. (1900) *Astrophys. J.* **11**, 251.
Tholen, D. J., and M. W. Baie (1990) *Bull Am. Astron. Soc.* **22**, 3, 1129.
Vergin, M., et al. (1991) *Planet Space Sci.* **39**, 131.
Watson, A. J., Donahue, T. M., and Walker, J. C. G. (1981) *Icarus* **48**, 150.

# 7

# Decametric Radioastronomy

## S. Ya. Braude
*Radio Astronomy Institute of the Ukraine, Kharkov*

The main results in experimental radio astronomy obtained at the Radioastronomy Institute of the Ukrainian SSR Academy of Sciences from 1958 to 1989 are presented. All observations of various cosmic objects have been conducted in the decameter band. The decameter radio telescope and interferometers used, UTR-2 and URAN, are briefly described. Data at these wavelengths radiated by both galactic and metagalactic objects are presented, namely: recombination lines, the Sun, HII regions, supernova remnants (SNR), pulsars, radio galaxies, and quasars. Some information on a discrete source catalogue and the cosmological conclusions obtained from a source count are presented.

## INTRODUCTION

Of the various bands of electromagnetic waves which are radiated by the cosmic bodies, only the longest, the decametric band at which ground observations can still be conducted will be considered in this paper.

The decametric band is of great interest to present radio astronomy. This interest is caused by the fact that in various cosmic objects both in the Galaxy and the Metagalaxy, the cosmic plasma density is such that it ceases to be transparent for decametric radio waves. These waves are partially absorbed and reradiated by the plasma, interacting with it. By considering this interaction, one can study the cosmic plasma and thus probe the interplanetary and interstellar media.

Using observational results at decametric wavelengths and similar data at the highest frequencies, some physical parameters for a number of cosmic objects can be determined, namely: ionized-hydrogen regions, SNRs, the solar corona, pulsars, radio galaxies, and quasars; the density and temperature of the electrons and other particles of the medium, the magnitude and direction of the cosmic magnetic fields, and the brightness distribution for the objects and other factors being studied.

As known, radio astronomical investigations were begun by Karl Jansky in 1931 using decameter waves (20 MHz). Nevertheless, for almost 25 years subsequently, radioastronomy used shorter waves, meter and decimeter ones. In spite of a great interest in the decameter wave band, radio astronomical observations at these frequencies only began in the late '50s. This delay was connected with a number of technical difficulties which were not so important at higher frequencies. We note some important difficulties which arise when conducting observations in the decameter band.

Firstly, there is the lower, compared to higher frequencies, resolution of decameter radio telescopes. Secondly, there are noises produced by various radio stations operating in the decameter band. These stations produce, near the radio telescope, energy flows of $10^5$–$10^6$ times more than those coming from the cosmic objects. Thirdly, the Earth's ionosphere has a hindering effect on the process of receiving cosmic signals, distorting the amplitude, phase, and polarization of the incoming signals. Fourthly, in the decameter band, radio signals from cosmic objects are received in the presence of very powerful nonthermal radiation of the Galaxy itself, whose brightness temperature achieves hundreds of thousands and millions of degrees which, in a number of cases, produces considerable difficulty in receiving weak cosmic signals.

These and other difficulties of the decameter band observations lead to a number of specific requirements which apply both to decametric radio telescopes and to observational methods. The difficulties mentioned above can be partially overcome by using radio telescopes of large size which allows one to obtain larger collecting areas, providing appropriate resolutions. Measurements at decameter wavelengths should be conducted as a rule by night and in fall-winter months using radiometers having the narrowest possibly bands over intermediate frequencies. Nevertheless, even observing the above requirements, a great number of repeated observations (about 10–20 measurements of the same object) should be conducted to obtain reliable data. One needs to take into account the fact that not more than 70% of the data obtained can be used for subsequent data reduction. Below we shall present the experimental results obtained at the Radioastronomy Institute of the Ukrainian SSR Academy of Sciences from 1958 to 1989. To consider these questions we shall begin by describing our experimental set-up.

## THE RADIO TELESCOPE AND RADIO INTERFEROMETERS

The main tools which are used for observations in the decameter band are the UTR-2 radio telescope (the Ukrainian T-shaped Radio telescope, type II) /1/ and the URAN radio interferometer (Ukrainian Radio Interferometer of the Academy of Sciences) /2/. We shall describe these tools.

To work successfully at decameter wavelengths, contemporary radio telescopes should have wide bands, high directivities in two planes with an optimum tuning between the sensitivity and the resolution, and maintain primary characteristics over the whole frequency operation band and over a wide angular range. They must provide for the quick operative change of beam orientation in two planes according to a set program,

# 7. DECAMETRIC RADIOASTRONOMY

being able to work simultaneously with several spaced beams; of course, they should be highly reliable, simple, and economical. An attempt to fulfill these requirements led to the construction of electrically-driven antennas having an optium cross- or T-form (or π-form). Starting from these requirements when constructing our decameter radio telescope, we chose a T-form with the directional pattern electrically driven.

The UTR-2 radio telescope operates at 10-25 MHz. It is situated near the village of Grakovo at 80 km from Kharkov. The ground area occupied by the instrument is 16 ha. The UTR-2 antenna area consists of two multi-element arrays arranged as a T-form. The N-S area is 1800 m long and 53 m wide, and the W-E area is 900 m long and 53 m wide. Both areas are filled with wideband horizontal resonators situated at a height of 3.5 m above the ground. The N-S area has 1440 resonators and the W-E area has 600 resonators. The resonators are directed along a parallel and are cylindrical forms with a diameter of 1.8 m and a total length of 8 m. The UTR-2 beam is pencil-like and is electronically driven in the sector of ±50° in right ascension and ±80° in declination. The beam drive is discrete and in the given coverage one can use 2,097,152 ($2^{21}$) beam positions for the celestial sphere. The change of the beam position is conducted both manually and using computers. To improve the efficiency of operation and to overcome noises, the UTR-2 has 5 beams spaced in declination. The drive of the antenna beam is performed using a timing principle by 39 phase-rotators with cable delay-lines that are switched. All the phases-rotators are placed in underground boxes which are situated just under the antenna areas. The observations of objects were conducted simultaneously by the 30-channel receiving equipment, and its signal is recorded and fed into the computer. The data reduction is conducted in a real time using a number of programs. To provide high reliability of the radio telescope operation, we envisage a system of automatic defense and search for malfunctions. In addition we use a system of high-frequency control allowing us to check conditions of the antenna as a whole using an accurate form of the directional pattern, to check the amplitudes and phases of the signals at the outputs of the phasing system, and to determine conditions of each of the resonators of the total antenna area. Because of the large size of the telescope, there are considerable losses in the communication cables for the phasing apparatus of its antennas. To compensate for these losses, a system of built-in amplifiers is used. To provide a high noise safeguard of the amplifiers against noise combinations, a special wideband-amplification system is used based on a planned division of the frequency band into a number of narrowband channels with a subsequent addition of them.

We present some parameters of the channels of the UTR-2. Its effective area is 150,000 $m^2$, the resolution at 25 MHz is about 30', the tracking sector over the hour angle is ±4 hours from the meridian, and the sensitivity is such that the recorded minimum density of the radiation flux for a signal-noise ratio equal to 3 is 20 Jy in the mid-frequency of the range. Fig. 7.1 shows I. S. Shklovskii in front of the south array of the UTR-2 and the W-E array of the UTR-2 is given in Fig. 7.2. At present the UTR-2

Figure 7.1. I. S. Shklovskii in front of the south array of the UTR-2 radio telescope.

Figure 7.2. The W-E array of the UTR-2 radio telescope.

## 7. DECAMETRIC RADIOASTRONOMY

radio telescope is one of the largest instruments in the world operating in the decameter band. The design and construction of the UTR-2 was conducted under the leadership of Professor A. V. Megn and myself, by a group of scientists, namely: L. G. Sodin, Yu. M. Bruk, N. K. Sharykin, G. A. Inyutin, and others.

Despite the large effective area and rather high resolution of the UTR-2, it is necessary to have still higher resolution. The URAN radio interferometer system, whose construction will be finished in 1991, has this resolution.

The maximum resolution which can be obtained when using decameter waves is limited by scattering due to ionospheric irregularities, and the interplanetary and interstellar plasmas; according to different estimations, it cannot be larger than 0.5–1". Therefore there is no sense in using a base with sizes exceeding a thousand kilometers when constructing decameter interferometers. The N-S antenna system of the UTR-2 radio telescope is the main component in the URAN system. The URAN-1 to URAN-4 radio telescopes are used as auxiliary instruments which form interferometers with the UTR-2.

The URAN-1 is situated in the suburb of Zmiev-town of the Kharkov region that is 42.6 km from the UTR-2. The radio telescope consists of 96 tourniquet wideband resonators forming 4 parallel rows oriented in the W-E direction. The size of the array is 172.5 x 22.5 m and each tourniquet element is situated at a height of 3.5 m above the ground and is formed of 2 wideband resonators perpendicular to each other; the design of the linear resonators are oriented at an angle of 45° with respect to the meridian. A front view of the tourniquets and an overall view of the whole antenna of the URAN-1 are given in Fig. 7.3 and Fig. 7.4. The antenna beam is driven electrically, providing control of the radiation sources over declination limits from –20° to 90°. The effective area of the antenna is 5,000 $m^2$. The UTR-2/URAN-1 interferometer has operated since 1975. The tourniquet resonmeters allow one to exclude the Faraday effect in the ionosphere by conducting observations in two polarizations. This interferometer provides a resolution of 30" at 25 MHz. The URAN-2 interferometer is situated 30 km from the city of Poltava at the distance of 144 km from the UTR-2. The radio telescope is being constructed and will be finished in 1991. It consists of an antenna area in which 512 tourniquet resonators of the same type as for the URAN-1 are placed (as for all other URANs). The array size is 116 x 240 m. The effective area is 27,000 $m^2$, and the resolution of the URAN-2/UTR-2 is about 9" at 25 MHz. The interferometer URAN-3 is situated in the vicinity of the town Shatsk in the Volyn' region at the distance of 960 km from the UTR-2. The radio telescope has already been constructed and was brought into operation in 1990. It consists of an antenna area in which 256 tourniquet resonators are placed. The array size is 58 x 240 m, the effective area is 13,500 $m^2$ at the mid frequency of the range, and the resolution of the UTR-2/URAN-3 is 1.3" at 25 MHz. The URAN-4 interferometer is situated 40 km from the city of Odessa at a distance of 616 km from the UTR-2. This radio telescope has operated since 1980. It consists of an antenna area in which 128 tourniquet resonators are placed. The array size is 29 x 240 m,

Figure 7.3. The tourniquet resonators of the URAN-1 radio telescope.

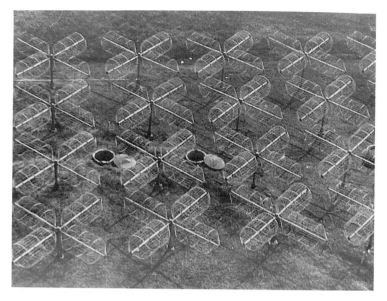

Figure 7.4. An overview of the URAN-1 radio telescope antennas.

the effective area is 7,250 m², and the resolution of the UTR-2/URAN-4 is 2.1" at 25 MHz. Because of the small effective area of the URAN radio telescopes it is not reasonable to use URAN interferometry between separate antennas of these instruments although it is suggested in the future to make the effective area of the antennas of the URAN-3 and URAN-4 two or three times larger and to increase the number of interferometer bases to 7 (instead of the present four). The URAN-system design was suggested by the author of this paper, and its further implementation was conducted under the leadership of Professor A. V. Megn with N. K. Sharykin (all the URANs), the late V. P. Tsesevich, V. V. Galanin (the URAN-4), A. N. Swenson, V. V. Koshevoy (the Uran-3), V. G. Bulatsen, A. I. Brazhenko (the URAN-2) participating.

The instruments mentioned above were and are used to conduct various radio astronomical observations which will be described briefly. We have studied some cosmic objects radiating decameter waves which are both in our Galaxy and outside it. We shall describe some results obtained for objects situated in our Galaxy.

## GALACTIC OBJECTS RADIATING DECAMETER WAVES. DECAMETER BAND SPECTRAL LINES.

One of the most important contributions of I.S. Shklovskii to astrophysics and radio astronomy is his pioneering works made in the 1940s and in the beginning of the 1950s was pointing out the possibility of using spectroscopy methods in radio astronomy. He was the first to analyze and provide a basis for discovering spectral lines of both neutral interstellar hydrogen and the superfine structure lines of some interstellar molecules. At the present, investigations of radio lines of atoms and molecules have become one of the main directions in radio astronomy. Though most of the lines are in bands of decimeter, centimeter, and millimeter radio waves, radio lines in the decameter band have appeared as well. I. S. Shklovskii has shown that superfine structure lines of neutral nitrogen should be observed in this band. These lines were very weak, having small temperature/continuum ratios (according to various previous calculations) of $10^{-4}$–$10^{-5}$; nevertheless I. S. Shklovskii believed that with such a powerful radio telescope as the UTR-2 it would be reasonable to try and discover such lines. In addition to these lines, N. S. Kardashev (I. S. Shklovskii's pupil) predicted in 1959 that one could observe the recombination radio lines of interstellar atoms; they were discovered in the USSR 5 years later at high frequencies. Although such lines should be very weak in the decameter band where the main quantum numbers should exceed 600, nevertheless the author in 1966 still suggested that such lines should be investigated. Work in this direction was developed at the beginning of the 1970s and conducted by L. G. Sodin and A. A. Konovalenko. At first, there were attempts to discover some hydrogen recombination-lines, but the result turned out to be negative. Then an attempt was made to discover neutral-nitrogen lines from $^{14}$N. For this purpose, the UTR-2 radio telescope beam was directed toward Cassiopeia A. It had been known that the radiation of this source goes through dense clouds of neutral hydrogen in the Galactic plane which concentrate both in the nearer Orion arm and in the farther Perseus arm. Therefore one could hope that in these arms neutral nitrogen $^{14}$N would appear with the nitrogen absorption line being observable. The observations were carried out at $\nu = 26.12$ MHz. In 1978, the decameter radio line was successfully found /3/. Nevertheless, its intensity appeared to be as large as about $10^{-3}$, and if this line were considered to be the nitrogen line $^{14}$N, then the nitrogen abundance in the Galaxy would be 15 times greater than the usual one. Calculations made by American radio astronomers /4/ showed that the parameters of the line observed could be explained by suggesting that it was not the nitrogen line $^{14}$N, but the carbon recombination-line C631$\alpha$. The experiments conducted by L. G. Sodin and A. A. Konovalenko /5/ some time after these calculations confirmed that, using the UTR-2, one could observe just the carbon recombination-lines with high quantum numbers. The discovery of such low-frequency lines resulted in a number of interesting conclusions. The main quantum numbers of the lines recorded are more than twice as large as the ones previously observed; the line frequencies are more than an order of magnitude lower than the frequencies of all other measured lines; for the first time, the lines were observed in absorption; a 10-times broadening of the lines was observed; and there is a clearly marked deviation from the conditions of balance. Since carbon and its ions play an important role in physical and chemical interstellar processes, an ionized-carbon study conducted in the ultraviolet and infrared regions using high-frequency radio lines can be enhanced considerably with the results obtained using the decameter lines. Such data are sources of new information about the physical conditions and phenomena in the interstellar medium.

The interstellar medium in the direction of the discrete source Cas A was studied in most detail using decameter radio spectroscopy methods. In this case we observed the carbon recombination-lines up to C768$\alpha$ at $\nu = 14.7$ MHz.[†] An analysis of the line-parameters obtained allowed us to determine the electron temperature of interstellar clouds. It is about 50 K and the cloud density is 0.1 cm$^{-3}$. Recombination lines appear in diffusive interstellar clouds and the ionization source is the disturbed ultraviolet radiation with a wavelength of more than 912 Å. Perhaps the mechanism of low-temperature dielectronic recombination has an effect on the population of the highly-excited carbon levels.

The unique parameters of both the UTR-2 radio telescope and highly-sensitive noise-stable decameter-radio spectroscopy methods developed by A. A. Konovalenko and his colleagues stimulated an investigation of various interstellar objects using decameter recombination-lines. The purpose of these investigations is to determine low-density interstellar medium parameters. Despite the low density, this plasma plays an important role in the energetics, dynamics, and evolution of galactic matter, and in a number of cases this plasma cannot be detected by other methods. In particular, such objects may be

---

[†] The carbon-lines measured are presented in Fig. 7.5.

# 7. DECAMETRIC RADIOASTRONOMY

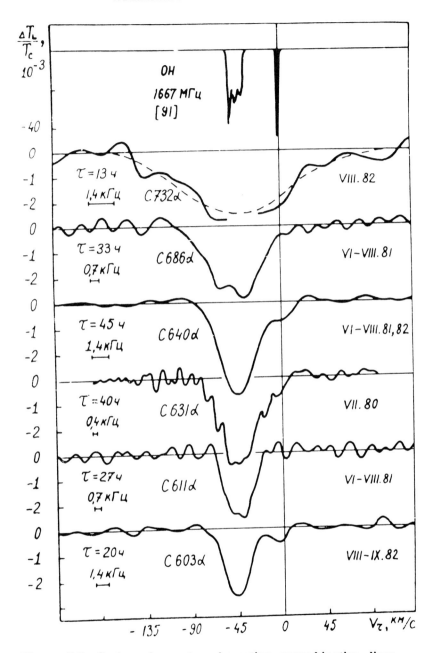

Figure 7.5. Carbon decameter absorption recombination lines. The main quantum number of the lines is marked.

diffusive and dark dust clouds, and partially-ionized media near emission nebulas. During recent years using the UTR-2 radio telescope, we reliably found decameter recombination-lines in some galactic objects, namely: 675.0+0.0; NGC 2024; DR 21; L 1407, and S 140. Such absorption-lines have an intensity of $10^{-3}$ against the background level and the line width is several kilohertz.

The results obtained demonstrate that considerable volumes of cold interstellar gas may contain ionized carbon with a density of 0.05–0.1 $cm^{-3}$ with practically neutral hydrogen. Thus, the corresponding model of the interstellar medium differs fundamentally from that forming the high-frequency recombination-lines. The results obtained give us hope that later on the circle of objects in which decameter recombination-lines can be observed will be widened.

Carbon recombination-lines with large quantum numbers over the meter-wave band were obtained in the USSR (Physics Institute of the USSR Academy of Sciences), USA, and India. I. S. Shklovskii was most interested in the investigations described. Kharkov radio astronomers honor his memory. Until now a nitrogen superfine-structure line has not been found. Therefore, it is planned to conduct corresponding experiments using the UTR-2 radio telescope with a probability of discovering this line that is considerably different from zero.

## THE INTERSTELLAR MEDIUM INVESTIGATED WITH DECAMETER RADIO ASTRONOMY METHODS

This section is concerned with using the decameter radio band to study the interstellar medium as a whole and in particular such objects as the emission nebulas of HII-regions and dust clouds. Since many of the objects noted are extended cosmic bodies and the original construction of the UTR-2 radio telescope was to observe point radio frequency-radiation sources, one had to develop new methods of observing. This was done by V. V. Krymkin; as a result it was possible to conduct measurements of the extended objects using for this purpose the five UTR-2 beams spaced in declination and the 6 frequencies of the decameter band. The use of decameter radio waves provides a practically unique possibility to obtain information about states and physical conditions in the outer layers of the cosmic plasma in emission nebulas. If the inner regions of these objects can be successfully studied using the recombination lines of the decimeter and centimeter bands, with increasing wavelength this medium becomes optically thick, the inner regions are screened, and it becomes possible to study the outer regions using both the decameter continuum and decameter recombination-lines.

Using the UTR-2, V. V. Krymkin and E. A. Abramenkov made a search of extended HII regions whose angular sizes are more than 1'. The number of regions, accessible to observations with the UTR-2, appeared to be 20. These were extended gaseous nebulas. Moreover, four molecular dust clouds were observed. The gaseous nebulas observed were found to have a number of important physical parameters, namely: the electron temperature ($T_e$), and the volume density of the nonthermal galactic background at 25 MHz ($\tau_{25}$). The data obtained are given in Table 7.1. All the objects were taken from the catalogue of Sharpless /6/. The parameters obtained for the first time with the UTR-2 radio telescope are marked with an asterisk.

As seen in Table 7.1, the parameters of different nebulas may vary over wide ranges. Thus, the electron temperature varies from 1,500 K to 11,000 K, the emission measure from 80 to 13,000, the optical thickness from 0.7 to 50, and the volume density from 7 to 30. In the 5 regions of HII in a linear approximation a radial electron-temperature distribution was found. Both these investigations and the results from other references showed that the thermal structures of gaseous nebulas have no analogs with respect to theoretical calculations, differing from them substantially. The values obtained in the direction of HII regions of the volume density of the nonthermal galactic background led to number of conclusions concerning the interstellar magnetic-field structure and the spiral structure of the Galaxy in nonthermal radio frequency radiation. We note some of the results obtained.

Until recently the primary information about the interstellar magnetic field was

Table 7.1. Physical Parameters of HII Regions

| No. | HII-region | $T_e$(K) | EM (pc cm$^{-6}$) | $\tau_{25}$ (kpc$^{-1}$) | $\eta_{25}$ (kpc$^{-1}$) |
|---|---|---|---|---|---|
| 1 | Sh 117 | 7,100 | 5,400 | 6.5 | 19* |
| 2 | Sh 119 | 5,200 | 500 | 0.9 | 19* |
| 3 | Sh 190 | 4,900 | 1,500 | 3.0 | 10 |
| 4 | Sh 199 | 8,300 | 3,150 | 3.0 | 10 |
| 5 | Sh 202 | 1,500* | 80 | 1.2 | 25 |
| 6 | Sh 205 | 4,000* | 550 | 1.2 | 23 |
| 7 | Sh 206 | 4,400 | 13,000 | 30.0 | 9* |
| 8 | Sh 209 | 2,900 | 13,000 | 50.0 | 8* |
| 9 | Sh 216 | 8,000 | 100 | 0.1 | — |
| 10 | Sh 220 | 4,400 | 1,500 | 3.3 | 26 |
| 11 | Sh 229 | 6,400 | 650 | 0.8 | 30 |
| 12 | Sh 236 | 11,000 | 8,000 | 8.0 | 7* |
| 13 | Sh 264 | 2,800 | 200 | 0.8 | 25 |
| 14 | Sh 273 | 4,100 | 275 | 0.7 | 21 |
| 15 | Sh 275 | 3,600 | 3,500 | 10.0 | 12 |
| 16 | Sh 276 | 5,900 | 220 | 0.4 | 12 |
| 17 | Sh 277 | 6,000 | 1,150 | 2.1 | 20* |
| 18 | Sh 280 | 2,000 | 1,400 | 6.4 | 12* |
| 19 | Sh 282 | 4,900* | 430 | 0.9 | 13* |
| 20 | Sh 284 | 2,000* | 1,550 | 9.7 | 4* |

obtained either from theoretical models or from laborious polarization-measurements. Decameter radio wave measurements provided a new independent method of studying the interstellar magnetic field, based on measuring absorption in the ionized gas at low frequencies. The use of the anisotropy discovered of the background volume-density resulted in successfully discriminating both the directed and isotropic components of the interstellar magnetic field over a scale of 1 kiloparsec in the direction of the Galatic anticenter. The ratio of their field intensities appeared to be about 0.6. The measured directed component of the magnetic field was found to be orthogonal to the galactic longitude $l = 160°$.

In the direction of the anticenter, we succeeded in discriminating the large-scale structure formations in nonthermal radio frequency radiation. These formations appeared to correspond to the local galactic arm, and the interarm space with respect to the Perseus arm. Estimations of the transverse size of the arms and interarm space were made at 25 MHz. It appeared that the local arm $\eta_{25} = 27$ kpc$^{-1}$, in the interarm space $\eta_{25} = 1$ kpc$^{-1}$, and in the Perseus arm $\eta_{25} = 10$ kpc$^{-1}$. Although according to the data of some papers the local galactic arm should not be observed in nonthermal radiation, a direct measurement of the volume density at 25 MHz does indicate a high intensity of the local-arm which is an order of magnitude larger than that in the interarm space.

One can estimate the intergalactic background intensity by measuring absorption of this component in HII-regions distant from the Galaxy center. From such measurements we obtained the brightness temperature of the isotropic metagalactic radiation which appeared to be about 31,100 K at 14.7 MHz. Using this value together with some data from the other papers, we succeeded in determining the spectral index of this component in the narrow limits between 2.75 and 2.90. Unlike most papers dealing with the metagalactic background estimation with the background parameters determined by considering various models, the measurements of HII region absorption for decameter waves allows one to obtain these parameters from experimental data. Together with the HII region measurements, we succeeded for the first time in discovering galactic background absorption in the direction of the extended molecular-dusty clouds of L1407 and Per OB-2, objects that are qualitatively different from gaseous emission nebulas. For these clouds, we obtained estimations of the ionization degree, the electron density, and the emission measure.

## OBSERVATION OF SNRS IN DECAMETER RADIO WAVES

Supernova remnants (SNRs) and their structures appear to be the main subjects of present astrophysics since these objects are connected with pulsars, black holes, cosmic-ray origin, stellar evolution, sources of X- and γ-rays, interstellar-medium parameters, and other important problems. I. S. Shklovskii paid great attention to supernovas and their remnants with two editions of his classical book devoted to these objects.

## 7. DECAMETRIC RADIOASTRONOMY

At present, 150 SNRs have been discovered in the Galaxy. Among these remnants there are objects whose angular sizes vary from several angular minutes to tens of degrees. Therefore, it has been possible to study SNRs in the decameter band using both the UTR-2 radio telescope and the URAN interferometer. Moreover, some data were obtained for the SNR in the Crab Nebula using scintillation techniques and the Lunar occultation method. Among old SNRs maps were obtained of the decameter radio frequency radiation in the Cygnus Loop, the Monoceros Loop in the direction of the source PRS 0607+17 which according to some data is part of the gigantic SNR called the Origen Loop, the SNRs in HB9, HB21, and CTA-1, and the supposed remnant of CTB13. Moreover, we also studied younger SNRs in such objects as the Crab Nebula and Cas A. The extended SNRs were studied by V. V. Krymkin and M. A. Sidorchuk. A radio map of the SNRs in the Cygnus Loop was obtained using UTR-2 at 25 MHz. One can see on this map a complex structure of source radiation where there are several maxima with different intensities. The N-E part of the source is separated from the main one by a lower-radiation region whose intensity decreases down to the cosmic background level. Comparing the radio brightness distribution for the decameter and decimeter waves, we succeeded in establishing that the sizes of this SNR from 25–430 MHz remained unchanged. The SNR frequency spectrum from 25–900 MHz is linear on a logarithmic scale and its spectral index is 0.25.

The SNR in the Monoceros Loop appeared to be an interesting object. It was measured by the UTR-2 at 14.7, 20, and 25 MHz. From these and other references a source frequency spectrum was constructed. It appeared that from 111 to 1400 MHz, the spectrum is linear and has a spectral index of 0.49 with a sharp break in the decameter band. It was found that radio maps of SNRs obtained at high and low frequencies are rather different. As shown by a detailed investigation, the main cause of the SNR frequency-spectrum break in the decameter band is the absorption of the nonthermal SNR radiation by ionized hydrogen. Nevertheless, the optical thickness of the absorbing matter is small and the 25 MHz absorption is about 40%. Using these and other data, the SNR location can be determined. It appears that the SNR lies between the Rosette nebula and NGC 2264, being genetically connected with the latter. Therefore, the valley in the SNR frequency spectrum at low frequencies is due to ionized hydrogen, manifested in two ways: in terms of HII absorption of the SNR nonthermal radiation and in terms of cosmic radio frequency radiation absorption generated along a line of sight in space behind the source. Thus the temperature drop accompanying the rest of the SNR-radiation produces an apparent valley in frequency spectrum. A similar situation was found for PSR 0607+17 as well. Near this SNR the sizes of which were very large (150') a new SNR called GR (0625+16) was discovered, having a spectral index of 0.47. Although both SNRs were situated so that a kind of ring-like structure was formed, the UTR-2 measurements gave no reason to consider that these SNRs were connected with the Origen Loop; the veracity of such a connection could be determined either by measuring the SNR in soft X-rays or by discovering a neutral-hydrogen envelope.

Investigations of SNR HB-9 in the decameter band together with data for high frequencies allowed us to determine the structure of this source. It was found that this SNR has a practically spherical form with a diameter of about 130', its distance is 1.1–1.4 kpc, its diameter is 41–53 pc, and its height above the galaxy plane is 53-64 pc. The frequency spectrum of HB-9 from 30–1000 MHz is linear with a spectral index $\alpha$ = 0.43±.005 and the electron density is 0.8 cm$^{-3}$. At lower frequencies we observed a spectral break connected with HII absorption; the optical thickness of the thermal component is small, $\tau$ = 0.05±0.03 and SNR HB-9 has an extended envelope which has a size in the decameter band of a half of its radius. The volume density of the radiation in the source center is about 2.5 x 10$^{-37}$ erg cm$^{-3}$ Hz$^{-1}$, and in the envelope it is 1.5–2 times lower. In the envelope the magnetic field is about 20 μOe and the total energy of the relativistic electrons and the magnetic field in the SNR is about 10$^{50}$ erg.

The rest of the extended SNRs studied in the decameter band such as HB-3, HB-21, and others /7,8/ are very similar to those described above. They are a mixture of thermal and nonthermal components. In different objects these component represents HII regions which may surround the SNR or be situated in front of the remnant. Therefore, the frequency spectra of the extended SNRs in the decameter band have a break whose magnitude and frequency at the flux maximum define the emission measure.

Together with the extended SNRs in the decameter band, we also measured compact details in SNRs. The SNR in the Crab Nebula belongs to these. Measurements were conducted for the first time in /9/ at 20 and 25 MHz. We used the URAN-1 interferometer. Subsequent URAN-1 measurements, made with more sensitive equipment, allowed us unlike /9/ to discover a dependence of the visibility function on the hour angle. From the data obtained and a number of other measurements, we considered the 3C461 frequency spectrum from 1 to 100 MHz in /10/; using the break of this spectrum we defined the emission measure of the ionized-hydrogen region with some absorption occurring in it. According to /10/ this region should have $N_e$ =1 cm$^{-3}$, $T_e$ = 5000 K, and EM = 100 pc cm$^{-6}$. Nevertheless, such a region was not discovered and therefore it was suggested in /10/ that the decameter radiation absorption is due to remnant ionization after the burst of the supernova. Compact detail in the Crab Nebula (3C144) was studied. This object was investigated by the methods of lunar-occultation, of scintillations, and of interferometry. Such investigations were conducted in the decameter band over 12.6 to 26.3 MHz /11,12/. It was shown that the compact detail coordinates ($\alpha$(1950) = 05$^h$31$^m$31$^s$.62±0$^s$.22; $\delta$(1950) = 21°58'52". 8±3".9) practically coincide with those of the PSR 0531+21 ($\alpha$(1950) = 05$^h$31$^m$31$^s$. 428±0$^s$.005; $\delta$(1950) = 21°58'54". 40±0".06). The apparent angular sizes of compact detail in 3C144 give $\theta$ = 2".2±0".25. The spectral index of this detail is 2.09±0.04. It appears that the dependence of the angular size $\theta$ on frequency is $\theta \sim \nu^{-1.95\pm0.25}$. Such a dependence is explained rather satisfactorily by radiation scattering in the turbulent interstellar medium when the dependence of $\theta$ = f($\nu$) has a form of $\theta$ =1.1 $\nu^{-2.2}$(sin b)$^{-0.6}$ with $\theta$ in degrees and b the galactic latitude /13/. Note that the numerical coefficient in this formula is approximate as the interstellar plasma parameters are not known exactly. If we take the data obtained for the pulsar in the 3C144 using the URAN-4 interferometer we shall, using |b| = 6° for it, obtain a value of 0.19 instead of 1.1 which is very favorable for decameter interferometry. It can be shown that if we apply the angular dependence θ = f(ν) obtained in the decameter band to the higher-frequency region, we obtain a θ-value that agrees with the visible angular size of the pulsar 0531+21. If the pulsed radiation of the pulsar did continue to the decameter band (as known, this radiation is not observed at ν < 70 MHz), the visible pulsar size would coincide with the angular size of the source; this means that the compact source has an extremely high brightness-temperature exceeding the so called Compton limit of $10^{12}$ K. Therefore, the radiation mechanism of this source cannot be described with only the incoherent synchrotron mechanism.

## DECAMETER RADIATION OF PULSARS

A review of pulsar radiation in the decameter band is presented in /14/. Therefore, we present here only those data which augment the review mentioned. We note that in the early 1970's the possibility of observing weak decameter pulsar radiation was doubted. This doubt was connected with the fact that with decreasing frequency the galactic background temperature increases and interstellar medium scattering becomes larger which in a number of cases, leads to pulse radiation disappearance as for example, in the cases of the Crab Nebula and the millisecond pulsar 1937±21. Only the experiments made in the early 1970s by Yu. M. Bruk and his colleagues showed that the pulsar radiation could be observed down to 10 MHz (PSR 0869+71). It appears that one can observe only near pulsars having a dispersion measure less than 20–30. Therefore, until now, we measured the parameters of only 8 pulsars, and upper limits of some parameters of 14 pulsars were also determined. It seems that the most basic difference of pulsar radiation in the decameter band from that at higher frequencies is explained by the fact that at low frequencies all observed pulsars have interpulse radiation rather densely filling the space between main pulses. Note that only a few per cent of the total number of pulsars has interpulses at high frequencies. This situation is caused by the fact that the interpulse spectral index is steep and considerably larger than that of the main pulse. The frequency spectra of the main pulse have radiation maxima from 40–60 MHz and this fact increases difficulties in observing decameter pulsar radiation. As shown in /14/, the pulse radiation of the pulsars is dependent on two oscillation modes coinciding in their forms with a high correlation coefficient of 0.92 and being shifted in their longitudes by 145°. It is possible that each of these modes is connected with its own magnetic pole. Though the longitudinal radiation pattern of the pulsar is rather complex, nevertheless, in a number of cases one can find the mean radiation profile with a number of stable maxima and minima. One can also observe a symmetry of the longitudinal radiation structure with respect to the main pulse. The low-frequency pulsar radiation is similar to the Jupiter decameter radiation in its structure; although these objects in their physical features are quite different, nevertheless, both the pulsars and Jupiter have magnetic fields not coincident with their rotation axes having a dipole character. It would be rather important to try and establish how far the analogy goes between these so different cosmic bodies because, if it appears that the processes of studying the pulsars and Jupiter are alike, the near position of the planet will allow us to carry out a study of these processes in detail.

## SOLAR-CORONAL DECAMETER RADIO FREQUENCY RADIATION

L. L. Bazelyan and E. P. Abranin carried out extensive investigations of solar-coronal radio frequency radiation in the decameterband using the UTR-2 radio telescope and the URAN-1 interferometer. To make these investigations, we worked out and constructed receiving system equipment consisting of an 8-channel spectrograph of a parallel coverage from 7.5–31 MHz and a 3-channel dynamic spectrograph of parallel-successive coverage with a passband of 10 KHz and a time constant of 0.01 s. This instrument could conduct measurements using any frequency of the UTR-2 and the URAN-1, the two-dimensional radio heliograph operating at 25, 20, and 16.7 MHz, and the circular-polarization polarimeter measuring the degree and sense of polarization of solar bursts. With this given equipment we made a number of investigations connected with the scattering features of the solar corona both for small $R_\odot$ distances (solar radii from 5 to 20) and for distances up to 200 $R_\odot$ /15–19/; we also made a search for local sources of the slowly varying component of the radiation which, according to some authors, exists in the decameter band and studied the sporadic decameter radio frequency radiation of the Sun. We present below the main results of the measurements made.

At angular distances from the Sun less than 10 $R_\odot$, we found a weaker (than that usually observed at the shorter wavelengths) quadratic increase in the scattering angle of the radio frequency radiation propagating within the corona depending on the wavelength. We created a theory of this phenomenon, based on plane electromagnetic wave propagation in a spherically symmetric inhomogeneous corona, taking account scattering and refraction, which explained the effect observed. The distribution law of the electron density in the corona at large distances ($R/R_\odot > 100$) was determined using pulsar radiation from the Crab Nebula which is translucent in the solar corona. Using a similar method, two-dimensional patterns of cosmic sources translucent through the solar wind were obtained both in the shadow region and in the illuminated one. To explain the results obtained, we used a previously unconsidered model of an elliptic solar corona with quasi-radial electron-density irregularities in the equatorial plane. Two-dimensional distributions of the quiet-Sun brightness were obtained at 25 MHz for the first time. Note that the observed variations of such distributions with a characteristic scale of several minutes are not only of ionosphere origin, but of solar origin as well.

At present to explain the distribution of quiet-sun brightness, computational models are used with small-scale isotropic irregularities in the upper corona penetrable by the solar wind. Such models cannot explain a number of experimental results. It is shown that, as in the case of the corona being sensed by radio frequency radiation of discrete sources, there is a better agreement between the calculations and experiment if one uses models with quasi-radial coronal structures having higher densities and temperatures.

The brightness temperature of the quiet Sun is found to increase for several tens of minutes after intensive type III bursts. This phenomenon is probably connected with the isotropization of some one-dimensional plasmons excited in the corona by electron beams.

Unlike high-frequency cases and the data of some authors, it is found that in the decameter band there are no slowly-changing components of the solar radio frequency radiation. This result is confirmed by measurements made in Australia for the long-wave part of the meter band.

It was shown that for sources of types III and IIIb at different heliolongitudes, they have an identical distribution of angular sizes, and the positions and senses of polarization of their radiation. These data demonstrate the common nature of these bursts.

To support a plasma generation mechanism of type III bursts, we have a number of multiple viewing angle measurements in two adjacent octaves at 25 and 12.5 MHz, and 12.5 and 6.25 MHz which confirm a harmonic correlation between the components of the double bursts of type IIIb-III. We found a decrease by about 17% of the mean velocity of the electron beams exciting type IIIb bursts in moving between the levels of 25 and 6.25 MHz. The threshold character of this decrease corresponds to a collisionless loss-mechanism. We have discovered diffuse stria-bursts and their chains of type-IIId which, as positional measurements show, are a superposition of the straight-line rays coming to the observer and the reflected ones from low coronal layers of a usual stria-burst which is generated at the second harmonic of the plasma frequency. We found that when the source of a decameter noise storm of type III is observed in the central part of the disc, its radiation is recorded near the limb at the second harmonic of the plasma frequency. This finishes the results obtained for objects situated in the Galaxy.

## EXTRAGALACTIC OBJECTS RADIATING DECAMETER RADIOWAVES. CATALOGUES OF DISCRETE SOURCES OF COSMIC RADIATION IN THE DECAMETER BAND

One of the main problems for whose solution the UTR-2 radio telescope was constructed was a catalogue compilation of discrete sources of the Northern sky. These observations have been conducted since 1972. The measurements are made in the transit regime. The five UTR-2 beams spaced in declination are fixed in a set direction and the Earth's rotation moves the beams over the sky, making a zone of about 2 degrees. The observations were conducted at the frequencies: 10, 12.6, 14.7, 16.7, 20, and 25 MHz. Measurements of the same zone are repeated 10 to 20 times. After finishing these measurements the instrument's beams are fixed at another declination so that the records at the extreme beams coincided in both experiments. Previously the radio telescope was calibrated using noise oscillators and low temperature equivalents. Thus, it is clear that absolute values of radiation flux densities can be measured. In the past, we obtained data about flux densities and coordinates of sources situated in zones over declinations of -13° to +20° and of 41° to 60°, and for right ascensions from 0 to 24 hours. About 4,000 sources were found, of which 4% had not been observed earlier. Using these data and the higher-frequency data from other references, we constructed the frequency spectra for more than 1,000 sources. We found that 70% of the frequency spectra are linear (on a logarithmic scale), 20% have a negative curvature at low frequencies, and 10% have a positive curvature /20–21/. Using data taken from the catalogue, it was possible, for the first time at such low frequencies, to investigate the cosmological evolution of cosmic radio sources. For this reason we made an analysis of experimental statistical dependence of a number of sources on the flux density, i.e. the value of $n_{25}(s)/n_o(s)$, where $n_v(s)$ is the number of sources at the frequency $v$ and $n_o(s)$ is the number of sources normalized to a Euclidian universe. At present such calculations are made at a number of frequencies from 5,000 to 178 MHz; moreover it appears that the relationships $n_v(s)/n_o(s)$ show the extent of the deviation of the source spacial distribution from an even one with decreasing frequency /22/. It appears that the lower the frequency, the narrower the region of source fluxes distributed evenly. It follows that the evolution should be most pronounced at low frequencies for extragalactic sources. This paradox has been discussed by many authors /23/. It follows from the data of some references that one should observe a discontinuity effect in the spacial distribution of the farthest radio sources having a value of their redshift which corresponds to the epoch of their formation. Nevertheless, as shown in /24–26/, the calculations of the sources at frequencies more than 400 MHz do not allow one to unambiguously find such a discontinuity because data obtained confirm both the presence and absence of this discontinuity. It appears that just calculations given of the sources at the lowest frequencies (25 MHz) allows one both to explain the paradox mentioned above and to discover the presence of a given discontinuity. In the papers by K. P. Sokolov /27-30/ an analysis was made of the physical causes defining the most pronounced manifestation of cosmological evolution effects in samples of extended sources at the lowest frequencies /27/. This is connected with the fact that one may find the correlation between the measured flux density and the distance to radio sources, A, only at the frequencies lower than 100 MHz. Assumptions such as these form the basis for all radio astronomical cosmological tests. Before presenting the results of this analysis we note the main features of the calculation of sources at 25 MHz /26–28/. It is here that one has narrowest interval of fluxes for which the dependence of $n_{25}(s)/n_o(s)$ has an inclination exceeding that of a Euclidian universe. A considerable deficiency of the radio sources is observed having S>80Jy and the integral surface density of the distribution of N < 30 sr$^{-1}$. Finally, there is a range of flux densities (10≤S≤20 Jy) for which one observes a flux density decrease of weak extended sources. Thus, data given at 25 MHz indicate a discontinuity in the spacial distribution of old extended radio sources. This agrees with the optical range discontinuity of the spacial distribution of the quasars for Z=3–4. We note that at 25 MHz the distribution discontinuity takes places for Z < 3. It is shown in /29/ that the character of the spacial source distribution is defined by the dependence of the fluxes on extended components and compact ones. If the extended component fluxes characterize the distance to the source, the compact component fluxes depend both on the distance and direction of the radiation, and on the Doppler enhancement or the weakening of the flux. Therefore, at frequencies higher than 1,000 MHz where compact component fluxes prevail, in principle one cannot unambiguously characterize the spacial source distribution and reliably find the presence of the above mentioned discontinuity. As a result, the observed spacial compact source distribution coincides with a Euclidian distribution over a wide range of flux densities.

## ANGULAR STRUCTURE OF EXTRAGALACTIC RADIO SOURCES IN THE DECAMETER BAND

The URAN-1 and URAN-4 interferometers were used to determine the angular structure of some extragalactic radio sources. About 20 radio galaxies and quasars were studied. As examples we shall consider the radio galaxy 3C123, the unidentified object 3C134, and the quasar 3C196. One very important problem which can be solved by interferometers is the variation of the angular size of the source with frequency. It is known from data obtained at high frequencies that most of discrete radio sources have compact details of small angular size. Reabsorption should be observed in these details and hence the frequency spectra of these sources should have a negative curvature at low frequencies. At the same time, our measurements show that less than 20% of the sources have such spectra. One of the possible explanations of this is connected with the fact that decameter and shorter waves are radiated by different source regions: the centimeter waves are radiated by compact details and the decameter ones are radiated by the outer envelope of the source. If such a model is valid, the angular source sizes could depend on frequency. The possibility of this dependence can be checked by interferometry.

The angular structure of one of the farthest galaxies considered to be the radio source 3C123 was studied over a wide frequency band. According to the measurements from 38 to 1425 MHz, the source has a one-component structure with an angular size of less than 25-30". At higher frequencies the total size of 3C123 does not exceed 30". Thus, from 38-8000 MHz the angular size does not change. The 3C123 source was measured by the URAN-1 interferometer at 25, 20, and 16.7 MHz /31/. It appears that for decameter waves as well, the source size is 23". Therefore, we may suppose that from 16.7 to 8000 MHz the main component of the 3C123 radio radiation is due to the same region of the source. It was pointed out in /31/ that the calculations for a two-component model of 3C123, which is used by some investigators from the data at greater than 2000 MHz, showed that the visibility function moduluses for such a model at the decameter waves do not correspond to the experimental results obtained with the URAN-1 interferometer. It appears that the unidentified source 3C134 has a structure different from that of 3C123. It is a typical two-component object which from measurements from 158 to 1425 MHz has the following angular sizes: it is 50 and 35" along the main axis; the total angular source-size is 138" and the position angle is 170°. At higher frequencies, in particular at 2695 MHz, it is stated that 3C134 should be representative of objects of the core-halo type. Decameter wave measurements were carried out with URAN-1 interferometer at 25, 20, and 16.7 MHz /32/. For all frequencies, a double-hump curve was obtained, characteristic of a two-component source with the interference-oscillations envelope depending on the hour angle. It appears that the ratio of the flux densities of both components is $1.22 \pm 0.15$ at 25 and 20 MHz. The conclusion is made in that paper that the two-component structure of 3C134 does not vary significantly from 16.7 to 1407 MHz, but the conclusion that at high frequencies this source is a core-halo type is ambiguous. It was further pointed out in /32/ that is one suggests that the components have equal isotropic Gaussian brightness distributions, then for the two components with angular sizes of 37" and a ratio of flux densities of 1.22, a rather satisfactory coincidence of the moduluses of the visibility function is obtained, depending on the interference bases (in terms of wavelengths). This coincidence occurs from 20 to 1425 MHz. This indicates that, over this frequency band, the angular sizes of source components practically do not change. Thus, in a number of the cases considered, the angular sizes of extended sources (such as SNR in the Cygnus Loop, the radio galaxy 3C123 and the unidentified object 3C134) remain unchanged over the wide range from 16.7 to 8000 MHz.

At the same time in some cases a different picture is observed. For example, let us consider the angular structure of the quasar 3C196. A number of papers /33–35/ and references therein are concerned with this source. It was shown in /33/ that in the whole frequency band for which measurements exist, the quasar 3C196 consists of two components from 38 to 408 MHz and of three-four components from 408 to 5000 MHz. These details have different brightnesses. Thus, two of them (S-W and N-E) are the brightest, have sizes of about 2 seconds, and are spaced about 5.4 arc seconds relative to each other for a position angle of about 26°. Measurements are presented in /34,35/ of the angular structure of 3C196 in the decameter band. The measurements were conducted by the URAN-1 and URAN-4 interferometers at 20 and 25 MHz. Before describing the results obtained, we note that at frequencies as low as 38 MHz it was shown in /33/ that 3C196 consists of two compact details with angular sizes of each less than 3.1", and the source does not contain extended components with a flux density of more than 5–10% of the total radiation intensity of the quasar 3C196. With the URAN-1 and URAN-4 interferometers we obtained values of the visibility-function modulus, but no hour dependence of this value was found either at 20 or at 25 MHz. We obtained values of the visibility-function modulus for the different interferometer bases; the absence of an hour dependence of this value and the frequency spectrum of the whole source 3C196 from 16.7 to 5000 MHz allowed us in /34-35/ to construct a source model which agrees with all experimental results. The model consists of a compact component and an extended one. A connection of this model with higher-frequency brightness distributions is shown in /34-35/. It appears that an extended detail exists for frequencies higher than 25 MHz, but the spectral index of this detail is high (1.25) while high-frequency compact details have an index of 0.83. For frequencies lower than 200 MHz for the frequency spectrum of the whole source, a break is observed connected with the fact that the spectrum of one of the compact details (N-E) begins to decrease and the flux maximum of this detail occurs at about 160 MHz; with a further frequency decrease, the flux density of this detail decreases as $v^{5/2}$. Reabsorption seems to take place in this source. The second compact component (S-W) has a frequency spectrum maximum from 40-25 MHz with a flux density decreasing as $v^{1.2}$. Thus, at frequencies lower than 20 MHz, the 3C196 radiation is determined only by the extended component and from 10-20 MHz the 3C196 quasar ia a one-component source. On the other hand from 30-100 MHz the two-component source has extended and compact details, and from 200-5000 MHz 3C196 is a multicomponent source with two, three, and even four compact details. Note that the accepted model gives sizes of the extended-component of about 18" and those of the compact component of 2-2.5". From preliminary data, the appearance of an extended component in 3C196 is also characteristic of some other quasars. Possibly the appearance of such extended components is connected with the fact that around quasars, even such as 3C196 (Z = 0.87), there are optical nebulae of the halo-type capable of radiating radio waves.

The results presented above were obtained by scientists of the Department of Decameter Radio Astronomy of the Radio Astronomy Institute of the Ukrainian SSR Academy of Sciences.

The author expresses his deep gratitude to all of them for the possibility of using the results of their investigations. The author is also grateful to E. A. Abramenkov, L. L. Bazelyan, Yu. M. Bruk, A. A. Konovalenko, A. V. Megn, and K. P. Sokolov for their materials that were presented.

## References

1. Braude, S. Ya., Megn, A. V., and Sodin, L. G. (1978) Antenny Sbornik No 26, Svyaz', Moscow.
2. Bobeyko, A. L., Bovkoon, V. P., Braude, S. Ya., et al. (1978) Antenny Sbornik No 26, Svyaz', Moscow.
3. Konovalenko, A. A., and Sodin, L. G. (1980) *Nature* **283**, 360.
4. Blake, D. H., Butcher, R. M., and Watson W. D. (1980) *Nature* **287**, 707.

5. Konovalenko, A. A., and Sodin, L. G. (1981) *Nature* **294**, 135
6. Sharpless, S. (1959) *Astron. Journ. Supp. Serr.***4**, 257.
7. Kabanova, T. N., Krymkil, V. V., and Sidorchuk, M. A. (1989) *Kinematika Fiz. Neb. Tel.* **5**, 344.
8. Krymkin, V. V., and Sidorchuck M. A., (1989) *Kinematika Fiz. Neb. Tel.,* in press.
9. Bovkoon, V. P., Braude, S. Ya., and Megn, A. V. (1982) *Astroph. Sp. Sc.* **81**, 221.
10. Vinyaikin et al. (1987) *Astron. Zh.* **64**, 987.
11. Bovkoon, V. V., and Zhuk, I. N. (1981) *Astroph. Sp. Sc.* **79**, 181.
12. Bovkoon, V. V., and Zhuk, I. N., and Sobolev, Ya. M. (1987) *Astron. Zh.* **64**, 734.
13. Dennison, B., et al. (1984) *Astron. Astrop.* **135**, 199.
14. Bruck, Yu. M., (1987) *Austr. J. Phys.* **40**, 861.
15. Abranin, E. P., Bazelyan, L. L., et al. (1979) *Sol. Phys.* **62**, 145.
16. Abranin, E. P., Bazelyan, L. L., et al. (1982) *Sol. Phys.* **78**, 179.
17. Abranin, E. P., Bazelyan, L. L., et al. (1984) *Sol. Phys.* **91**, 383.
18. Abranin, E. P., Bazelyan, L. L., and Tsybko, Ya. G. (1984) *Sol. Phys.* **92**, 293.
19. Braude, S. Ya., Megn, A. V., et al. (1989) *Doklady AN SSSR,* No 10,70.
20. Braude, S. Ya., Sharykin, N. K., Sokolov, K. P., and Zakharenko, S. M. (1985) *Astrop. Sp. Sc.* **111**, 1.
21. Braude, S. Ya., Sharykin, N. K., Sokolov, K. P., and Zakharenko, S. M. (1985) *Astrop. Sp. Sc.* **111**, 237.
22. Wall, J. V. (1978) *M.N.R.A.S.* **182**, 381.
23. Kellerman, K. I., and Pauliny-Toth, I. I. K. (1981) *Ann. Rev. Astron. Astroph.* **19**, 373.
24. Wall, J. V., Pearson, T.J., and Longair, M. S. (1980) *M.N.R.A.S.* **193**, 683.
25. Peacock, J. A., and Gull, S. F. (1981) *M.N.R.A.S.* **196**, 611.
26. Condon, J. J. (1984) *Astroph. J.* **287**, 461.
27. Sokolov, K. P. (1988) *Pis'ma Astron. Zh.* **14**, 202.
28. Sokolov, K. P. (1986) *Astron. Zh.* **63**, 426.
29. Sokolov, K. P. (1988) *Astron. Zh.* **65**, 236.
30. Sokolov, K. P. (1986) *Pis'ma Astron. Zh.* **12**, 254.
31. Megn, A. V., Braude, S. Ya., Rashkovsky, S. B., et al. (1987) *Pis'ma Astron. Zh.* **13**, 751.
32. Megn, A. V., Braude, S. Ya., Rashkovsky, S. B., et al. (1985) *Astr. Zh.* **62**, 38.
33. Megn, A. V., Braude, S. Ya., Rashkovsky, S. B., et al. (1987) *Izvest. VUZ Radiofizika* **30**, 474.
34. Megn, A. V., Braude, S. Ya., Rashkovsky, S. B., et al. (1990) *Izvest. VUZ Radiofizika* **33**, 523.
35. Megn, A. V., Braude, S. Ya., Rashkovsky, S. B., et al. (1990) *Izvest. VUZ Radiofizika* **33**, 534.

# 8

# New Developments in Cosmic Gas Dynamics: Galactic Shocks, Hot Protogalaxies, and Galactic Superwinds

L. S. Marochnik
*Space Research Institute, Moscow*
A. A. Suchkov
*Rostov State University, Rostov*

This review highlights recent developments in two areas of cosmic gas dynamics: gas flowing through the density waves in spiral galaxies and the evolution of a gaseous protogalaxy interacting with a strong galactic wind. The discussion is preceded by a short overview of a controversy regarding steady-state and supernova-dominated models of the interstellar medium (ISM). We conclude that the former one seems to be more compatible with current observations. We also argue that magnetic fields in the ISM are plausibly subjected to turbulent diffusion which strongly reduces their effect upon the dynamics of the ISM. The review presents an illustration of how gas flows through density waves produce large-scale galactic shocks that generate small-scale gas clouds. It also provides a short account of the "hot" model of galaxy formation whose key idea is the formation of a "hot" protogalaxy as a result of interaction of an early galactic "superwind" with a massive gaseous component surrounding the region of an initial burst of star formation. This scenario is illustrated by results of numerical integration of gas dynamic equations for a model protogalaxy. They show the capability of the model to explain the stellar dimensions of giant galaxies, the mass and iron abundance of the gas in clusters of galaxies, and the origin of X-ray coronae around giant ellipticals.

## I. INTRODUCTION

Over the last decade, there has been a tremendous advance in all-wavelength observations of cosmic gas in and outside galaxies. A wealth of new phenomena involving various dynamical processes in rarefied gas have been discovered, which has stimulated rapid development of new areas in cosmic gas dynamics studies, such as currently vary popular cooling flows of X-ray emitting gas in giant galaxies and clusters of galaxies. Also, the classical problems of interstellar gas dynamics have not been left intact. An example is provided by the theory of the interstellar medium where the widely accepted supernova-dominated model of McKee and Ostriker (1977) has been questioned as not being compatible with the data obtained on the base of a new generation of radio astronomy tools (cf. Heyles 1986).

The aim of the present review is mainly to provide a short account of two areas in cosmic gas dynamics which seem to us very promising. The one is related to gas flows in density waves of spiral galaxies, and the other one bears upon the evolution of protogalactic gas clouds and their interaction with the early galactic winds. This choice is motivated mainly by the fact that these topics have been the focus of our own activity over the last years. At the same time we are quite aware that they are related to the most fundamental problems of modern astrophysics such as the structure and evolution of spiral galaxies, star formation history, and galaxy formation.

In Section II we discuss the current understanding of the ISM in disk galaxies, including the role played by the magnetic field in the dynamics of the ISM. Section III highlights some results in studies of gas flows through the density waves in spiral galaxies. In section IV we discuss the dynamics of a protogalactic gas interacting with an early galactic wind produced by an initial burst of star formation.

## II. THE STATE OF THE ISM: STEADY-STATE VERSUS SUPERNOVA-DOMINATED MODELS

The ISM is recognized to include components covering a wide range of density and temperature. However, its most fundamental features are usually discussed in the framework of two models, both of which ignore to a greater or lesser extent relatively small variations in physical parameters within major components (phases) of the ISM as opposed to large contrasts between the phases. The two-phase steady-state model (Pikel'ner 1967; Field et al. 1969) assumes that the ISM represents a rarefied gas layer within the galactic disk, supported against gravitation along the z-direction by the thermal pressure at temperatures of about 600 to 800 K; this is the warm phase of the ISM. Embedded in it are clouds of much more condensed and cold gas that are in pressure equilibrium with the warm intercloud medium; they are referred to as the cold phase of the ISM and are commonly identified with observed diffuse HI clouds. The pressure equilibrium between the phases is believed to be maintained by the balance of radiative and cosmic ray (or X-ray) heating.

McKee and Ostriker (1977) have developed an essentially different model of the ISM, based on the idea that the ISM is governed by supernova explosions. In this model the ISM is envisioned as being largely in the form of a very tenuous hot gas ($T \sim 10^5$–$10^6$ K) which represents in fact the overlapping supernova remnants; only a small portion of the interstellar space is allowed to be occupied by warm- and cold-phase gas.

It is recognized that the observational evidence for either model remains controversial. The supernova-dominated model has been certainly favored by most authors, evidently due to its apparent unavoidability from the physical point of view once the conventional rate and energetics of Type II supernovae (SNe) is adopted. However, the current progress in radio observations of interstellar hydrogen, especially of its small-scale structure in nearby spiral galaxies, indicates that it is the cold and warm gas that dominates the disks of spirals rather than the hot gas (cf. Heyles 1987). A recent study of Richter and Rose (1988) seems to provide an explanation for the discrepancy between the predictions of the SNe dominated ISM and the latter observations. These authors have found evidence that (as also suggested by Heyles (1987)) the SNe rate has been strongly overestimated for the majority of galaxies (while just as strongly underestimated for a small number of galaxies producing the major part of all SNe).

The dominance of the hot phase has been questioned occasionally also on several other grounds. Turner (1981) has noted that "A hot tenuous medium cannot be compressed by nor develop a spiral density wave (sdw) shock when it passes through the spiral gravitational field. Nevertheless, there is observational evidence for a density enhancement in the HI (factor of ~ 10) in the sdw shocks in external galaxies (van der Kruit and Allen 1978)". Turner's major impression is that the filling factor of the hot component must be small, at least "somewhere, sometime".

The same view is supported by Heyles (1987). He argues that both the low value of the soft X-ray background and the small number of hot supercavities in the galactic disk is not compatible with a large filling factor for the hot phase in our Galaxy.

Thus, there are currently good reasons to turn again to a two-phase model of the ISM of the kind envisioned by Pikel'ner (1967) and Field et al. (1969). In the following discussion we shall proceed from the assumption that the major understanding of the ISM implied by this model is true. The ISM will be considered as composed of warm rarefied intercloud gas and dense cold clouds with diffuse HI clouds being almost in pressure equilibrium with the intercloud gas due to the balance of radiative cooling and cosmic ray heating. Each of the two phases may actually cover a range of density and temperature due to variations of basic parameters governing the two-phase state within a galaxy. The two-phase state can certainly be only quasistationary rather than strictly stationary, at least because of the spiral density waves.

When a true supernova-dominated ISM occurs, we believe that it happens in spiral galaxies, but it appears only as a consequence of powerful bursts of star formation that strongly increase the Type II supernovae rate (cf. Ikeuchi et al. 1984).

Now we remark on the role of interstellar magnetic fields. The most important dynamical property of the interstellar gas is its enormous compressibility caused by radiative energy losses. It allows the gas to develop very large density contrasts and this is believed to occur within the galactic disks when cold dense clouds are formed from a tenuous warm gas. The corresponding physics is rather well understood. However, a lucid physical picture for the dynamical behavior of such a gas becomes immediately blurred once magnetic fields are involved. Magnetic fields, being "frozen" in to the gas and behaving as an elastic medium, resist efforts to compress the gas across the force lines. Since the interstellar magnetic fields possess a high entangled component (together with a regular one), the corresponding magnetic pressure acts isotropically in fact (on scales larger that the homogeneity scale of the field). When the magnetic pressure within the galactic disk is comparable to the thermal pressure, the medium as a whole, i.e., gas plus field, is expected to behave essentially adiabatically with its dynamics being governed mainly by the interstellar field.

However, in this case any strong compression necessary to produce a dense cloud in a rarefied medium is impossible. On the other hand, gas contraction along the force lines of a presumably regular magnetic field can hardly be accepted as a universal mechanism for cloud formation because the real interstellar fields do contain a strong random component. Nonetheless, the clouds are somehow formed in gaseous disks of galaxies.

Similarly, a regular component of a large-scale magnetic field directed along a spiral arm, being "frozen-in", must prevent the interstellar gas from forming a highly compressed region known as a galactic shock along the arm; yet such shocks are observed.

All this taken together suggests that magnetic fields can be carried away very efficiently from a contracting gaseous element.

A mechanism for efficient transporting the embedded magnetic fields in cosmic gas bodies has been proposed over two decades ago; this is the rapid diffusion of the field caused by turbulence (cf. Parker 1973). The ISM in the galactic disk is known to be turbulent over a very large range of scales. So the enhanced turbulent diffusion of the field is expected to operate on the scale of the galactic disk as a whole (Parker 1972) and down to scales of the smallest clouds.

In case of strong turbulence, the turbulent diffusion is believed to occur on a time scale of the order of the time required for sound to travel through a given gaseous element. If so, on scales larger than this time the embedded magnetic fields would not be able to prevent this element from contracting to high density. On the other hand, the same sound travel time characterizes the rate at which a gaseous element can change its density due to a dynamical process. This implies that the capability of a rarefied, radiatively cooling gas to develop large density contraction would not be drastically changed by allowing for magnetic fields if the gas is turbulent; the fields would only somewhat slow down this process rather than totally inhibit it.

These considerations suggest that ignoring magnetic fields in describing the dynamical behavior of the ISM is in certain respects much more adequate than involving the approximation of fully "frozen-in" fields. The latter approach is hardly capable of accounting for the most important processes in the ISM, namely, formation of dense clouds and strong galactic shocks while the former one predicts and excellently describes them.

## III. GALACTIC SHOCKS AND CLOUD FORMATION

Let us consider the initially circular motion of gas in a galactic disk where the gravitational field of a stellar density wave gradually arises. In the case of a tightly wound two-armed spiral pattern rotating at a speed $\Omega_p$, the equations governing the motion are as follows (cf. Roberts 1969; Marochnik et al. 1983):

$$\frac{du}{dt} - 2\Omega(v - v_o) = -\frac{1}{\rho}\frac{\partial p}{\partial x} - \frac{\partial \varphi_s}{\partial x},$$

$$\frac{dv}{dt} + \frac{\kappa^2}{2\Omega}(u - u_o) = 0,$$

$$\frac{dx}{dt} = u,$$

$$\frac{d\rho}{dt} + \rho\frac{\partial u}{\partial x} = 0,$$

$$\frac{3}{2}\frac{d}{dt}\frac{P}{\rho} + \frac{p}{\rho}\frac{\partial u}{\partial x} = -L(\rho, T, \zeta_i),$$

$$\varphi_s = A(t)\cos\left(\frac{2x}{r \sin i}\right),$$

So the following discussion, which does not take explicitly into account the interstellar magnetic fields, may be regarded as based on a conviction that the fields can very efficiently withdrawn from any contracting gas volume due to turbulent diffusion.

where u and v are velocity components directed perpendicularly and tangentially to the spiral arm respectively, x is the spiral coordinate orthogonal to the direction along the arm, L $(\rho,T,\zeta_i)$ is the rate of internal energy variation due to radiative cooling and cosmic ray heating, $\varphi_s$ is the gravitational potential of the density wave, A(t) is its amplitude and $\kappa$ is the epicyclic frequency. The coordinate x is related to the polar galactocentric coordinates r and $\theta$ as

$$x = r[\cos i \ln(r/r_o) + (\theta - \Omega_p t) \sin i],$$

where i is the pitch angle of the arm. The reference frame rotates at a speed $\Omega_p$, and $u_o = r(\Omega - \Omega_p) \sin i$, $v_o = r(\Omega - \Omega_p) \cos i$ are the velocity components of the background circular motion (the subscript "o" refers to the initial values of gas parameters).

Marochnik et al. (1983) have numerically integrated the above equations for a set of models differing in $n_o$, $\Omega_p$, A, r, $\Omega$, $\kappa$, and $\zeta_i$. To illustrate the most conspicuous features of the solutions, we present the one corresponding to $n_o = 0.05$ cm$^{-3}$, $\Omega_p =$

23 km s$^{-1}$ kpc$^{-1}$, r = 5 kpc, A = –265 km$^2$ s$^{-2}$ (which is a stationary value after t ≅ 3 x 10$^8$ years), i = 7°, Ω = 46 km s$^{-1}$ kpc$^{-1}$, and κ = 73 km s$^{-1}$ kpc$^{-1}$.

Figure 8.1 shows density profiles across the spiral arm at three different moments of time. It can be seen that the gas pressure and density grow in the potential well of the density wave and a galactic shock gradually forms. When the density at the shock front reaches a value corresponding to the upper critical point of the equilibrium curve, $n_{cr}^{up}$, a certain part of the gaseous mass there rapidly cools and contracts with the two contact discontinuities being formed at its boundaries; the density and temperature jump at the boundaries by about two orders of magnitudes, the pressure varies much less, and the velocity remains initially almost continuous. The cooled mass displayed as a density

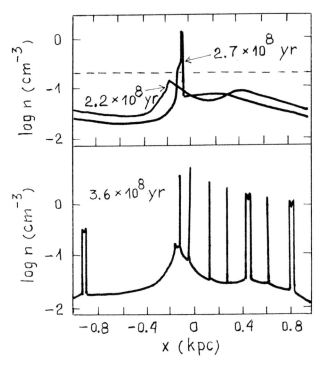

Figure 8.1. Development of density profiles across the spiral arm. The time elapsed after the gravity of the density wave is turned on is indicated. The zero point on the X-axis corresponds to the center of the gravitational well of the density wave. The dashed line indicates the critical density corresponding to the upper critical point of the equilibrium curve in the pressure-density plane. Note the formation of a density peak on the shock front as the latter reaches the upper critical point.

peak in the upper panel of Fig. 8.1 is carried away as a whole out of the shock front and moved down-stream with the main flow. It is very natural to interpret this process as the formation of a cool dense cloud from initially warm gas.

Formation of the cloud reduces the pressure and density at the shock front and some time is required for these parameters to reach the upper critical values again. After that the process repeats and a second cloud is formed, then a third one, etc. (the lower panel of Fig. 8.1). So the solution obtained envisions that a galactic shock can really be a "machine" which "manufactures" clouds in galaxies.

The density peaks seen in Fig. 8.1 are in fact unresolved regarding their true spatial structure and dimensional on scales of Fig. 8.1. Their true size can be seen in Fig. 8.2 where we show on a much smaller scale the structure of a "cloud" corresponding to one of the peaks from Fig. 8.1.

We have chosen for this illustration one of those "clouds" which reveal rather specific behavior. As seen in Fig. 8.1, the gas flow exhibits essentially non-stationary behavior with gas parameters varying significantly both in the clouds and the ambient warm gas. In the case of the cloud shown in Fig. 8.2 the density in the cloud center at some moment has dropped below the value corresponding to the lower critical point of the equilibrium curve. As a result, the temperature rapidly jumps up here which entails a

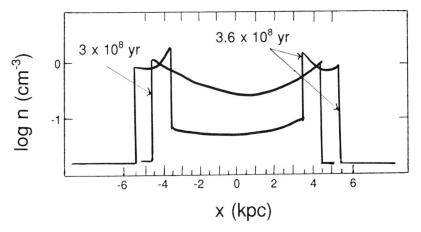

Figure 8.2. Density profiles across the first density peak formed as shown in Fig. 8.1. The peak is interpreted as a cloud formed on the front of the galactic shock. The cloud inflates because the central density overshoots above the critical density, $n_{cr}^{up}$. The inflation forms a shock on the inner boundary of the dense shell; the outer boundary represents a contact discontinuity slowly propagating into ambient warm gas.

rapid pressure increase. This leads to a shock propagating outwards. The cloud inflates, forming a bubble of hot gas inside a dense, cold shell.

It is important that formation of separate density peaks requires rotation, or, to be more exact, the action of the Coriolis force. The above equations with $\Omega = 0$ have yielded a density profile with an infintely growing peak on the shock front; in this case the "cloud" cannot leave the gravity potential well (Berman et al. 1986). This result may be pertinent to the general problem of cloud formation.

Summarizing this section, we see that an initially uniform gas flow in the presence of a stellar density wave easily develops a large-scale galactic shock as well as small-scale contact discontinuities and shocks.

It is to be expected that the pressure of magnetic fields would smooth these processes. At the same time, non-stationary shocks and contact discontinuities would undoubtedly generate miscellaneous waves, thus generating turbulence, and the turbulent diffusion of magnetic fields would reduce the effect of magnetic pressure on the development of density contraction in the gas. Therefore it may be suggested that magnetic fields (and cosmic rays) do not influence as strong as they could (if fully frozen into the gas) the formation of shocks and clouds, and that the variation of magnetic field energy (and cosmic ray density) in the ISM is significantly smaller than that of the gas density.

## IV. HOT PROTOGALAXIES AND GALACTIC SUPERWINDS

A large new area for gas dynamics applications in astrophysics has been opened by recent progress in understanding how gaseous protogalaxies turn into stellar systems. There is little doubt that galaxy formation and early evolution involve enormous energy input into large gaseous masses that leads to various shocks, hot outflows, thermal instabilities, etc. It is widely recognized that galaxy formation is accompanied by the occurrence of strong galactic winds and subsequent mass loss which in fact determines the global characteristics of resulting stellar systems (Bookbinder et al. 1980; Dekel and Silk 1986; Arimoto and Yoshii 1987). The corresponding gas dynamics was also examined in relation to the origin of hot X-ray coronae and stellar shells in giant ellipticals (Lowenstein and Mathews 1987; Umemura and Ikeuchi 1987). Below we give an illustration of how these gas dynamics may operate to form giant galaxies, X-ray coronae, and hot, metal-rich intergalactic gas.

As argued by Suchkov et al. (1987), several lines of evidence suggest that the main star formation in giant galaxies is preceded by an evolutionary phase at which a protogalaxy becomes overheated so that the thermal energy overcomes the binding energy of its own gravity and violent outflow of gaseous material occurs. This phase is supposed to be due to a large number of Type II supernova appearing at the initial burst of star formation which is believed to produce mainly very massive stars. The major star formation is supposed to occur long afterwards from a material enriched by supernovae with metals when the inner part of the expanding hot protogalaxy, after being decelerated, reverses its motion into radical contraction and rapidly cools down as the density surpasses some critical value.

This scenario was first suggested by one of us (A.S.) and his colleague over a decade ago to explain our discovery that the galactic components (the halo, the intermediate component, and the disk) are decoupled in respect of their metallicity; the very existence of quite distinct galactic components, regarding their kinematics, spatial configuration, and age, was also attributed to star formation characterized by bursts and subsequent lulls (the theory of "active phases of evolution" of the Galaxy; Marsakov and Suchkov 1976; Suchkov 1981).

The later development has added new pieces of evidence in support of our conviction that some dramatic events occur in gaseous protogalaxies before they turn into predominantly stellar systems. Thus, the famous mass-metallicity relation for galaxies is universally regarded at present as an indication for a heavy mass loss at early stages of galaxy evolution. The same is inferred from the large iron abundance and iron mass in the intergalactic gas of clusters of galaxies. The observations of distant galaxies and quasars straightforwardly reveal violent star-formation events and the presence of large mass of extremely hot gas showing violent dynamical behavior.

These facts has led us to "a hot model" of galaxy formation (Suchkov 1988; Suchkov et al. 1987). Its key notion and key object of study is "a hot protogalaxy", whose origin, evolution, and relation to observational manifestations of galaxy formation are thought of as follows (Berman and Suchkov 1990).

We assume that a burst of initial star formation in a giant galaxy, covering mainly its central part, leads to the development of a galactic "superwind", enriched with metals, interacts with the rest of the protogalactic gas representing itself cold gas clouds, possibly embedded into a much hotter intercloud gas, that move under the overall gravity of the galaxy toward the center of the system. As the highly supersonic galactic wind encounters the clouds, strong shocks heat the latter up to high temperatures that make them disperse and mix up with the wind material. A similar effect upon the clouds is produced by a shock induced by the superwind in the intercloud medium as well as by the hot post-shock gas. The result is two-fold. First, the massive remainder of the initial protogalaxy becomes overheated and turns into expansion: Second, in this model the wind is capable of enriching the most remote outskirts of the protogalaxy, so its outer parts, being expelled into the intergalactic space, would produce the hot, metal-rich intracluster gas.

The subsequent evolution of the retained part of the "hot" protogalaxy is deceleration and a reverse from expansion to contraction which finally results in rapid cooling and a major burst of star formation responsible for the dominant metal-rich stellar population in the giant galaxies (dwarf galaxies lose all their remaining gas together with metal-enriched wind during an early active phase so they are left only with the initial metal-poor stellar population).

The dynamics of a gaseous protogalaxy interacting with a galactic superwind is illustrated well in Fig. 8.3. It displays the behavior of one of a series of spherically-symmetric models examined by numerical integration of the gas dynamic equation (Berman and Suchkov 1990). In this particular model the galactic wind blows for $10^8$ yr at a rate $\dot{M} = 100$ M$_\odot$ yr$^{-1}$, its initial velocity $u_o = 1500$ km s$^{-1}$, and the temperature $T_o = 5 \times 10^7$ K. The wind interacts with the gas clouds of total mass $M_{cl} = 10^{11}$ M$_\odot$, and the intercloud gas of mass $M_{int} = 10^{10}$ M$_\odot$ located initially within a sphere of radius $R_o = 100$ kpc. The gas moves towards the center under its own gravity and that of the dark halo whose mass, $M_v = 2 \times 10^{12}$ M$_\odot$ is distributed within 20 kpc. The equations describing the model are as follows:

$$\frac{d\rho}{dt} + \frac{\rho}{R^2}\frac{\partial R^2 u}{\partial R} = \dot{\rho}_{cl},$$

$$\frac{du}{dt} = -\frac{1}{\rho}\frac{\partial p}{\partial R} - \frac{GM(R)}{R^2} - f(R) - \dot{\rho}_{cl}\frac{u - u_{cl}}{\rho},$$

$$\frac{3}{2}\frac{d}{dt}\left(\frac{P}{\rho}\right) = \frac{P}{\rho R^2}\frac{\partial uR^2}{\partial R} + \frac{\dot{\rho}_{cl}}{\rho}\left[\frac{(u-u_{cl})^2}{2} - \frac{3}{2}\frac{P}{\rho}\right] - L(\rho,T),$$

$$\frac{du_{cl}}{dt} = -\frac{GM(R)}{R^2} - f(R),$$

$$\frac{d\rho_{cl}}{dt} + \frac{\rho_{cl}}{R^2}\frac{\partial R^2 u_{cl}}{\partial R} = -\dot{\rho}_{cl},$$

where f(R) is the gravitational acceleration due to the dark halo, L ($\rho$, T) is the rate of energy loss due to radiative cooling, and $\dot{\rho}_{cl}$ is the rate at which the mass of continuous gas medium is growing due to cloud evaporation. The latter quantity for this particular model is taken in the form

$$\dot{\rho}_{cl}(R) = \beta\rho(R)\left[\frac{3}{2}\frac{P}{\rho} + \frac{1}{2}(u - u_{cl})^2\right],$$

where

$$\beta = M_{cl}\left[M_w\left(\frac{3}{2}\frac{P_{wo}}{\rho_{wo}} + \frac{u_{wo}^2}{2}\right)t_{cl}\right]^{-1} \quad \text{if } \dot{\rho}_{cl} > 0,$$

$$\beta = 0 \quad \text{if } \dot{\rho}_{cl} = 0,$$

$M_w = \int_0^{t_w} \dot{M}_w \, dt$, and $t_{cl}$ is the time during which a gaseous mass with a total energy $M_w \left( \dfrac{3}{2} \dfrac{P_{wo}}{\rho w_o} + \dfrac{u_{wo}^2}{2} \right)$ is capable of evaporating the cloud mass $M_{cl}$; the subscript "o" refers to the boundary values of the parameters. For the model displayed in Fig. 8.3, $t_{cl}$ is taken to be $2.5 \times 10^8$ yr.

The upper panel in Fig. 8.3 shows a strong shock in the intercloud gas at R ~ 60 kpc, a hot, shocked gas between 20 and 60 kpc, an outer boundary of the wind at 20 kpc, and a region from 5 to 20 kpc occupied at that time by the wind. By the time displayed in the mid-panel the shock has moved far beyond 100 kpc while within 100 kpc and even farther away the hot gas has already developed a subsonic flow directed towards the center. At that time the gas density in a layer at R ~ 15 kpc has surpassed the value at which radiative cooling becomes the dominant effect. The result is the rapid formation of an extremely dense, cold shell. The latter is undoubtedly subjected to various instabilities and must fragment into separate clouds that may be regarded as material ready for star formation. So, at a certain stage the layer is assumed to be fragmented and is excluded from further gas dynamic calculations.

At times $t > 1.5 \times 10^9$ yr the system is approaching a steady-rate and a kind of a quasistationary cooling flow settles. At the moment displayed in the lower panel, nearly half (48%) of the initial gaseous mass already resides within ~ 5 kpc, forming a dense cloud nucleus; in fact this gas must be essentially converted there into stars so the sharp boundary at R ~ 5 kpc separating the nucleus from the extended tenuous hot envelope is to be considered as determining the dimension for a stellar system formed from this gas. The other half (49%) of that mass is expelled into the intergalactic space. It carries away 40% of metals contained in the wind. Taking into account that the nucleus has captured 56% of metals, we conclude that the model leads to approximately equal portions of barionic mass and nearly equal iron abundance in giant galaxies and in the intergalactic gas that originates from ejecta of hot protogalaxies and fills the corresponding cluster of galaxies. This is a very encouraging result since, on the one hand, the X-ray observations have yielded just this proportion of mass and iron abundance, and, on the other hand, the model seems to be quite reasonable from the point of view of both the galactic parameters and galactic wind parameters (as deduced, e.g., from far-infrared galaxies).

The tenuous envelope containing about 3% of the initial gaseous mass might well be identified with X-ray coronae of the giant ellipticals; it emits on the levels typical for observed giant ellipticals, only slightly decreasing its X-ray luminosity on a time-scale of about $10^{10}$ yr.

Summarizing this section we conclude that the idea of early galactic superwinds interacting with gaseous protogalaxies is capable of providing a self-consistent picture for galaxy formation that accounts for various fundamental facts such as the sizes of giant galaxies, the mass and iron abundances of the intracluster X-ray gas, X-ray coronae of giant ellipticals, etc.

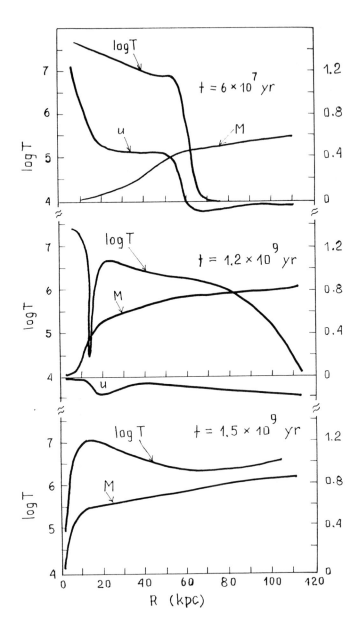

Figure 8.3 (facing page). The dynamics of a "hot" protogalaxy. The right scale is for mass, M, and velocity, u, normalized to $5 \times 10^{10}$ M$_\odot$ and 1150 km s$^{-1}$, respectively. M is the total mass of galactic wind, intercloud gas, and gas clouds currently dissolved, and R is the distance from the galactic center. At times $t > 1.5 \times 10^9$ yr the gas velocity is almost zero everywhere excluding the central part, R < 5 kpc, so the system reaches a quasistationary state.

## REFERENCES

Arimoto, N., and Yoshii, Y (1987) *Astron. Astrophys.* **173**, 23; *Astron. Ap.* **188**, 13.
Berman, V. G., Marochnik, L. S., Mishurov, Yu. N., et al. (1986) *Soviet Astronomy* **30**, 1, 20.
Berman, V. G., and Suchkov, A. A. (1988) *Astrofizika* **28**, 85 (*Astrophysics* **28**, 50).
Berman, V. G., and Suchkov, A. A. (1989) *Astrofizika* **30**, 48 (*Astrophysics* **30**, No. 1).
Berman, V. G., and Suchkov, A. A. (1990) *Astrophys. Space Sci.*, submitted.
Bookbinder, J., Cowie, L. L., Krolik, J. H., et al. (1980) *Ap. J.* **237**, 647.
Dekel, A. and Silk, J. (1986) *Ap. J.* **303**, 39.
Field, G. B., Goldsmith, D. W., and Habing, H. J. (1969) *Ap. J.* **155**, L149.
Heckman, T. M., Armus, L., and Miley, G. K. (1987) *Astron. J.* **93**, 276.
Heyles, C. (1987) *Ap. J.* **315**, 555.
Ikeuchi, S., and Habe A., and Tanaka, J. D. (1984) *MNRAS* **207**, 909.
Marochnik, L. S., Berman, V. G., Mishurov, Yu. N., and Suchkov, A. A. (1983) *Astrophys. Space Sci.* **89**, 177.
Marsakov, V. A., and Suchkov, A. A. (1976) *Pisma Astron. Zh.* **2**, 381.
McKee, C. F., and Ostriker, J. P. (1977) *Ap. J.* **218**, 148.
Parker, E. N. (1973) *Astrophys. Space Sci.* **22**, 279.
Pikel'ner, S. B. (1967) *Astron Zh.* **44**, 1915.
Richter, O.-G., and Rosa, M. (1988) *Astron. Astrophys* **206**, 219.
Suchkov, A. A. (1981) *Astrophys. Space Sci.* **77**, 3.
Suchkov, A. A. (1988) *Astron. Zh.* **65**, 1 (*Sov. Astron.* **32**, 1).
Suchkov, A. A., Berman, V. G., and Mishurov, Yu. N. (1987) *Astron. Zh.* **64**, 708 (*Sov. Astron.* **31**, No. 4).
Turner, B. E. (1981) in The Phases of the Interstellar Medium, J. M. Dickey, ed. NRAQ, Green Bank, p. 69.
Umemura, M., and Ikeuchi, S. (1987) *Ap. J.* **319**, 601.

# 9

# Observational Evidence for Magnetic Fields in the Galaxy and Galaxies

## Richard Wielebinski
*Max Planck Institute fur Radioastronomie, Bonn*

Considerable data has accumulated over the years giving definite evidence of the existence of magnetic fields on various scales in galaxies. Many galactic objects like stars, pulsars, supernova remnants, etc. are known to possess magnetic fields. In external galaxies large-scale ordered magnetic fields have been detected. The jet phenomenon in radio galaxies requires a magnetic field for its alignment. Finally recent detections of magnetic fields in clusters of galaxies have been reported.

The magnetic field strength in astrophysical objects spans some 20 orders of magnitude. The weakest fields ($\sim 10^{-9}$ Gauss) are expected in intergalactic space. Fields of $\sim 10^{-6}$ Gauss have been detected in the Coma cluster of galaxies. The magnetic field of the Earth lies in the middle of the range with a field strength of $B \sim 0.5$ Gauss. Magnetic stars have been observed with fields of $10^4$ Gauss. Pulsars and X-ray sources possess fields of $\sim 10^{12}$ Gauss or more.

In this article the observational evidence about magnetic fields will be presented. The theoretical treatment of the magnetic fields is beyond the scope of this short review. The everlasting contributions of J. S. Shklovsky and S. B. Pikel'ner to the theory of magnetic fields and the interstellar medium must be remembered.

## INTRODUCTION

The existence of magnetic fields was realized in antiquity since both iron and lodestone were available. The fact that the Earth possesses a dipole field was used by ancient mariners to navigate around the globe. Experiments with magnets were performed by William Gilbert (1540–1603). Soon afterwards experiments with static electricity were suggesting similar cases of both phenomena. The first proof of a real connection between electric and magnetic phenomena came from H. C. Oersted (1777–1851) who observed a deflection of a magnetic needle when placed close to a conductor carrying an electric current. Further studies by A. M. Ampere (1775–1836) gave us the laws which connect electricity and magnetism and which are the basis for our modern civilization. The final synthesis came through the electromagnetic theory formulated by J. C. Maxwell (1831–1879) which is still of great importance today.

The search for magnetic fields in astronomical objects dates back to the end of the last century when a number of techniques of measurement became available. A connection between polarization and magnetic fields was realized quite early and hence various polarizers were placed in optical telescopes and a variety of objects studied. The Felspar crystal was a typical early polarization analyzer. Later other techniques, in particular a half-wave plate placed in front of a polarizer was introduced to improve the accuracy of measurement. The exact mechanism for polarized light is still not finally proven. One of the accepted theories suggests that starlight is polarized by preferential absorption in one plane of polarization by nonspherical dust grains aligned in magnetic fields (the Davis-Greenstein effect). However it must be noted that polarization can be caused by scattering only, which complicates the issue.

Observations of the Zeeman effect, (doubling the spectral lines due to the splitting of a wave into two opposite circular polarizations in a magnetic field), is a very powerful tool for studying magnetic fields. Observations of sunspots were made by G. E. Hale using the Zeeman effect in 1908 giving us a new insight into the importance of magnetic fields in solar physics.

The first substantiated measurement of polarized light in a galactic object is due to W. F. Meyer from the Lick Observatory who observed the Hubble's variable nebula NGC2261 in 1920. The first discovery of a magnetic field in an external galaxy was reported by Y. Öhman in 1942 who used first a Felspar polarimeter and later a Wollaston prism to observe the polarized emission in the Andromeda nebula (M31). Observations of magnetic stars by H. W. Babcock followed in 1946.

The progress in our knowledge about magnetic fields in the universe changed dramatically after radio techniques were developed for astronomical observations. In fact it turned out that the radio emission discovered by Karl Jansky in 1934 (synchrotron emission) was due to relativistic electrons being braked in interstellar magnetic fields. However the acceptance of this theory took a long time. It was J. S. Shklovsky who in 1953 played a major role in the interpretation of the origins of the non-thermal radio emission. The radio synchrotron emission which is emitted up to 75% linearly polarized gives us another observational possibility. On its way to the observer it is Faraday rotated due to the longitudinal component of the magnetic field in the thermal interstellar medium. Hence we can get information about the magnetic field in the line of sight and the thermal electrons. At radio wavelengths the Zeeman effect has been detected in HI complexes and in OH, $H_2O$ and CCS molecular clouds. The contributions of S. B. Pikel'ner to the theory of interstellar medium were anticipating these recent discoveries.

In this contribution the present situation of our knowledge about magnetic fields will be reviewed—from the Solar system to the most distant reaches of the universe.

## 9. OBSERVATIONAL EVIDENCE FOR MAGNETIC FIELDS

## THE MAGNETIC FIELD IN THE SOLAR SYSTEM

The magnetic field in the planets (including the Earth) come mostly from the in-situ observations. Various spacecrafts have measured the magnetic fields of planets and their moons out to Neptune. The field strength seems to be correlated with the planet's size (its molten core mass) and rotation. The Earth with B ~ 0.5 Gauss lies in the middle range of magnetic field values. Not surprisingly Jupiter has the most intense magnetic field of B ~ 10 Gauss. Venus, with its orbital period of 244 days, shows hardly any field at all. There is still some dispute about the field of Mercury. The magnetic field structure of the inner planets is essentially dipolar but with the polar axis usually not quite aligned with the rotation axis. The most extreme misalignments are in Uranus and Neptune where an angular difference of ~50° was measured by the Voyager spacecraft. There are local differences in the magnetic fields some of which are attributable to the interaction of the planet's magnetic field with the solar wind. We have a great amount of data on the Earth's magnetic field variations.

The data on solar magnetic fields is obtained by astronomical measurements. As we already mentioned G. E. Hale measured the strength of the magnetic fields in sunspots using the Zeeman effect in 1908 and found that B ~ 4000 Gauss. Later measurements showed that the Sun has fields of opposite polarity at its two poles, and that this polarity changes at the start of a new sunspot cycle. Magnetic fields are also found in the prominences and there is a large-scale field. Radio studies of solar emission gave us many clues about the emission processes in particular in the solar corona. X-ray studies of the Sun have also shown their connection with the magnetic phenomena. In general the magnetic field changes are closely related to solar activity. The Sun is responsible for an interplanetary magnetic field which is connected to the solar wind.

Comets, the irregular visitors of the Earth's neighborhood, possess magnetic fields. The return of Halley's comet increased the interest in magnetic phenomena in these objects since a definite detection of a magnetic field was reported (e.g. Yeroshenko et al. 1987).

## MAGNETIC FIELDS OF STARS (AND PULSARS).

Strong magnetic fields have been detected in Ap (A type peculiar) stars with field strengths of B ~ $10^4$ Gauss or more. These stars rotate very rapidly which again shows that there is a definite correlation between rotation and field strength. The fields are dipolar with the field axis often not aligned with the rotation axis. Also white dwarfs have magnetic fields with field strengths of $10^6 <$ B $< 10^8$ Gauss. Pulsars are probably the sources of the most intense fields in the universe with B ~ $10^{12}$ Gauss. Also X-ray sources near neutron stars have magnetic fields similar to pulsars.

# MAGNETIC FIELDS IN OUR GALAXY

## Discrete Objects

Magnetic fields have been found in a variety of galactic objects. Reflection nebulae, molecular clouds, Bock's globulae all have confirmed magnetic fields (e.g., see a series of papers in Dickmann et al. 1988). Perhaps the bi-polar action is due to alignment by a magnetic field. The hardest evidence for magnetic fields in Galactic objects is found in supernova remnants (SNR'S) for which a lot of data was recently collected by Lozinskaya (1986) and Kundt (1988).

The Crab nebula was the object in which brilliant suggestion of Shklovsky, made in 1953, namely that synchrotron radiation is responsible for all the emission, could be tested. Optical polarization observations of Dombrovsky (1954) and Vashakidze (1954) confirmed this suggestion. The radio observations of Mayer et al. (1957) added further proof. Now we also know that the X-ray emission of the Crab nebula is due to magnetic fields. The association of the 30 millisec pulsar with the Crab nebula was a further milestone of astronomy as a result of studies of this remarkable object. In all the new breakthroughs the magnetic fields played a major role.

Supernova remnants are observed as a number of types. Shell-type supernova remnants show a radial magnetic field in the early phases of their development. Older shell-type remnants invariably have a tangential magnetic field. Plerions, amorphous types of SNR's have a highly tangled magnetic field. In Figures 1a and 1b the two types of supernovae with different magnetic field configurations are shown. The young SNR Cassiopea A has a radial field while the older shell of CTB1 shows a predominantly tangential magnetic field.

## THE LARGE-SCALE MAGNETIC FIELDS IN OUR GALAXY

One of the methods to study the magnetic fields in the Galaxy is to measure starlight polarization. The earliest results are due to Hiltner (1949) and Hall (1949). Many results for the northern hemisphere were collected by Behr (1959). The southern stars were observed by Mathewson and Ford (1970) who then combined the whole set and produced an all-sky distribution of starlight polarization. A re-analysis of this data was made by Ellis and Axon (1978). The general conclusion is that the data give good information only about the local neighborhood. The field is aligned along the Galactic plane, but in a direction of $l=45°$. Beyond a circle of 600 pc the magnetic field is directed towards $l=70°$. There are only a few stars observed beyond a circle of 1.2 kpc. Radio methods offer possibilities to study more distant magnetic fields.

The final confirmation for the existence of synchrotron emission (and hence of ordered magnetic fields) in the Galaxy came as a result of the observations of radio polarization

## 9. OBSERVATIONAL EVIDENCE FOR MAGNETIC FIELDS 121

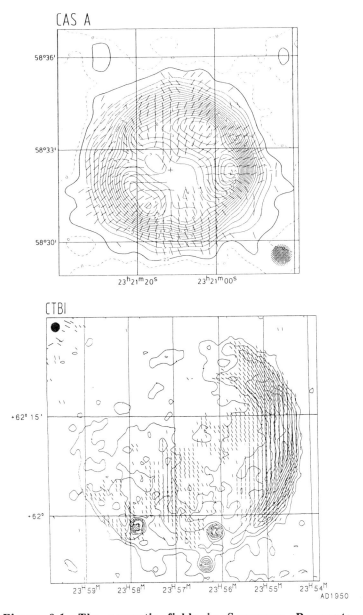

Figure 9.1. The magnetic fields in Supernova Remnants. (a) Cassiopea A has a radial magnetic field. (b) CTB1 shows a predominantly tangential field. (Courtesy of Prof. E. Fürst)

by Westerhout et al. (1962) and Wielebinski et al. (1962). Further large-scale surveys at a number of frequencies in the northern hemisphere were supplemented by southern observations of Mathewson and Milne (1965). The synchrotron emission is emitted with the 'E' vector perpendicular to the orientation of the magnetic field. Thus the radio measurements, after correction for Faraday rotation, give us the orientation of the magnetic field in the line of sight. the lower frequency results suggest that the local field is in the direction of $l = 50°$. Higher frequency surveys (e.g. Spoelstra, 1984; Junkes et al., 1987) show that more distant magnetic fields (possibly in the whole Galaxy) can be traced by these direct measurements.

The studies of Rotation Measures (RM) of extragalactic radio sources enable us to trace the magnetic field of the Galaxy. Multi-frequency surveys of thousands of sources are necessary to enable a reasonable delineation of the field structure. The present state of this important work is given in Simard-Normandin et al. (1981) who based their studies on 560 sources. There is a large-scale field along the galactic plane with many local features. Field reserves are observed (as reversals of rotation measure) on scales of a few degrees. Continuation of these studies with data on thousands of sources should give us a better understanding of the magnetic field of the Galaxy.

Pulsars offer the most direct method of determining $B_{\parallel}$. The reason for this is the fact that we can measure both the RM and the Dispersion Measure (DM). The combination of RM and DM gives us $B_{\parallel}$. Recent reanalysis of all the available data by Lyne and Smith (1989) gave the mean field $B_{\parallel} \sim 3$ μG, directed towards $l = 90°$. However sudden field reversals (indicated by high positive and negative RM) have also been observed. The maximal values of the observed magnetic field are $B_{\parallel} \sim 8$ μG.

There has been a long controversy about Zeeman effect measurements in HI clouds. After final success (e.g. Verschuur, 1979) numerous observations were made. The limiting field values that can be measured in HI clouds at present are $B > 10$ μG. Zeeman effect observations in OH clouds have been made where the fields are $B > 50$ μG. The CCS molecule can probe denser and warmer clouds where fields of $B \sim 100$ μG exist. In regions with $H_2O$ masers magnetic fields of $B \sim 10$ mG were measured (Fiebig and Güsten 1989).

All the data presented so far suggested that the magnetic fields in the Galaxy are in the galactic plane. A recent analysis of all the available data by Vallee (1988) indeed confirmed this. The magnetic fields seem to follow spiral arms with small deviations from the pitch angle. There may be reversals of the field direction from one arm to the next but the data for this are not very significant.

Radio observations of polarized emission towards the galactic center suggest that the magnetic fields in the center of our Galaxy are in the Z-direction. This has been shown in multifrequency polarization observations of the central 100 parsecs (e.g. Seiradakis et al., 1989). This magnetic field in the central area is seen perpendicular to the plane, along filaments, with $B \sim 100$ μG or more. Even stronger magnetic fields are indicated in the central 1 parsec.

# 9. OBSERVATIONAL EVIDENCE FOR MAGNETIC FIELDS

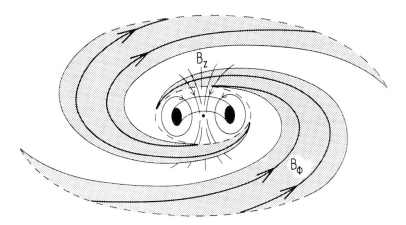

Figure 9.2. A model of the global magnetic field in the Galaxy.

A model of the magnetic fields in the Galaxy is shown in Figure 2. The fields in the disc have a uniform component $B_{u\parallel} \sim 3$ μG. This means that the mean uniform field is $B_u \sim 5$ μG. In addition, a turbulent (random) component or $B_r \sim B_u$ is also present. The mean total magnetic field in the spiral arms attains a strength of $B_t \sim 7$ μG, possibly around 10 μG.

## MAGNETIC FIELDS IN GALAXIES

As often happens in astronomy we derive much of our knowledge about the magnetic fields by studying nearby galaxies. The same holds as for our Galaxy;- we can measure optical polarization and/or radio synchrotron emission. The advent of powerful new instruments and of new detectors made extragalactic studies possible. A great collection of recent data on galactic and extragalactic magnetic fields can be found in Sofue et al. (1986), Beck and Gräve (1987), and in Beck, Kronberg, and Wielebinski (1990).

The original work on M31 in optical polarization by Öhman (1942), Muliarchik (1957) and Hiltner (1958) led to further observations of other nearby galaxies. However the sensitivity was a problem (e.g., Appenzeller 1967; Elvius 1978). The optical observations have become recently attractive with the advent of sensitive CCD detectors. We do not need the largest telescopes, because the angular resolution is not a problem.

At radio wavelengths large single-dish radio telescopes (Effelsberg, Germany; Parkes, Australia) have been used with broad-band polarimeter to map galaxies at high frequencies. The synthesis arrays (Westerbork, Holland; V.L.A., New Mexico) have been used to give complementary results, in particular at lower frequencies. The Australia

Telescope in Narrabri will soon add new data on southern galaxies.

Practically all the larger galaxies (disc size 10' or more) have now been studied in radio or in the optical range (or both). In Figures 3 to 5 some nearby galaxies are shown for which the global magnetic field is known as a result of multi-frequency observations.

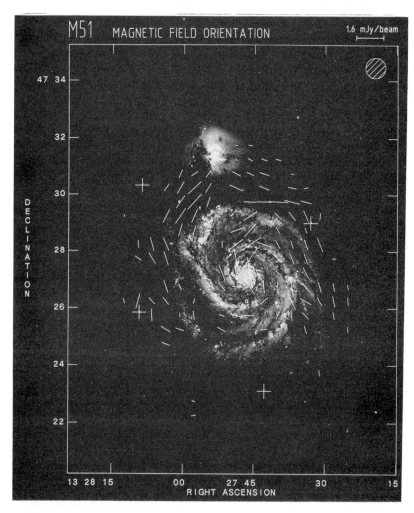

Figure 9.3. The magnetic field orientation in M51. (Courtesy of Cathy Horellou)

Figure 9.4. The magnetic field orientation in M81. (Courtesy of Dr. M. Krause)

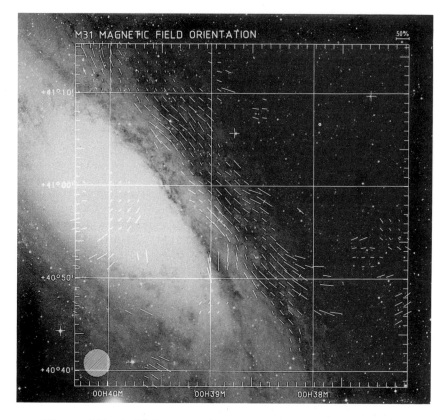

**Figure 9.5. A high resolution map of the magnetic field in M31. (Courtesy of Dr. R. Beck)**

One general conclusion is that the magnetic fields seem to follow the arms in spiral galaxies. There is observational evidence that some spirals (M51 or M81) have bisymmetric spiral (BSS) patterns. A few other galaxies, like M31 or IC342, have an axisymmetric spiral (ASS) field. Irregular galaxies (like the Large Magellanic Cloud) show filamentary fields which in the case of the LMC originate in the 30 Doradus nebula. Also dwarf galaxies have been detected in polarization (e.g. the Small Magellanic Cloud). The few galaxies for which both optical and radio polarization data exist indicate that the same fields that align the grains are also responsible for the generation of the non-thermal emission.

Most of the observations made so far were for the brighter face-on galaxies. The fields are obviously in the disc, along the spiral arms. Some recent studies of edge-on galaxies show a continuation of the field into a halo. The present results suggest that the halo fields are nearly normal to the disc as shown in Figure 6 for NGC4631. Observations of NGC891 show a less regular field structure.

In a number of mildly active galaxies (e.g., M82, NGC1808, NGC4945, etc.) the nuclear region shows optical filaments aligned normal to the disc. In these galaxies rotating rings of CO were observed. Also radio continuum, Hα, and HII co-exist in these rings. In a scenario of Lesch et al. (1989) this rotating molecular ring, which is also partially ionized, is thought to carry a current. This ring current can produce a dipole seed field which in turn can be amplified further by the dynamo action. Are magnetic fields at the root of galactic structure?

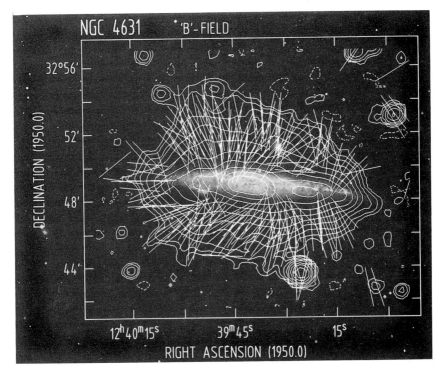

Figure 9.6. The magnetic field orientation in the halo of NGC4631. (Courtesy of Dr. E. Hummel)

## RADIO GALAXIES

Radio galaxies are objects of enormous extent with radio emission extending over Megaparsecs while the optical galaxy is seen coincident with the radio nucleus. From the very beginning of the studies of radio galaxies it was clear that magnetic fields played a decisive role in these objects. Already the single-dish maps of Centaurus A by Cooper et al. (1965) showed polarized synchrotron emission in the outer lobes. The increase of angular resolution, which started with the Cambridge 1-mile telescope, was continued with the Westerbork synthesis telescope, and reached its zenith in the Very Large Array, allowed detailed studies of magnetic fields in radio galaxies. A lot of data on magnetic fields in radio galaxies is to be found in Heeschen and Wade (1982). The jets which apparently transport the enormous energy from the nucleus to the lobes are polarized. In fact it seems that the magnetic fields provide the mechanism for jet confinement. An example of a radio galaxy is shown in Figure 7. Initial studies using Very Long Baseline Interferometry have given evidence of the existence of magnetic fields on milliarcsecond scales in active nuclei.

## CLUSTERS OF GALAXIES

The discussion on the possible strength of the intergalactic magnetic field must look first to the results for clusters of galaxies. Many theories have postulated magnetic fields of $10^{-9}$ Gauss in the intergalactic space but no observational confirmation of such fields has been made so far. Recently Kim et al. (1990) have studied rotation measures of point sources in the direction of the Coma A cluster of galaxies and near the cluster. Their result suggests that a magnetic field of ~2 µG (r.m.s.) exists in the cluster between the galaxies. This result certainly adds new reasons to search for intercluster magnetic fields.

## REFERENCES

Appenzeller, I. (1967) Polarimetric observations of spiral galaxies. *Publ. Astron. Soc. Pacific* **79**, 600.

Beck, R., and Gräve, R. (1987) *Interstellar Magnetic Fields*, Springer-Verlag, Berlin.

Beck, R., Kronberg, P. P., and Wielebinski, R. (1990) *Galactic and Intergalactic Magnetic Fields*, Kluwer Academic Publishers, Dordrecht.

Behr, A. (1959) *Die interstellar Polarisation des Sternlichts in Sonnenumgebung*, Veröff. Univ. Sternw. Göttingen No. 126.

Cooper, B. F. C., Price, R. M., and Cole, D. J. (1965) *Austr. J. Phys.* **18**, 589.

Dickmann, R. L., Snell, R. L., and Young, J. S. (1988) *Molecular Clouds in the Milky Way and External Galaxies*, Springer-Verlag, Berlin.

Dombrovsky, V. A. (1954) *Doklady Akad. Nauk USSR* **94**, 1021.

## 9. OBSERVATIONAL EVIDENCE FOR MAGNETIC FIELDS

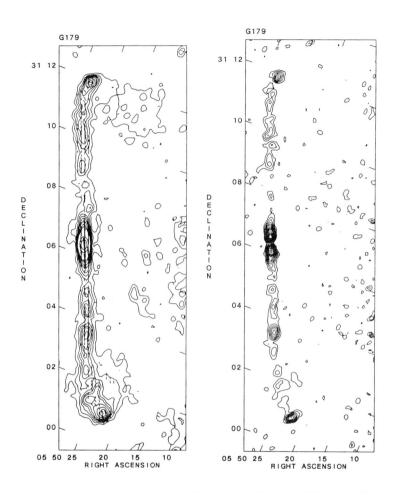

Figure 7.9. The polarization of radio galaxy G179. (a) Total intensity with polarization vectors. (b) Polarized intensity. (Courtesy of Prof. E. Fürst)

Ellis, R. S., and Axon, D. J. (1978) *Astrophys. Space Sc.*, **54**, 425.
Elvius, A. (1978) *Astrophys. Space Sc.* **55**, 49.
Fiebig, D., and Güsten, R. (1989) *Astron. Astrophys.* **214**, 333.

Hall, J. S. (1949) *Science* **109**, 166.
Heeschen, D. S., and Wade, C. M., (1982) *Extragalactic Radio Sources*, Reidel Publishing Company, Dordrecht.
Hiltner, W. A. (1949) *Science* **109**, 471
Hiltner, W. A. (1958) *Astrophys. J.* **128**, 9.
Hummel, E., Lesch, H., Wielebinski, R., and Schlickeiser, R. (1988) *Astron. Astrophys.* **197**, L29.
Junkes, N., Fürst, E., and Reich, W. (1987) *Astron. Astrophys. Suppl.* **69**, 451.
Kim, K-T, Kronberg, P. P., Dewdney, P. F., and Landecker, T. L. (1990) *Astrophys. J.* **355**, 29.
Kundt, W. (1988) *Supernova Shells and their Birth Events*, Springer-Verlag, Berlin.
Lesch, H., Crusius, A., Schlickeiser, R., and Wielebinski, R. (1989) *Astron. Astrophys.* **217**, 99
Lozinskaya, T. A. (1986) *Supernova Remnants and Stellar Winds*, Nauka, Moscow (in Russian).
Lyne, A., Smith, Graham F. (1989) *Mon. Not. Roy. Astr. Soc.* **237**, 533.
Mathewson, D. S., Ford, V. L. (1970) *Mem. Roy. Astr. Soc.* **74**, 139.
Mathewson, D. S., and Milne, D. K. (1966) *Austr. Soc.* **18**, 635.
Mayer, C. H., McCullough, T. P., and Sloanaker, R. M. (1957). *Astrophys J.* **126**, 468.
Muliarchik, T. M. (1957) *Izvest. Astrophys. Inst. Alma Ata.* **5**, No. 7.
Öhmann, Y. (1942) *Stockholm Obs. Bull.* **10**, 68.
Seiradakis, J. H., Reich, W., Wielebinski, R., Lasenby, A. N., and Yusef-Zadeh, F., (1989) *Astron. Astrophys. Suppl.* **81**, 291.
Simard-Normandin, M., Kronberg, P. P., and Button, S. (1981) *Astrophys J.* **45**, 97.
Sofue, Y., Fujimoto, M., and Wielebinski, R. (1986) *Ann. Rev. Astron. and Astrophys.* **24**, 459.
Spoelstra, T.A.-Th. (1984) *Astron. Astrophys.* **135**, 238.
Vashakidze, M. A. (1957) Astronom. Circ. No. 127, p. 11.
Verschuur, G. L., (1979) *Fund. Cosmic Phys.* **5**, 113.
Westerhout, G., Seeger, Ch. L., Brouw, W. N., and Tinbergen, J., (1962) *Bull. Astron. Inst. Neth.* **16**, 187.
Wielebinski, R., Shakeshaft, J. S., and Pauliny-Toth, I. I. K. (1962). *Observatory* **82**, 187.
Wielebinski, R. (1990), in *The Interstellar Medium in Galaxies*, D. Hollenbach and H. Thronson (eds.), pp. 349–369, Kluwer Academic Publishers, Dordrecht
Yeroshenko, Y. G., Styashkin, V. A., Riedler, W., Schwingenschuh, K., and Russell, C. T. (1987). *Astron. Astrophys.* **187**, 69.

# 10

# Recombination Radio Lines

R. L. Sorochenko

*Astro Space Center, Lebedev Physical Institute, Moscow*

## ABSTRACT

Recombination radio lines (RRL) detected more than 25 years ago became an effective tool for astrophysical studies. Presently they are observed in a wide range from 1 mm to 20 m; radio lines are recorded from levels of excited atoms up to n = 768. They were used to determine the temperature, density, and velocity of internal motions in HII regions, to obtain the distribution of ionized hydrogen and the helium abundance in the Galaxy, to the estimate the intensities of galactic cosmic rays, etc. RRL have begun to be used to study extragalactic objects.

## 1. INTRODUCTION

In 1959 N. S. Kardashev, a student of I. S. Shklovskii, came to the conclusion that it was possible to form highly excited atoms in the interstellar medium in the process of recombination of charged particles, electrons and ions, with a definite probability. During the subsequent cascade transitions due to the small difference between energies of highly excited levels, the emission of radio quanta should occur, i.e., the emission of radio lines (Kardashev 1959).

The frequencies of radio lines are determined by the well-known Balmer equation:

$$\nu = R\left(\frac{1}{n_1^2} - \frac{1}{n_2^2}\right) \cong \frac{2R\Delta n}{n^3} \quad \text{(for } n \gg 1\text{)}, \tag{1}$$

where $n_2$ and $n_1$ are the numbers of upper and lower levels, respectively, $\Delta n = n_2 - n_1$, and R is the Rydberg constant.

For different elements the same transitions occur whose frequencies are somewhat shifted relative to each other due to the dependence of the Rydberg constant on the atomic mass:

$$R = R_\infty\left(1 - \frac{m_e}{m_a}\right), \qquad (2)$$

where $m_e$ is the electron mass, $m_a$ is the atom mass, and $R_\infty = 3.2898422 \times 10^{15}$ Hz.

Kardashev showed that transitions with $\Delta n = 1$ should have the highest probability and the radio lines formed in this manner, firstly the radio lines of excited hydrogen, could be observed with radio astronomy instruments.

In 1964 the predicted radiation was measured. Sorochenko and Borodzich (1964) using the 22-meter radio telescope in Pushchino of the Lebedev Physics Institute recorded the radio line of hydrogen $n_{91} \to n_{90}$ ($\nu = 8872$ MHz) in the spectrum of the Omega Nebula. At the same time at the Pulkovo Astronomical Observatory, Dravskich et al. (1964) measured the still longer wavelength line $n_{105} \to n_{104}$ ($\nu = 5008$ MHz). Somewhat later Höglund and Mezger (1965) observed the hydrogen radio line $n_{110} \to n_{109}$ at Green Bank and at Harvard, Lilley et al. (1966a) measured the lines $n_{159} \to n_{158}$ and $n_{157} \to n_{156}$. Soon the radio lines of helium, the second most abundant element in the interstellar medium, were measured (Lilley et al. 1966b; Gudnov et al. 1968) whose observation was also predicted in the pioneering work of Kardashev (1959).

Beginning with these studies, in many observatories research began of radio lines of highly excited atoms for recombination radio lines (RRL), as they were called, because the emission of these lines accompanies a recombination act.

## 2. PHYSICAL PROPERTIES OF RECOMBINATION RADIO LINES

### 2.1. Width and Form of the RRL Profile

It was assumed before the beginning of experimental research on recombination lines that the lines should be significantly broadened due to the Stark effect due to the interaction of highly excited atoms with charged particles. This conclusion was obtained both according to the estimates made by Kardashev (1959) and in subsequent more detailed calculations of RRL widths and intensities (Sorochenko 1965) carried out using a complete theory of spectral line broadening in plasma (Griem 1960).

However, the first results of RRL observations were in disagreement with the theory. It was clarified that the lines have no broadening and disperse with an increase of the level n as expected (Sorochenko and Borodzich, 1965; Höglund and Mezger, 1965; Lilley et al., 1966a). All the observed radio lines up to $n = 166$ to within the accuracy of measurement had a constant ratio of line width to frequency which indicates pure Doppler broadening. The line width is naturally explained by thermal and turbulent motions of the radiating atoms:

$$\Delta \nu_D = \frac{1.66\nu}{c}\sqrt{\frac{2kT_e}{m_a} + V_t^2}, \quad Hz \tag{3}$$

where $\Delta \nu_D$ is the Doppler width at half intensity, $T_e$ is the kinetic (electron) temperature of the HII regions, k is Boltzmann's constant, $m_a$ is the atomic mass, c is the velocity of light, and $V_t$ is the velocity of internal turbulent motions, all in cgs. Meanwhile, according to the theory, Stark broadening rapidly increases with the growth of n and for n = 100 should already be comparable with Doppler broadening and exceeds the latter by ten times at n = 166.

A new theoretical examination (Griem 1967, Minaeva et al. 1967) showed that in interstellar medium conditions having a low density and high velocity of perturbing particles, the impact approximation is correct in determining Stark broadening. In this case for an elastic interaction of excited atoms both with electrons and with ions a consistent compensating Stark effect occurs: neighboring levels close to each other are perturbed similarly and the difference in energies between them changes significantly less than the energy of the levels themselves. As a result the weak inelastic interaction becomes dominant; in this case inelastic collisions with electrons play the basic role in Stark broadening.

The correctness of the theory considered was verified by studies of Stark broadening carried out in a combined program on two radio telescopes: the 22-meter radio telescope at Pushchino and the 100-meter radio telescope at Eiffelsberg (Smirnov et al. 1984). On the 22-meter radio telescope at 8.2 mm ($\Delta \varphi_A = 1.9'$) the form of the H56$\alpha$ line was measured very accurately in the direction of the central part of the Orion Nebula with which the lines Doppler contour was determined. In the same direction using the 100-meter telescope at $\lambda = 3.3$ cm the forms of lines H90$\alpha$, H114$\beta$, H128$\gamma$, H141$\delta$, H152$\varepsilon$, and H161$\eta$ were measured[†] with close angular resolution.

With the combined action of Doppler and Stark broadening the RRL profile has the form of a Foigt function and is given by the equation of packing:

$$I(\nu) = \frac{1.66\nu_D}{\pi^{3/2}(\Delta \nu_D c)} \int_{-\infty}^{\infty} \frac{2\Delta \nu_{st} \exp\left\{-\left(\frac{1.66\nu_0 u}{c\Delta \nu_D}\right)^2\right\}}{\Delta \nu_{st}^2 + 4\left(\nu - \nu_0 + u\frac{\nu}{c}\right)^2}, \tag{4}$$

where $\nu_0$ is the central line frequency, $I(\nu)$ is the normalized spectral density distribution,

---

[†]The following designation of recombination radio lines is used: the name of the element according to Mendeleev's table, the level number at which the transition occurs, and the line order (Greek letter). For example, the line H114$\beta$ indicates a hydrogen line due to transitions between the 116th and 114th levels).

$\int I(\nu)d\nu = I$, and $\Delta\nu_{st}$ is the Stark width at half intensity. The total line width at the half intensity level in this case is determined with good accuracy by:

$$\Delta\nu = 0.53\Delta\nu_{st} + \sqrt{0.22\,\Delta\nu_{st}^2 + \Delta\nu_D^2} \qquad (5)$$

Equation (5) provides a means of determining the Stark broadening for a known total width and Doppler width of a line. As a result of studies performed a good agreement was obtained of the dependence of the Stark broadening of RRL with the theory considered on the level number and electron density:

$$\Delta\nu_{st} = 8.3\,N_e \left(\frac{n}{100}\right)^{4.4}, \text{ Hz} \qquad (6)$$

This dependence was further used for determining the electron density of the interstellar medium.

## 2.2. Intensity of Recombination Radio Lines

Under thermodynamic equilibrium (TE) of atomic level populations, the optical depth in RRL are (Gulyaev and Sorochenko 1974):

$$\tau_L^{TE}(\nu) = \frac{1.05 \times 10^7 N_e N_x^+ l f_{n+\Delta n,n} I(\nu)}{T_e^{5/2} n} \qquad (7)$$

where $N_e$ and $N_x^+$ are the densities of electrons and ions of a given material, respectively, $l$ is the length along the line of sight, pc, $T_e$ is the electron temperature, and $f_{n+\Delta n,n}$ is the oscillator strength (for $\Delta n = 1$, $f_{n+\Delta n,n} = 0.19\,n$). The distribution of spectral density, $I(\nu)$, for the general case of the Foigt profile at line center is (Shaver 1975):

$$I(\nu_o) = \frac{1}{1.06\Delta\nu_D(1 + 1.48\,\Delta\nu_{st}/\Delta\nu_D)} \qquad (8)$$

In the case of a nonequilibrium population

$$\tau_L(\nu) = b_n \beta_n \tau_L^{TE}, \qquad (9)$$

where $b_n$ is the coefficient characterizing the degree of deviation of the population of the n-level from TE (for equilibrium $b_n = 1$), and $\beta_n = \left[1 - \frac{kT_e}{h\nu}\frac{d\ln b_n}{dn}\Delta n\right]$ is a quantity taking into account the change of the optical thickness due to the nonequilibrium population (h is Planck's constant).

One can see from Eq. (9) that the nonequilibrium population may considerably change the optical depth in the line, make it 0, and even negative, i.e., in a medium amplification will occur at the line frequency, i.e., the maser effect (Goldberg 1966).

For radio astronomical observations the RRL intensity can be most conveniently expressed in units of brightness temperature in the line $T_L$. A solution of the transfer equation for the case $\tau_L < \tau_c \ll 1$, where $\tau_c$ is the optical depth in the continuum gives:

$$T_L(\nu) = T_e \tau_L^{TE}(\nu) b_n \left(1 + \frac{\tau_c}{2} \frac{kT_e}{h\nu} \frac{d \ln b_n}{dn} \Delta n \right) + T_{B.gr} e^{-\tau_c} \tau_L^{TE}(\nu) b_n \beta_n, \qquad (10)$$

where $T_{B.gr}$ characterizes the radiation incident on the region of RRL formation from the side opposite the observer.

## 2.3. Range Investigations of a Recombination Radio Line

*RRL in the high frequency part of the radio range.* RRL studies, initially carried out at centimeter and decimeter wavelengths, were rapidly extended to the millimeter range. The observation of RRL in this range has a series of advantages: 1) even for the most dense HII regions radio lines are broadened insignificantly due to the Stark effect. For this reason the RRL profiles are determined by the purely Doppler line widths; 2) the optical depth of HII regions is small and the maser effect is correspondingly small which considerably simplifies the interpretation of observations; and 3) a single-dish instrument has a higher angular resolution in the millimeter range.

At the same time in the millimeter range due to the increase with frequency of the Doppler line width its brightness temperature is weaker than at longer wavelengths. Because difficulties of creating an instrument with a high spectral sensitivity grow simultaneously with an increase of frequency, considerably fewer observations are carried out in millimeter RRL than in centimeter and decimeter ones.

The first RRL observations in the millimeter range were carried out at Pushchino on the 22-meter radio telescope. The radio line H56$\alpha$ ($\lambda$ = 8.2 mm, Fig. 10.1a) was recorded in the Omega Nebula (Sorochenko et al. 1969). This radio line was later detected in a series of sources; in one of them, the Orion Nebula, shorter wavelength lines were detected: H42$\alpha$, $\lambda$ = 3.5 mm (Waltman et al. 1973), H40$\alpha$, $\lambda$ = 3 mm, and H39$\alpha$, $\lambda$ = 2.8 mm (Wilson and Pauls 1984). In a recent series of observations Gordon (1989) detected the radio line H40$\alpha$ in the seven brightest HII regions.

The shortest wavelength RRL detected currently is the radio line H29$\alpha$ ($\lambda$ = 1.17 mm) recorded on the 30-meter IRAM radio telescope in the source MWC349 (Martin-Pintado et al. 1989). The line is considerably enhanced by maser amplification and has the two-humped profile shown in Fig. 10.1b.

RRL of the millimeter range turned out to be very useful in determining the electron temperature of HII regions (see below). It was also possible using them to determine the population of highly excited levels and define the quantities $b_n$. For n < 60 ($\lambda$ < 1 cm) for $N_e = 10^2–10^4$ cm$^{-3}$ typical for HII regions, impact processes are already not dominant in determining the population of levels. The quantities $b_n$ deviate significantly from 1 which facilitates their measurement. It was found (Sorochenko et al. 1988) that the dependence of $b_n$ is well described by the theory of Salem and Brocklehurst (1979).

*RRL at low frequencies. Gigantic atoms in the cosmos.* At the beginning of the '80s a considerable expansion of the range of RRL studies toward the long wavelength side occurred. In Pushchino several RRL lying in the meter range were detected. However, these were not lines of hydrogen which were initially sought and which were not detected, but lines of carbon: C427α ($\nu$ = 84.2 MHz), C486α ($\nu$ = 57.1 MHz), C538α ($\nu$ = 42.1 MHz), and C612β ($\nu$ = 57.1 MHz). The lines were detected in absorption in the spectrum of Cassiopia A on the DKR-1000 radio telescope (Ershov et al. 1984).

Carbon lines with still higher transition numbers, C630α ($\nu$ = 26.2 MHz), C631α ($\nu$ = 26.1 MHz), and C640α ($\nu$ = 25.0 MHz) were detected by Konovalenko and Sodin (1981) using the UTR-2 radio telescope of the Radioastronomy Institute of the Ukrainian Academy of Sciences in Kharkov. Historically, the decameter RRL of carbon were

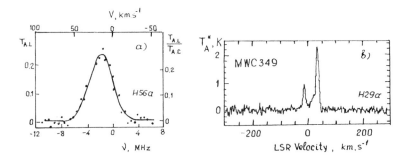

Figure 10.1. a) Spectrum of the H56α ($\lambda$ = 8.2 mm) line in the Omega Nebula obtained on the 22-meter radio telescope at Pushchino. The abscissa is the radial velocity or frequency. The ordinate on the left is the antenna temperature and on the right it is the ratio of antenna temperatures in the line and in the continuum. b) Spectrum of the H29α ($\lambda$ = 1.17 mm) line in the source MWC349 obtained on the 30-meter radio telescope in IRAM (Martin-Pintado et al. 1989).

## 10. RECOMBINATION RADIO LINES

detected prior to the meter ones, and also after unsuccessful searches of hydrogen lines. Subsequently directing this radio telescope towards Cas A, a whole series of carbon radio lines were recorded in the range 87.9–16.7 MHz up to the line C732α (Konovalenko 1984). At the same time all attempts to record higher frequency radio lines for $\nu \cong 100$ MHz (n = 380–400) were unsuccessful (Ariskin et al. 1982; Ershov et al. 1987). The spectra of low-frequency carbon RRL which were obtained at Pushchino and Khrakov are shown in Fig. 10.2.

The detection of low-frequency carbon radio lines was quite unexpected, extremely interesting, and immediately posed a series of questions:

1. Why for the highest excitation levels (n > 400) is it possible to reliably observe carbon lines and not to observe the lines of hydrogen which is significantly more abundant in the interstellar medium?
2. Why do radio lines for excitation levels n = 530–732 ($\nu$ = 44–17 MHz) turn out to be more intense than lines of less excited atoms for n = 420–480 ($\nu$ = 88–59 MHz) while lines for n = 380–400 ($\nu \cong 100$ MHz) are in general impossible to record?
3. If one succeeded in recording lines corresponding to the 732nd level and in this case their intensities were not decreased with an increase of n, then where is the limit to radio line formation and how is it determined?

As shown by the analysis presented for the formation of radio lines with n > 400 it is necessary that the interstellar plasma: a) have a low electron density; and b) be cold (Sorochenko and Smirnov 1987). The first requirement is connected with the fact that at high density lines will be dispersed due to Stark broadening. Therefore, for each level there is some limiting $N_e \sim n^{-9}$. The second requirement is due to the optical depth dependence of RRL: $\tau_L \sim N_e T_e^{-5/2}$ (see Eq. (7) setting $N_c^+ = N_e$). Since the density $N_e$ should be very low, a significant line intensity may occur only at low temperature.

Both conditions in the interstellar medium, low $N_e$ and $T_e$ are fulfilled far from hot stars in HI–CII regions. In these regions hydrogen, whose ionization is accompanied by strong heating of the medium ($T_e \cong 10^4$ K), is neutral. At the same time the less abundant elements having a lower ionization potential than hydrogen are ionized. Among them carbon is the most abundant. The medium in this case is weakly heated, $T_e < 100$ K. Therefore, for the highest excitation levels carbon lines are observed and hydrogen lines are not observed.

The answer to the second question can be found in the strong dependence of the intensity of low-frequency radio lines and the character of the atomic level population. In spite of the high cross sections of highly excited atoms, in conditions of small electron density the population of levels is a nonequilibrium one. For n < 500 stimulated radiation, decreasing the optical depth of the line, begins to be important. This explains the decrease of line intensities for n = 420–480 and the unsuccessful attempts to measure

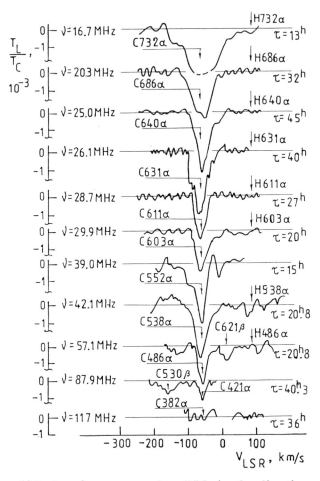

Figure 10.2. Low-frequency carbon RRL in the direction of Cas A. ν < 30 MHz from Khrakov (Konovalenko 1984) and ν > 30 MHz from Pushchino (Ershov et al. 1984; 1987). The ordinate is the contrast in the line relative to the source radiation in the continuous spectrum. On the left is the frequency and on the right is the integration time in hours. The arrows indicate the calculated positions of lines of carbon and hydrogen which have not been detected yet.

lines for n = 380–400. For $\nu \cong 100$ MHz for physical conditions existing in the RRL formation region in the direction of Cas A, $\tau \cong 0$, absorption lines are converted into emission lines (see Eq. (9)). This fact was verified recently by observations of Payne et al. (1989) who measured in the Cas A spectrum in emission the lines C300α–C303α, C272α, and C273α.

What determines the maximum number of distinct excited levels and the limiting sizes of atoms? It was clarified in our Galaxy that this is the background nonthermal radio emission. The brightness temperature of the background due to this radiation increases with a decrease of frequency. For the same reason the density of quanta capable of causing induced transitions in highly excited atoms grows rapidly with the level number:

$$N_q = 0.6 \times 10^{-9} T_B \nu = 0.1 \left(\frac{n}{100}\right)^{4.65} \text{q cm}^{-2} \text{ Hz}^{-1} \text{ s}^{-1} \tag{11}$$

Simultaneously, with an increase of the excitation level the cross section of induced transitions increases:

$$\sigma_{n \to n+1} = 0.5 \left(\frac{n}{100}\right) \text{ cm}^{-2} \tag{12}$$

Both reasons lead to a very rapid ($\sim n^{5.65}$) growth of the rate of induced transitions with the growth of n. As a result for n ~ 1000 the lifetime in this level becomes so small that we do not see distinct lines (Ershov et al. 1982).

Present experimental studies of RRL are approaching this limit. An averaged spectrogram is presented in Fig. 10.3 of the carbon radio lines C764α–C768α in the direction of Cas A obtained recently at Kharkov (Konovalenko 1990). The signal integration time for one line was 50 h. These are the lowest frequency radio lines ($\nu = 14.7$ MHz) corresponding to the highest exitation levels of atoms observed until now. One can see from a comparison with the spectrogram of the C603α line shown on the same figure how much the C768α line is broadened due to the shortened lifetime of highly excited levels. For n ~ 1000 the width should be so large that neighboring lines blend with each other and with the continuous spectrum.

## 3. RECOMBINATION RADIO LINES AS AN EFFECTIVE TOOL OF ASTROPHYSICAL STUDIES

RRL turned out to be quite an effective tool for carrying out astrophysical studies. They are unique in the number of transitions and the width of the range in which they can be observed. On the electromagnetic wave scale RRL cover more than four orders of magnitude which allows one to study objects differing considerably in their physical properties. We consider the basic possibilities of RRL and illustrate them with some examples.

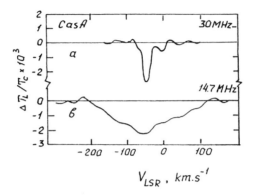

Figure 10.3. Profile of carbon RRL with the highest level of atomic excitation: a) spectrogram of the line C603α and b) average spectrogram of the lines C764α – C768α.; (Konovalenko 1990)

### 3.1. Physical Conditions in HII Regions. Distribution of Ionized Hydrogen in the Galaxy

The emission of RRL is the most important criterion in the classification of radio sources, dividing them into thermal HII regions and nonthermal ones. As shown by the studies performed, the majority of sources of continuum radio emission at centimeter wavelengths located close to the plane of the Galaxy are HII regions. Of 500 sources situated in ±1° from the galactic plane and studied for the presence of recombination radio lines, a positive result was obtained in 462 (Lockman 1989). In this case of the remaining 38 sources all of them are not nonthermal. In a series of them it is possible that radio lines were not detected due to insufficient sensitivity.

The information contained in RRL provides a means of finding all basic parameters of HII regions: electron temperature, density, velocities, internal motions, helium abundance, etc.

*Electron Temperature*

RRL measurements provide the simplest and most accurate method of determining $T_e$ of HII regions. The possibility of these measurements is connected with the probability of recombination processes leading to RRL radiation and the probability of free-free processes with which radiation in the continuum occurs depend differently on temperature. As a result a measurement of the relative brightness temperatures in a line and in a part of the spectrum near to it allow one to determine $T_e$.

The most accurate $T_e$ values, to a few percent, are provided by RRL measurements of hydrogen at high frequencies for $\lambda \lesssim 1$ cm. In this range one can neglect Stark broadening of radio lines and the maser effect. As a result the electron temperature of HII regions is determined by the simple equation:

$$T_e = \frac{27.6\times10^{-6} b_n \nu^2}{(T_c/T_L) \Delta\nu \left[\frac{3}{2}\log T_e + 1.7 - \log \nu\right] [1 + (N_{He}/N_e)]}, \quad (13)$$

where $T_L$ and $T_c$ are the temperature in the line and in the continuum, $\nu$ is the line frequency, $\Delta\nu$ is its width at the half-intensity level in MHz, $N_{He}/N_e$ is the ratio of helium ion and electron densities, and $b_n$ is a coefficient close to 1.

The most accurate and most extensive measurements of electron temperature in HII regions were carried out using RRL. The measurements encompassed about 200 HII regions of the Northern and Southern hemisphere, including cold and weak ones, and those obscured by absorbing material. It was established that in the Galaxy the electron temperature of HII regions lies in the range $(4-10)\times10^3$ K. An important dependance appeared in the $T_e$ distribution: the temperature of HII regions decreases upon nearing the galactic center (Churchwell and Walmsley 1975; Churchwell et al. 1978).

According to the most complete and careful measurements carried out with the lines H76$\alpha$ and H110$\alpha$ in the Southern Hemisphere, the electron temperature gradient is $+433\pm40$ K kpc$^{-1}$ (Shaver et al. 1983). The results of these measurements are presented in Fig. 10.4. The dispersion of $T_e$ at each halocentric distance is not due to measurement errors, but reflects the actual dispersion of electron temperature of HII regions connected with differences in electron density and the effective temperature of the exciting star.

The data obtained in the $T_e$ gradient are most probably explained by the assumption that upon approaching the galactic center the intensity of star formation processes increases during which light elements are transformed into heavy ones, C, N, O, etc., cooling the interstellar medium (Churchwell and Walmsley 1975). This suggestion found its confirmation in recent years in the indirect appearance of a gradient of heavy element abundance with measurements in the lines NII, OII, OIII, etc. (Shaver et al. 1983). It should be noted that in these measurements RRL played a significant role providing accurate electron temperature measurements which are necessary to determine abundances with optical lines.

*Electron Density of HII Regions*

The Stark width dependence on the level number established in RRL studies provides a means of determining the electron density $N_e$. The equation for determining $N_e$ has the form (Smirnov 1985):

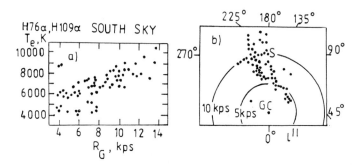

Figure 10.4. Electron temperature gradient of HII regions in the Galaxy according to observations on the lines H76α and H110α (Shaver et al. 1983). a) Electron temperature as a function of galactic radius and b) distribution in the plane of the Galaxy of HII regions on which measurements were performed. GC and S are the galactic center and sun, respectively.

$$N_e = \frac{\Delta\nu_{st}}{8.2\left(\frac{n}{100}\right)^{4.5}\left(1 + 2.25\frac{\Delta n}{n}\right)} \text{ cm}^{-3}, \qquad (14)$$

where $\Delta\nu_{st}$ is the Stark component of the line width in Hz.

To determine $\Delta\nu_{st}$ with the line width one must measure lines for different transitions including transitions with relatively small n, where the width is a purely Doppler one. The electron density in several HII regions was determined using this method. For the central part of the Orion Nebula it was $N_e = (1\pm0.3)\times10^4$ cm$^{-3}$ (Smirnov et al. 1984). For the compact detail in the DR-21 source, $N_e = (4.3\pm0.4)\times10^4$ cm$^{-3}$, and for shell structures and sources W3A and W3B, $N_e = (1.5\pm0.3)\times10^4$ cm$^{-3}$ (Smirnov et al. 1985).

It should be noted that Stark broadening of RRL gives the local, i.e., true value, of $N_e$ determined by collisions. It is an advantage of the Stark broadening method over the other radio astronomic method of determining density, with radio brightness in a continuous spectrum, which gives an average $N_e$ over the line of sight. Comparing data on density obtained with RRL and with radio brightness led to the conclusion about the nonuniformity of the $N_e$ distribution in the Orion and Omega Nebulae, and the presence in them of $N_e$ condensations (Sorochenko and Berulis 1970).

## Velocities of Turbulent Motions

In the millimeter range where Stark broadening is practically absent, RRL widths are determined only by the electron temperature and velocities of turbulent motions $V_t$. Because RRL measurements allow one to determine $T_e$ with great accuracy, one can determine with these lines simultaneously the quantities $V_t$. Substituting in Eq. (3) the values of the constants, we obtain for measurements with hydrogen RRL:

$$V_t = 12.8 \sqrt{1.96 \times 10^8 \left(\frac{\Delta v_D}{v}\right)^2 - \frac{T_e}{10^4 K}} \text{ km s}^{-1} . \tag{15}$$

It was found according to data in the line H56α that in HII regions $V_t = 8-15$ km s$^{-1}$.

## Distribution of Ionized Hydrogen in the Galaxy

At the present time as the result of reviews presented recombination radio lines are measured in more than 600 well-identified HII regions which basically lie close to the galactic plane. About 160 lie in the Southern hemisphere and in them the radio lines H76α (McGee and Newton 1981), H109α and H110α (Wilson et al. 1970; Shaver et al. 1983) are recorded. In the Northern hemisphere according to the most complete survey using the RRL H85α, H87α, H100α, H101α, H110α, H125α, and H127α, 462 HII regions were identified (Lockman 1989).

RRL were also detected in many directions in the galactic plane where HII regions were absent. Gottesman and Gordon (1970) were the first to show this after observing the rather weak (~0.05 K), but quite distinct line H157α in some of these directions. It was supposed initially that the observed radio lines are emitted by the diffuse, partially ionized hydrogen of the interstellar medium homogeneously distributed in the plane of the Galaxy. However, subsequent studies did not support this hypothesis. A diffuse nature of RRL cannot explain the significant changes of the spectrum of radio lines at small angular scales as well as the absence of correlations between large scale distributions of ionized hydrogen obtained in the line H166α and the HI distribution (Lockman 1989).

On the basis of a survey of the galactic plane in the line H272α (v = 325 MHz) Anantharamaiah (1986) concluded that all or almost all of the radiation of decimetric RRL in the galactic plane is caused by external rarified envelopes of usual HII regions. For $l < 40°$ in any direction in the plane of the Galaxy the line of sight will pass through at least one Hu regions. According to measured data electron density in these envelopes is 1–10 cm$^{-3}$, their sizes are 30–300 ps, and temperatures 3000–8000 K.

In Fig. 10.5 the longitude-velocity diagram is presented confirming this conclusion for RRL H110α, H166α, and H272α for $l = 50°$ (Anantharamaiah 1986). One can see that

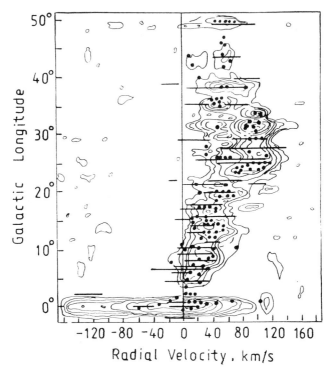

Figure 10.5. Longitude-velocity diagram of RRL H272α (horizontal lines), H166α (contours), and H110α (points). The length of the horizontal lines is the observed half-width of the line H272α (Anantharamaiah 1986).

there is a good correlation between the position of points corresponding to the radiation of dense well-identified HII regions and data in the lines H166α and H272α whose radiation corresponds to the external layers of HII regions and their rarified envelopes.

RRL measurement showed that the majority of HII regions are located in inner parts of the Galaxy and basically at distances 4–8 kps from the galactic center. In this case in the distribution of ionized hydrogen for the Northern and Southern Hemispheres there is some asymmetry. Data with the line H166α evidences that in the fourth quadrant of the galaxy ionized hydrogen is approximately 20% larger than in the first quadrant (Cersosimo et al. 1989). The distribution of the integral intensity in the line H166α per kiloparsec is shown in Fig. 10.6 for the first and fourth quadrants. Both of them, in the

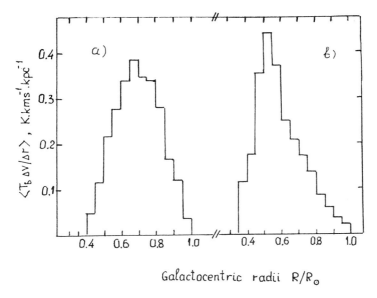

Figure 10.6. Distribution of line intensity at kiloparsecs for a) the fourth quadrant and b) the first quadrant of the Galaxy (Cersosimo et al. 1989).

Northern and Southern distributions indicate a sharp drop of the mass of ionized hydrogen for R < 4 kps; at the same time for R > 6 kps the ionized hydrogen in the fourth quadrant is significantly larger than in the first one.

*Relative Abundance of Chemical Elements in HII Regions*

The helium abundance can be reliably determined using RRL in many HII regions. It was found that the average helium abundance relative to hydrogen is ~8% (Mezger 1980). Dependences of helium abundance on heliocentric distributions were not established (Shaver et al. 1983).

## 3.2. Physical Conditions of the Interstallar Medium Outside HII Regions

The atoms excited to levels n > 400 are quite sensitive indicators of the physical conditions in the rarified cold interstellar medium outside HII regions. Using them one can determine the electron density for $N_e \sim 1$ cm$^{-3}$ and less.

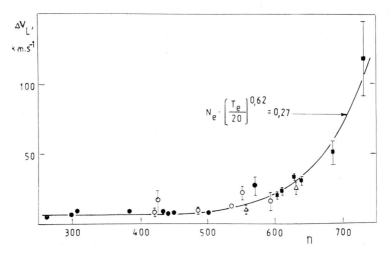

Figure 10.7. Dependence of the carbon RRL width ($\Delta n = 1$ transition) in the direction of Cas A on the level number. Different designations of experimental points correspond to observations by different authors (Sorochenko and Smirnov 1990).

The dependence of the RRL carbon widths are presented in Fig. 10.7 in the direction of Cas A on the level number n. With the carefully observed width of radio lines one can determine the density $N_e$, more precisely the quantity $N_e T_e^{0.62}$. According to this data, $N_e = 0.15$–$0.3$ cm$^{-3}$ depending on the possible values of $T_e = (18$–$50)$ K. Because the carbon abundance is $3.3 \times 10^{-4}$ of hydrogen, the density of atomic hydrogen should be $N_H \cong 10^3$ cm$^{-3}$ (Sorochenko and Smirnov 1990).

A series of arguments indicate the CII regions where carbon RRL are formed are connected with molecular clouds. Goss et al. (1984) identified in the solid angle of Cas A in the Perseus arm with observations in the line at 6 cm with a resolution 10" at least 16 H$_2$CO clouds, with average sizes 0.3 pc and density $N_{H_2} = 10^4$ cm$^{-3}$. These clouds have HI envelopes with an average thickness 0.2 ps and density $N_H = 300$ cm$^{-3}$. The lines of each cloud are narrow, 1–2 km s$^{-1}$ and have different radial velocities. As a result the formaldehyde line integrated over the solid angle of Cas A has a width 14 km s$^{-1}$ in near agreement with the Doppler width $D = 15.9$ km s$^{-1}$ obtained with the carbon line. The radial velocities of the integrated H$_2$CO line and the carbon line also nearly coincide.

The data obtained indicate that carbon RRL in the direction of Cas A are formed most probably in two or several formations "HI envelope-molecular cloud." In this case the

CII regions are located in the surface layers of these formations and the ionization occurs by external UV radiation. One can suppose that the pattern of RRL formation in the direction of Cas A is also typical for other regions of the Galaxy. The regions where low-frequency carbon RRL recorded are CII envelopes around molecular clouds. Thus, RRL of carbon at low frequencies provides quite valuable information on the physical conditions in the transition region between the rarified interstellar gas and the denser molecular clouds.

### 3.3. Use of RRL for Estimating the Intensity of Galactic Cosmic Rays

Assuming that in the cold interstellar medium hydrogen RRL should be formed in the same volume of gas as the 21 cm line, Shaver (1976) obtained the dependence for determining the hydrogen ionization rate by cosmic rays $\zeta_H$ in relation to the optical depths of the 21 cm line and hydrogen recombination lines:

$$\zeta_H = 5.7 \times 10^{-15} \phi_2 \left[ \frac{\tau_{H_n}}{\tau_{HI}} \right] \left[ \frac{\nu}{100 \text{ MHz}} \right] \left[ \frac{T_e}{T_s} \right] \left[ \frac{T_e}{(b_n \beta_n)_H} \right] \text{ s}^{-1}, \qquad (16)$$

where $\tau_{H_n}$ and $\tau_{HI}$ are the optical depths of hydrogen RRL and the 21 cm line, respectively, $T_s$ is the hydrogen spin temperature, and $\phi_2$ is a coefficient weakly depending on temperature ~3.

The detection of carbon RRL made it possible to estimate the hydrogen ionization rate by still another method, by the ratio of optical depths of RRL of carbon and hydrogen (Sorochenko and Smirnov 1987):

$$\zeta_H = \frac{6.8 \times 10^{-15} \phi_2 \left[ N_e T_e^{0.62} \right]}{T_e^{1.12}} \frac{\tau_{H_n}}{\tau_{C_n}}, \qquad (17)$$

where $\tau_{C_n}$ is the optical depth of carbon lines. In this case there is no necessity to postulate an electron density on which the quantities $b_n$ and $\beta_n$ in Eq. (16) depend because the quantity $N_e T_e^{0.62}$ is immediately obtained from the measurements. At the same time in the second method it should be kept in mind that the populations of excited levels of hydrogen and carbon may deviate from equilibrium values for various reasons. To take into account this circumstance it is necessary to correct the ratio of optical depths of RRL of hydrogen and carbon according to the equation:

$$\left[ \frac{\tau_{H_n}}{\tau_{C_n}} \right] = \left[ \frac{\tau_{H_n}}{\tau_{C_n}} \right]_{meas} \cdot \frac{[b_n \beta_n]_C}{[b_n \beta_n]_H}, \qquad (18)$$

where $[b_n\beta_n]_C$ and $[b_n\beta_n]_H$ are the calculated coefficients taking into account variations of RRL intensity due to the nonequilibrium population of levels for carbon and hydrogen, respectively.

The upper estimate obtained by both methods of the hydrogen ionization rate in the direction of Cas A is $\zeta_H < 10^{-16}$ s$^{-1}$. This upper limit is an order of magnitude lower than the quantity $\zeta_H \cong 10^{-15}$ s$^{-1}$, obtained from theoretical calculations of an equilibrium interstellar medium heated by cosmic rays (Kaplan and Pikel'ner 1979). Because in the heating mechanism of the interstellar medium by cosmic rays a series of difficulties were previously noted, the question about alternate heat sources without hydrogen ionization is quite timely.

### 3.4. RRL in Extragalactic Sources

In the second half of the seventies RRL studies were extended to extragalactic sources. The first of these source (discounting the Magellanic Clouds) was the galaxy M82 in which Shaver et al. (1977) detected the radio line H166$\alpha$. Later, using aperture synthesis at Westerbork, Roelfsema (1987) carried out a mapping of M82 in this line with an angular resolution of 13". The data obtained showed that the ionized gas in M82 is situated in a disk or ring which rotates around the galactic center as a solid body.

Up to now besides M82 recombination radio lines were measured only in the galaxy NGC 253 (Seaquist and Bell 1977). Efforts to observe RRL in the direction of some other galaxies and quasars have not yet led to positive results (Bell et al 1984). However, it is clear that with an increase of the sensitivity of radio telescopes success in these observations will be reached and the study of extragalactic sources will become a quite promising direction for RRL.

## CONCLUSION

The studies presented show that in galactic conditions atoms, as quantum systems, may exist to excitation levels n ~ 1000. In this case they reach gigantic sizes ~0.1 mm. The factor preventing the existence of still higher excited atoms is the nonthermal galactic radio radiation.

The spectral lines radiated (absorbed) by highly excited atoms, recombination radio lines, may be observed at the Earth in a wide range of radio waves from millimeter to decimeter ones. They are an effective tool of astrophysical studies with a large diversity in their possibilities. RRL provide a means of measuring the physical properties of the interstallar medium, both in HII regions and in HI-CII regions, the electron temperature, density, and velocity of internal motions. The information contained in RRL also provide a means of estimating the hydrogen ionization rate by cosmic rays, of obtaining the distribution of ionized hydrogen in the Galaxy, of finding the helium content, etc.

RRL are quite promising for studying extragalactic objects.

## References

Anantharamaiah, K. R. (1986) *J. Astrophys. Astr.* **7**, 131.
Ariskin, V. I., Kolotovkina, S. A., Lekht, E. E., et al. (1982) *Astron. Zh.* **59**, 38.
Bell, M. B., Seaquist, E. R., Mebold U., et al. (1984) *Astron. Ap.* **130**, 1.
Cersosimo, J. C., Azcarate, I. N., Hart, L., and Colomb, E. P. (1989) *Astron. Ap.* **208**, 239.
Churchwell, E., and Walmsley, C. M. (1975) *Astron. Ap.* **38**, 451.
Churchwell, E., Smith, L. F., Mathis, J., et al. (1978) *Astron. Ap.* **70**, 719.
Dravskich, A. A., Dravskich, Z. B., and Kolbasov, V. A. (1964) *Trans. IAU* **12**, 360.
Ershov, A. A., Ilyasov, Yu. P., Lekht, E. E., et al. (1984) *Pis'ma Astron. Zh.* **10**, 838.
Ershov, A. A., Lekht, E. E., Rudnitskii, G. M., and Sorochenko, R. L. (1982) *Pis'ma Astron. Zh.* **8**, 694.
Ershov, A. A., Lenht, E. E., Smirnov, G. G., and Sorochenko, R. L. (1987) *Pis/ma Astron. Zh.* **13**, 19.
Goldberg, L. (1966) *Ap. J.* **144**, 1225.
Gordon, M. A. (1989) *Ap. J.* **337**, 782.
Goss, W. M., Kalberla, P. M., and Dickel, M. R. (1984) *Astron. Ap.* **139**, 317.
Gottesman, S. T., and Gordon, M. A. (1970), *Ap. J.* **162**, L93.
Griem, H. (1960) *Ap. J.* **132**, 883.
Griem, H. (1967) *Ap. J.* **148**, 547.
Gudnov, V. M., Zotov, V. V., Nagornykh, L. M., and Sorochenko, R. L. (1968) *Astron. Zh.* **45**, 942.
Gulyaev, S. A., and Sorochenko, R. L. (1974) *Astron. Zh.* **51**, 1237.
Höglund, B., and Mezger, P. G. (1965) *Science* **150**, 339.
Kaplan, S. A., and Pikel'ner, S. B. (1979) *Physics of the Interstellar Medium*, Nauka, Moscow.
Kardashev, N. S. (1959) *Astron. Zh.* **36**, 838.
Konovalenko, A. A. (1984) *Pis'ma Astron. Zh.* **10**, 912.
Konovalenko, A. A. (1990) *IAU Coll N 125: Radio Recomb. Lines: 25 Years of Investigation*, M. A. Gordon and R. L. Sorochenko (eds.), Kluwer, Dordrecht.
Konovalenko, A. A., and Sodin, L. G. (1981) *Nature* **294**, 135.
Lilley, A. E., Menzel, Q. H., Penfield, H., and Zuckerman, B. (1966a) *Nature* **209**, 468.
Lilley, A. E., Palmer, P. Penfield, H., and Zuckerman, B. (1966b) *Nature* **211**, 174.
Lockman, I. J. (1989) *App. J. Supp.* **71**, 469.

Martin-Pintado, J., Bachiller, R., Thum, C., and Walmsley, M. (1989) *Astrom. Ap.* **215**, L13.
McGee, R. X., and Newton, L. M. (1981) *Mon. Not. R. Astr. Soc.* **196**, 889.
Mezger, P. G. (1980) in *Radio Recombination Lines*, B. A. Schaver (ed.), Reidel, Dordrecht.
Minaeva, L. D., Sobel'man, I. I., and Sorochenko, R. L. (1967) *Astron. Zh.* **44**, 995.
Payne, H. E., Anantaramaiah, K. R., and Erikson, W. C. (1989) *Ap. J.* **341**, 890.
Roelfsema, P. R. (1987) Ph.D. Thesis, University of Groningen, The Netherlands.
Salem, M., and Brocklehurst, M. (1979) *Ap. J. Supp.* **39**, 633.
Seaquist, E. R., and Bell, H. B. (1977) *Astron. Ap.* **60**, L1.
Shaver, P. A. (1975) *Pramana* **5**, 1.
Shaver, P. A. (1976) *Astron. Ap.* **49**, 149.
Shaver, P. A., Churchwell, E., and Rots, A. H. (1977) *Astron. Ap.* **55**, 435.
Shaver, P. A., McGee, R. X., Newton, L. M., et al. (1983) *Mon. Not. R. Astr. Soc.* **209**, 53.
Smirnov, G. G. (1985) *Pis'ma Astron. Zh.* **11**, 17.
Smirnov, G. T., Sorochenko, R. L. , and Pankonin, V. (1984) *Astron. Ap.* **135**, 116.
Sorochenko, R. I., and Borodzich, E. V. (1964) *Trans. IAU* **12**, 360.
Sorochenko, R. L. (1965) *Tr. Fiz. In-ta AN SSSR* **28**, 90.
Sorochenko, R. L., and Berulis, I. I. (1970) *Astron. Zh.* **47**, 850.
Sorochenko, R. L., and Borodzich, E. V. (1965) *Dokl. AN SSSR* **163**, 603.
Sorochenko, R. L., and Smirnov, G. T. (1987) *Pis'ma Astron. Zh.* **13**, 191.
Sorochenko, R. L., and Smirnov, G. T. (1990) *IAU Coll. N 125: Radio Recombination Lines: 25 Years of Invest.*, M. A. Gordon and R. L. Sorochenko (eds.), Kluwer, Dordrecht.
Sorochenko, R. L., Puzanov, V. A., Salomonovich, A. E., and Shteinschleiger, V. B. (1969) *Astrophys. Lett.* **3**, 7.
Sorochenko, R. L., Rydbeck, G., and Smirnov, G. T. (1988) *Astron. Ap.* **198**, 233.
Waltman, W. B., Waltman, E. B., Schwartz, P. R., et al. (1973) *Ap. J. Lett.* **185**, 135.
Wilson, T. L., and Pauls, T. (1984) *Astron. Ap.* **138**, 225.
Wilson, T. L., Mezger, P. G., Gardner, F. F., and Milne, P. K. (1970) *Astr. Ap.* **6**, 364.

# 11

# Cosmic Masers: Yesterday, Today, and Tomorrow

## V. S. Strelnitskii
*Institute for Astronomy, Moscow*

Both researchers with which this review is connected, S. B. Pikel'ner and I. S. Shklovsky, made contributions to the study of cosmic masers. These unusual radio sources are interesting both for the variety of physical phenomena and the means of astrophysical research connected with these objects, very young and very old stars. S. B. Pikel'ner was more interested by the physics of the natural maser while I. S. Shklovsky was occupied to an equal degree with the physical and astrophysical aspects of the problem.

This article is not a systematic review of the problem of cosmic masers. Its first section (Yesterday) is a memory about Pikel'ner and Shklovsky on the background of the problem which occurred to me with these two contributing people. In the section Today a series of current problems will be briefly considered connected with cosmic masers. In the section Tomorrow some directions of future research will be noted.

One can find more complete details on cosmic masers in the reviews of Reid and Moran (1981), Elitzur (1982; 1987), Reid and Moran (1988), and Moran (1989) as well as in several articles of the Proceedings of IAU Symposium 129 (Diamond 1988; Kylafis 1988; Strelnitskii 1988).

## YESTERDAY

In 1963 the well-known monograph of S. A. Kaplan and S. B. Pikel'ner, "The Interstellar Medium," appeared (Kaplan and Pikel'ner 1963). Discussing in the Appendix the contribution of induced radiation, $1 - (\tilde{\omega}_{n'}N_n/\tilde{\omega}_n N_{n'})$, to the absorption coefficient (here $N_{n'}$ and $N_n$ are the populations of levels, and $\tilde{\omega}_{n'}$ and $\tilde{\omega}_n$ are their statistical weights), the authors proposed that in the interstellar medium "in transitions between lowly excited levels the quantity $\tilde{\omega}_{n'}N_n/\tilde{\omega}_n N_{n'}$ may be both smaller and larger than unity (i.e., here a laser effect is possible)."

This prediction was made several years before the discovery of cosmic masers and, apparently, independently of the earlier remark of D. Menzel that "in the absence of thermodynamic equilibrium it is possible to think of a situation when the value of the integral (energy absorbed in a spectral line) turns out to be negative" (Menzel 1937). It is true that Menzel considered that "in practice this possibility, apparently, is never realized" (Menzel 1937). The prediction of Kaplan and Pikel'ner was more optimistic, but they proposed that, in the best case, the maser effect in the interstellar medium could only be quite weak.

All this explains why interstellar masers on OH molecules discovered in 1965 (Weaver et al. 1965) and $H_2O$ masers detected in 1969 (Cheung et al. 1969) excited researchers not so much by the fact of a population inversion in natural conditions (it, as we saw, was actually predicted) as by the large brightness (brightness temperatures to $10^{16}$ K) and, consequently, the large, completely unexpected, value of maser amplification ($>10^{12}$ for the brightest sources).

The cause of such a strong amplification was soon understood: the condensations detected in maser radiation were not the "usual" interstellar clouds; they had much smaller sizes ($\cong 1$–10 AU), but much larger densities ($n_H \cong 10^6$–$10^{11}$ cm$^{-3}$). In these dense condensations with a relatively high molecular content, the optical depth in a molecular radio line, $\tau$, can have a modulus much larger than unity. In an unsaturated maser (i.e., in the case where the maser radiation itself does not influence the population of levels) the amplification is approximately $e^\tau$, i.e., it can become very large.

The line-of-sight concentrations of OH, $H_2O$, and SiO molecules are sufficiently large for maser amplification also in envelopes of old cold stars of high luminosity. These stellar masers were discovered quickly after the detection of masers in regions of star formation. Their brightness temperatures are somewhat lower, but also very high, to $\sim 10^{12}$ K.

Here it is appropriate to recall the timing of masers; their discovery almost exactly coincided in time with the emergence of a technological breakthrough, with the invention of very-long baseline interferometry (VLBI). It was only with this method that it was possible to measure the small angular sizes of maser condensations (to $10^{-3}$–$10^{-4}$") and with the observed flux density and angular size to determine radiation brightness temperatures.

In 1966 soon after the discovery of the first OH masers in regions of star formation, J. S. Shklovsky put forward the hypothesis that maser radiation is generated in the outer layers of protostars and proposed an actual pumping mechanism for by these masers, the infrared radiation of protostars (Shklovsky 1966). The "protostar" hypothesis of Shklovsky played the role of a catalyst in the study of cosmic masers. It attracted the attention of astrophysicists who understood that, due to these strong compact radio sources, they had in their hands a unique observational tool for studying the earliest stages of star formation. As a rule, the spectra of maser sources consisted of several narrow details. If one interprets the differences in frequencies of details as the result of Doppler shifts due to the spatial motion of gas clumps (each of which Shklovsky considered a separate protostar), then the dispersion of the corresponding radial velocity of OH clumps is usually of order 10 km/s. This agreed well with the hypothesis about the formation from the protostar cloud of a stable stellar aggregation with a negative total energy.

The "protostellar" hypothesis for OH masers became so popular that one initially again began to associate the discovered $H_2O$ masers with protostars. Following Shklovsky, Litvak (1969) put forward the proposition that $H_2O$ masers are also pumped by the IR radiation of protostars.

However, it was soon detected that the spectra of some $H_2O$ sources were very broad, to several hundred km/s. It became clear that a system of bodies with such a velocity dispersion, independent of their individual masses, should have a positive total energy, i.e., should break up (Strelnitskii and Sunyaev 1972). Being convinced that stars do not form as the result of the breakup of superdense bodies (Byurakan conception) but by the gravitational condensation of gas, I. S. Shklovsky rejected his protostar hypothesis and took the concept proposed by us of $H_2O$ masers as light clouds driven to the high observed velocities by the stellar wind of a very young star. He even was attracted to the idea that our estimates of cloud masses gave "planetary" values and soon he spread our "protoplanetary" hypothesis with the same enthusiasm and sincerity as his own "protostellar" one. The truth (especially if it contained a good dose of romanticism) was undoubtedly for him more valuable than any personal scientific prestige although he also valued the latter to some extent.

S. B. Pikel'ner also supported the idea of the driving of $H_2O$ maser condensations by a stellar wind. In conversations with him we worked through the idea that such a strong driving of condensations certainly should be accompanied by strong gas dynamic effects and this could become a key to the explanation of the amplification mechanism of the $H_2O$ maser (see the next section).

Concerning OH masers, in contrast to the "protostar" hypothesis which did not stand the test of time, the idea of I. S. Shklovsky about the radiative IR pumping of these masers turned out to be fruitful. IR radiation is the inevitable satellite of compact molecular objects in regions of star formation. A considerable fraction of their flux is radiated in the IR also by old cold stars. The polar molecules such as OH or $H_2O$ have a rich spectrum of resolved IR transitions, vibrational and rotational. Therefore, the populations of molecular energy levels are often controlled by just this IR radiation. Following Shklovsky, models of IR pumping of OH masers were developed by many other researchers. For example, it is now generally accepted that OH masers in the 1612 MHz line observed in envelopes of old cold stars of high luminosity are pumped by the IR radiation of the star (e.g., Elitzur et al. 1976). It is quite probable that OH masers, associated with very young hot stars or "cocoons" (their well-known prototype is the source W3-OH) are also pumped by the IR radiation of the dust cloud ("cocoon"), remaining around the star after its formation from the gas-dust cloud (e.g.: Burduzha & Varshalovich 1973).

The problem with $H_2O$ sources, the brightest of known maser sources, became more complicated. Multiple attempts to explain the strong radiation of $H_2O$ masers by the pumping mechanism including as the source or sink of energy radiative processes did not meet with success. The point is that all these mechanisms are limited in rate by effects of "trapping" of the resonance radiation in IR transitions whose large optical depth automatically follows from the requirement of the large amplification in the maser transition if one does not assume a quite special, strongly elongated or plain source geometry. The problem of pumping of strong $H_2O$ masers cannot be considered solved to this day. Some contemporary approaches to its solution will be discussed in the following section.

In finishing this section of memories, I would like to consider still another property of cosmic masers, immediately attracting the attention of J. S. Shklovsky, the polarization of their radiation.

The radiation of many spectral OH details is very strongly circularly polarized (to 100%). Because OH is a paramagnetic radical having one unpaired electron, it is natural to connect the radiation polarization with the Zeeman effect. Zeeman splitting for the 1665 MHz line is, in radial velocity units, about 0.3 (km/s)/mG. Consequently, the Zeeman shift, exceeding the observed line width ($\lesssim 1$ km/s) is expected in a magnetic field of order several mG. In OH maser spectra one often observes spatially coincident pairs of oppositely polarized details which are interpreted as the $\sigma$-components of the Zeeman triplet in a longitudinal magnetic field. Their separation and frequency indicates magnetic fields of several mG. The absence of the linearly polarized $\pi$-component of the triplet may be explained by depolarization caused by Faraday rotation in the partially ionized gas of the maser condensation. Those cases when only one $\sigma$-component is observed remain a puzzle. Shklovsky proposed a refined model explaining this phenomenon (Shklovsky 1969): the rotating protostar with a radial gradient of the magnetic field wound up in it is observed on the background of a nearly situated HII region. He showed that the oppositely situated edges of this protostar in the presence of a population inversion will amplify the circularly polarized radiation with the same sign of rotation as observed. And although the protostar hypothesis of the nature of OH masers turned out to be erroneous, the idea of the combination of gradients of velocity and magnetic field for explaining features of OH circular polarization maintain their value to this day.[†]

## TODAY

We consider some problems that are being intensively developed at the present.

---

†This idea was independently given previously by Cook (1966).

## OH Masers

These are the most understood sources of maser radiation at the present time. The maser emission in several OH radio lines from the regions of star formation is observed from the vicinity of very young OB stars, not having reached the main sequence, but emitting sufficient ultraviolet for the ionization of a compact ($\leq 0.1$ pc) HII region. OH emission is generated in the neutral gas-dust envelope (cocoon) surrounding the compact HII region. Active molecules amplify the continuum radio emission of the HII region in narrow lines; therefore, OH maser condensations are most often projected on the corresponding HII region. It is quite probable that the maser is pumped by the far IR emission of the dust contained in the cocoon. The basic energetic condition for radiative pumping is that the number of pumping photons should be at least as large as the number of radiated maser photons which is necessarily satisfied here.

There is no longer a large uncertainty for the several hundred OH masers observed in the 1612 MHz line from expanding shells of cold old stars of high luminosity (see, e.g., Engels 1979). The shells of these stars provide the excess IR radiation from which the short name of these objects has emerged, OH/IR stars. The typical two-humped OH spectrum of a OH/IR star is well modeled by the radiation from an optically thick shell expanding with a constant velocity ~10–50 km/s. Many point sources from the IRAS catalog in the OH radio lines have this characteristic two-humped radiation and are classified as OH/IR stars even if in the visual spectral region they are completely invisible due to the very strong absorption by dust in the shell (Lewis et al. 1985). All OH/IR stars are regular or semi-regular variables with a characteristic period of several hundred days. The maser radiation varies with the same period as IR radiation with a small phase delay.

One of the clearest accomplishments in the study of OH/IR stars is the measurement of the delay time for maser emission coming from the rear half of the shell (red peak) in comparison with emission from the front half (blue peak) (see, e.g., the study of Diamond et al. (1985) with the references to previous work). The observed delay time gives a light diameter of the shell ~10 days, i.e., ~$10^3$ AU. Because the angular diameter of the shell emitting in the maser radio line may be measured directly with an interferometer, the possibility of rises of determining the distance to the star by means of a comparison of the angular and linear diameters of the shell. The relatively small statistics of these measurements (15 stars) was sufficient for Herman et al. (1985) with the newly determined distance to these stars to redetermine the distance to the center of the galaxy (9.2±1.2 kpc) in the framework of the galactic rotation model of Schmidt.

## $H_2O$ Masers

One of the most significant accomplishments in the study of $H_2O$ masers in the last ten years was the measurement of their proper motions by the VLBI method, first carried out by Genzel et al. (1981). The prediction was verified about the expansion of maser condensations in sources with wide spectra. The predicted value of the speed of the expansion was verified corresponding to the dispersion of radial velocities in the spectrum.

It was no less important that these measurements became a new method of determining the distance to regions of star formation. If one supposes that maser condensations fly off from the central source isotropically, then the dispersion of radial velocities and velocities in the plane of the figure should be equal. The distance to the source is then determined by the classical method of statistical parallax as the ratio of the dispersion of velocities in linear scale (determined by radial velocities) to their dispersion in angular scale (by their proper motions). Variations of this method were applied to those cases when one could suspect a regular character to the motion, for example, an expansion with constant velocity. In this way the distance to the $H_2O$ masers in Sgr B2 was determined which is a radio source close to the galactic center, $7.1\pm1.5$ kpc (Reid et al. 1988), which is somewhat lower than previous estimates of distances to the center.

In a personal conversation, I. S. Shklovsky once told me that he considered his best work the new method he proposed for determining distances to planetary nebula. During his life only the first measurements of the proper motions of $H_2O$ masers were carried out and the new method of determining intragalactic distances had still not been verified. Now, I think when the method described has already determined distances to five galactic regions of star formation (Ori A, W51M, W51N, Sgr B2, and W49; Moran 1989) and in turn about the measurements of intergalactic distance (see the next section) that I. S. Shklovsky would have esteemed highly this new application of cosmic masers for solving one of the fundamental problems of astronomy.

The spectra of $H_2O$ masers in regions of star formation have, as a rule, quite irregular structures. However, in recent years an ever greater number of examples of regular spectra have been published, in particular, triplets maintaining a stable structure for a time of at least several years. One of the examples is shown in Fig. 11.1. Elmergreen and Morris (1979), Grinin and Grigor'ev (1983a,b), and Cesarony (1989) developed the hypothesis according to which the triplet spectrum is formed in the near-stellar Keplerian disk, observed edge on. In the framework of this hypothesis, Cesarony (1989) was able to also explain variability features of the triplet component in the source S255. If this hypothesis is true, then valuable information on the structure of the disk, its size, its angular moment, and mass of the central star may be given by VLBI measurements of distances between hot spots emitting triplet components and a measurement of the angular sizes of spots.

Recently great attention was given to the problem of observing $H_2O$ line widths. The typical width of a single, clearly not blended component is several tens of kHz which corresponds to a kinetic temperature $\sim 10^2$ K. Sometimes quite narrow lines are observed, to 20–30 kHz. The corresponding temperatures of <100 K cannot provide a sufficient population of working $H_2O$ levels whose height above the ground state is, in temperature units, about 700 K. It is true that the line can be narrowed by several times

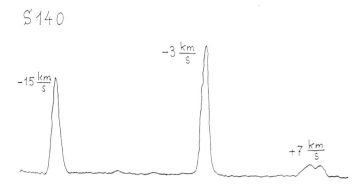

Figure 11.1. The spectrum of the $H_2O$ maser source S 140 obtained by summing spectra for ten years of observations from 1981–1990 (Sorochenko et al. 1991).

(in comparison with the thermal Doppler profile of the absorption coefficient) if the maser is not saturated because with nonsaturated (exponential) amplification the intensity in the center of the line grows faster than in the wings. In addition, processes of cross relaxation of the profile of the absorption coefficient with the trapped resonant IR radiation or with elastic collisions with surrounding molecules may delay the process of secondary line broadening at the onset of saturation (Goldreich and Kwan 1974; Elitzur 1990). In principle, this provides a means of reconciling the observed line narrowness with the supposition that $H_2O$ masers work in the saturated regime (without this supposition it is difficult to explain the observed linear polarization of $H_2O$ masers and the high strength of their emission).

However, greater difficulties arose with the quantitative explanation of the observed dependence in many spectral details of their width on intensity: the observed widths change with intensity according to a power law: $\Delta v \propto I^{-\alpha}$ with a quite large (in modulus) index $\alpha$ from 0.3 to 1 at the same time as the unsaturated narrowing is much weaker: $\Delta v \propto (\ln I)^{-0.5}$. One of the most carefully studied cases of the variability of line width as a function of its intensity is shown in Fig. 2 (Lekht 1991). Qualitatively, the change $\Delta v$ reminds one of what is expected in the transition from the unsaturated regime to the saturated one. However, it is easy to be convinced that as with many other details, the variation of line width with the variation of its intensity occurred here much more rapidly than would be possible with unsaturated amplification (solid line in Fig. 11.2).

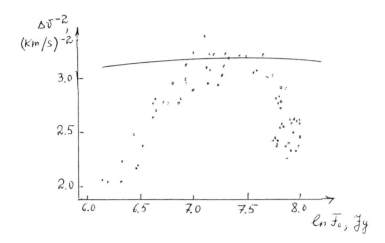

**Figure 11.2.** Points: the dependence of the square of the inverse width of the H$_2$O line with the radial velocity 42.2 km/s in the source G 43.8–0.1 as a function of the logarithm of the flux density in the center of the feature (Lekht 1991). Solid line: the dependence between the same quantities in a numerical model of a two-stream cylindrical maser for its transition from a nonsaturated regime to a saturated one.

One of the possible explanations of such strong variations of line width is the variation of kinetic temperature in the region of amplification (Strelnitskii 1986, 1988). Still another possibility was recently proposed by Ishankuliev (1990). It is connected with the supposition that in H$_2$O masers line broadening by the maser radiation itself, $\Delta v_R$, is comparable with the Doppler width of the absorption coefficient, $\Delta v_D$ (it is necessary for this that masers with an observed brightness temperature $10^{14}$–$10^{16}$ K radiate in a wide solid angle, $\Omega \sim 1$). For $\Delta v_R \gtrsim \Delta v_D$ it is necessary to take into account the evolution of not only the amplitude, but also the phase of the signal in the process of amplification and, according to the calculations of Ishankuliev, the phase fluctuations give line broadening varying according to $\Delta v \propto I^{-0.5}$ as observed. This mechanism requires serious further study.

The observed linear polarization of H$_2$O maser radiation provides a means of estimating the magnetic field strength in the source. A comparison of the observed degree of polarization with theoretical models usually leads to values of order tens to hundreds of mG, i.e., by one or two orders of magnitude higher than in OH masers. Qualitatively this agrees with the conclusion from models of pumping and amplification that the $H_2O$ condensations should be denser by 3–5 order than OH condensations and, consequently, the magnetic field there should be stronger for the condition of its partial freezing into the gas.

An important step was the first measurement of the quite weakly circularly polarized $H_2O$ radiation (Fiebig and Güsten 1988). Values of field strength to 0.1 G were obtained. An attempt at the interpretation of one case of a measurement of rotation of the polarization plane in the brightest $H_2O$ maser line in Orion (Matveenko 1984) gave an even higher value of the field strength, to ~10 G (Strelnitskii 1987).

We note two different approaches to the problem of explaining extremely high strengths of $H_2O$ maser emission following from the observed values of flux density for some assumptions on the magnitude of the solid angle in which the maser emission is concentrated. The first approach is a search for effective pump mechanisms (Strelnitskii 1980, 1984; Bolgova 1981; Kylafis and Norman 1986, 1987). In all the works cited the collision-collision mechanism of pumping was discussed with particles of two sorts with different kinetic temperatures. This mechanism has no limitation in principle on the gas density (in contrast to radiative mechanisms) and, for densities greater than $10^{11}$ cm$^{-3}$, apparently, can provide the necessary rate of pumping even in the brightest of known sources.

The second approach is a search of those geometrical source models which provide very small values of the emission solid angle, reducing correspondingly the requirement on the pump power. In one of the latter models (Deguchi and Watson 1989) to explain bright bursts of maser emission the possibility was studied of randomly forming two cylindrical masers along the line of sight. For a sufficient distance between cylinders the solid angle of the maser beam may be very small and the high brightness temperature at the output of the maser does not require a large pump rate.

For a long time only one $H_2O$ radio line, $\lambda = 1.35$ cm (the rotational transition $6_{16}$–$5_{23}$) was observed in maser emission. Recently radiation was recorded in several more lines: the rotational transitions $4_{40}$–$5_{33}$ and $5_{50}$–$6_{43}$ in the excited vibrational state $v_2$ from old cold stars of high luminosity (Menten and Melnick 1989), and the rotational transitions $10_{29}$–$9_{36}$ and $5_{15}$–$4_{22}$ in the ground vibrational state from several regions of star formation (Menten et al. 1990a,b).

## Protostars

The desire to detect stars at the earliest stage of their formation, the free-fall stage, has always been very large for researchers. Despite the unsuccessful hypothesis of Shklovsky with OH masers, attempts to find something among maser sources which is directly connected with protostars continues. One of the candidates for this role may be masers on methanol molecules ($CH_3OH$) (Strelinskii 1981). The sizes (~$10^{16}$ cm), masses (~10

M⊙), and temperatures (~ $10^2$ K) of gas clumps displaying themselves in the weak maser emission of methanol at λ = 1.25 cm in Orion and other sources in any case supports the assumption about the gravitational instability of these clumps. Their spatial position in Orion is quite indicative: they are around a compact aggregation of very young hot stars of high luminosity, still surrounded by their cloud of origin (and thus visible only in the IR). Aren't we dealing here with the "contagious" growth of a stellar cluster whose nucleus has already formed and whose stellar wind induces gravitational instability and formation of new stars from clumps of the surrounding gas?

Ten years ago we knew about methanol maser lines only with frequencies near 25 GHz ($J_2$–$J_1$ rotational transition series) and only in the single source Ori A (OMC 1). During the intervening years many new lines of this molecule were discovered in many sources. The majority of the lines show clear signs of maser line amplification, but there are also clear "thermal" ones. Batrea et al. (1987) divided all observed methanol maser lines into two classes. Class B are lines observed in the direction of compact HII regions. They have a close spatial connection with the OH masers observed there. The lines with frequencies 12.1, 19.9, and 23.1 GHz belong to Class B. The sources of lines of Class A are not connected with compact HII regions. The prototype of lines of this class are the already noted lines $J_2$–$J_1$ at 25 GHz. One often associates the lines at 44 and 36 GHz with it, being very close to it in the form of spectra and in the spatial localization of sources. Menten and Batrea (1989) also considered that the sources of these lines may be connected with protostars.

The results of observations of the region of active star formation NGC 6334 from Menten and Batrea (1989) are shown in Fig. 11.3 in the lines of methanol, hydroxyl, and water vapor. One can see well the difference in the spatial localization of methanol sources of Class B connected with the compact HII region (F) and sources of Class A projected on the peak of far IR radiation. In the region of this peak there are no compact HII regions or sources of near IR radiation usually connected with the more advanced phases of star formation. It is possible that long-waited protostars will be identified here. The fact of the protostar's contraction could be maintained by a careful comparison of the radial velocities in the spectra of different molecules.

## Masers in Hydrogen Recombination Lines

Calculations of standard populations of excited levels of hydrogen in HII regions show that transitions between these levels, as a rule, are inverted. However, the actual maser effect in hydrogen recombination lines had not been observed until recently. The cause is the very small optical depth of HII regions in these lines. Recently Martin-Pintado et al. (1989a,b) detected the first case of an actual maser effect in the recombination lines H35α–H29α in the Be-star MWC 349. The emission spectra are shown in Fig. 11.4. The arrow indicates the radial velocity of the center of mass of the system determined by

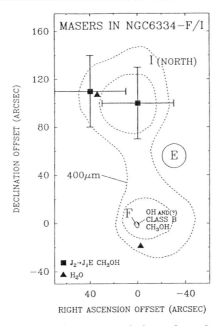

Figure 11.3. Sources of maser emission of methanol of Class A (squares) and water vapor (triangles) in NGC 6334. The probable position of OH masers and $CH_3OH$ masers Class B are also noted of at the edge of the compact HII region which is a source of radio continuum with the designation F. The dashed lines indicate isophots of far IR radiation at 400 μm (after Menten and Batrea 1989).

optically thin recombination radio lines. An attempt to explain the observed spectra in the framework of a model of an expanding ionization shell (Martin-Pintado et al. 1989) did not give satisfactory agreement of the observed and theoretical spectra. Ponomarev et al. (1989) proposed another model: the emission of components shifted to the red arise in a disk-shaped shell near the star in which the gas accretes onto the star almost radially to distances ~10 AU from the star (where maser radiation is generated); the emission of a "blue" component arises in a jet of bipolar outflow. Further observations of this unique object are necessary in recombination and other lines.

## Megamasers

One of the accomplishments of the last ten years in this domain is the discovery of masers in near-nuclei regions of other (active nuclei) galaxies. The extragalactic OH and

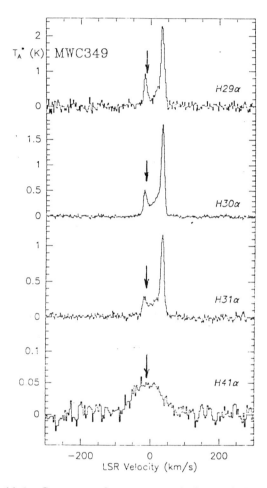

Figure 11.4. Spectra of maser emission of MWC 349 in hydrogen recombination lines. The arrows indicate the radial velocity of the center of mass of the system determined by thermal recombination lines (after Martin-Pintado et al. 1989).

$H_2O$ masers turned out to be much brighter than galactic ones and have obtained the name "megamasers." It is understood that we are here dealing with the effect of observational selection. The existence of megamasers in no way excludes the presence of ordinary maser sources in quiet galaxies which have not yet been detected due to the relatively small brightness. Active galaxies with OH emission differ by especially strong IR excesses correlated with the strength of the OH maser emission. It is assumed that this IR radiation provides the pumping of the maser.

To explain the emission of $H_2O$ megamasers (and $H_2O$ masers in the Galaxy) is more complicated. If the near-nuclei $H_2O$ megamasers do not have a very high emission directivity (which is not very probable), then to explain their high strength requires a very effective pump mechanism. It is possible that here as in the brightest galactic $H_2O$ masers collisional-collisional pumping occurs (Moran 1989).

## TOMORROW

We will attempt to note directions of research which can lead in the near future to answers to problems posed by the science of masers today.

1. Interferometric studies of $CH_3OH$ masers at different spatial scales. A comparison of the spatial and kinematic structure of $CH_3OH$ masers at different transitions between themselves and with masers at transitions of other molecules. These studies may lead to a solution of the problem of searching for protostars.

2. VLBI studies of $H_2O$ masers with triplet spectra should be carried out. They are necessary for verifying the hypothesis on the generation of maser radiation in a Keplerian disk and, in the case that this hypothesis is supported, for determining the size of the disk, its angular momentum, and the mass of the central star.

3. Searches and observations of masers and lasers at low levels of hydrogen in shells of old and new stars should be performed. There is hope for success in the positive results for MWC 349 in transitions of the millimeter range and results of numerical modeling of the excitation of hydrogen for different densities and temperatures showing the possibility of inversion of almost all transitions in a wide range of conditions.

4. Observations of proper motions of $H_2O$ masers in other galaxies and a determination of distances to them by statistical methods should be carried out. These time-consuming experiments have already begun. Recently the first synthesized interferometric maps of $H_2O$ masers in the M33 galaxy that are the first step for determining proper motions of maser condensations were obtained (Greenhill et al. 1990). In the domain of extragalactic interferometry we will have more hopeful results in the near future of an experiment with cosmic interferometry (Kardashev 1991).

## References

Batrea, W., Matthews, H. E., Menten, K. M., and Walmsley, C. M (1987) *Nature* **326**, 49.
Bolgova, G. T. (1981) *Scientific Inform. Astron. Council AN SSSR* **47**, 9.
Cesarony, R. (1989) *Astron. Astrophys.*, in press.
Cheung, A. C., Rank, D. M., Townes, C. H., et al. (1969) *Nature* **221**, 626.
Cook, A. H. (1966) *Nature* **211**, 503.
Deguchi S., and Watson, W. D. (1989) *Astrophys. J.* **340**, L17.
Diamond, P. J., Norris, R. P., Rowland, P. R., et al. (1985) *MNRAS* **212**, 1.
Diamond, P. J. (1988) in The Impact of VLBI on Astrophysics and Geophysics, IAU Symp. 129, M. J. Ried and J. M. Moran (eds.), Kluwer Academic Publishers, Dordrecht.
Elitzur, M. (1982) *Rev. Mod. Phys.* **54**, 1225.
Elitzur, M. (1987) in Interstellar Processes, D. J. Hollenbach and H. A. thornson (eds.), Reidel, Dordrecht.
Elitzur, M. (1990) *Astrophys. J.*, in press.
Elitzur, M., Goldreich, P., and Scoville, N. (1976) *Astrophys. J.* **205**, 384.
Elmegreen, B. G., and Morris, M. (1979) *Astrophys. J.* **229**, 593.
Engels, D. (1979) *Astron. Astrophys. Suppl.* **36**, 337.
Fiebig, D., and Güsten, R. (1988) *Astron. Astrophys.*
Genzel, R., and Downes, D. (1977) *Astron. Astrophys. Suppl* **30**, 145.
Genzel, R., et al. (1981) *Astrophys. J.* **244**, 884.
Greenhill, L. J., Moran, J. M., Reid, M. J., et al. (1990) *Astrophys. J.*, in press.
Goldreich, P., and Kwan, J. (1974) *Astrophys. J.* **190**, 27.
Grinin, V. P., and Grigor'ev, S. A. (1983a) *Sov. Astron.* **27**, 298; (1983b) *Sov. Astron. Lett.* **9**, 244.
Herman, J., Baud, B., Habing, H. J., and Winnberg, A. (1985) *Astron. Astrophys.* **143**, 122.
Ishankuliev, D. (1990) *Astron. Circl.* **1542**, 11.
Kaplan, S. A., and Pikel'ner, S. B. (1963), The Interstellar Medium, Nauka, Moscow.
Kardashev, N. S. (1991), this volume.
Kylafis, N. D. (1988) in The Impact of VLBI on Astrophysics and Geophysics, IAU Symp. 129, M. J. Reid and J. M. Moran (eds.), Kluwer Academic Publishers, Dordrecht.
Kylafis, N. D., and Norman, C. (1986) *Astrophys. J.* **300**, L73; (1987) *Astrophys. J.* **323**, 346.
Lekht, E. (1991), in preparation.
Lewis, B. M., Eder, J., and Terzian, Y. (1985) *Nature* **313**, 200.
Litvak, M. (1969) *Science* **165**, 855.

Martin-Pintado, J., Bachiller, R., Thum, C., and Walmsley, M. (1989a) *Astron. Astrophys.* **215**, L13.

Martin-Pintado, J., Thum, C., and Bachiller, R. (1989b) *Astron. Astrophys.* **229**, L9.

Matveenko, L. M. (1984) *Pis'ma Astron. Zh.* **10**, 199.

Menten, K. M., and Batrea, W. (1989) *Astrophys. J.*, **341**, 839.

Menten, K. M., and Melnick, G. J. (1989) *Astrophys. J.* **341**, L91.

Menten, K. M., and Melnick, G. J., and Phillips, T. G. (1990a) *Astropohys. J.* **350**, L41.

Menten, K. M., Melnick, G. J., Phillips, T. G., and Neufeld, D. A. (1990b) *Astrophys. J.* **363**, L27.

Menzel, D. (1937) *Astrophys. J.* **85**, 330.

Moran, J. M. (1989) in Molecular Astrophysics, T. Hartquist (ed.), Cambridge University Press, Cambridge.

Moran, J. M. (1989) Handbook of Laser Science and Technology, M. J. Weber (ed.), CRC Press, Boca Raton, Florida.

Ponomarev, V. O., Smirnov, G. I., Strelnitskii, V. S., and Chugai, N. N. (1989) *Astron. Circl.* **1540**, 5.

Ponomarev, V. O., Strelnitskii, V. S., and Chugai, N. N. (1990), *Astron. Circl.* **1545**, 37.

Reid, M. J., Schneps, M. H., Moran, J. M., et al. (1989) in The Galactic Center, IAU Symp. 136, M. Morris (ed.), Kluwer Academic Publishers, Dordrecht.

Reid, M. J., and Moran, J. M. (1981) *Ann. Rev. Astron. Astrophys.* **19**, 231.

Reid, M. J., and Moran, J. M. (1988) in Galactic and Extragalactic Radio Astronomy, G. L. Verschuur and K. I. Kellermann (eds.), Springer-Verlag, New York.

Shklovsky, I. S. (1966) *Astron. Circl.* **372**, 1.

Shklovsky, I. S. (1969) *Astron. Zh.* **46**, 3.

Sorochenko, R. L., Lekht, E., and Logvinenko, S. (1991), in preparation.

Strelnitskii, V. S. (1980) *Pis'ma Astron. Zh.* **6**, 354.

Strelnitskii, V. S. (1981) *Pis'ma Astron. Zh.* **7**, 406.

Strelnitskii, V. S. (1984) *MNRAS* **207**, 339.

Strelnitskii, V. S. (1986) *Astron Circl.* **1465**, 1.

Strelnitskii, V. S. (1987) *Astron Circl.* **1490**, 4.

Strelnitskii, V. S. (1988) in The Impact of VLBI on Astrophysics and Geophysics, IAU Symp. 129, M. J. Reid and J. M. Moran (eds.), Kluwer Academic Publishers, Dordrecht.

Weaver, J,. Williams, R. W., Dieter, N. H., and Lum, W. T. (1965) *Nature* **208**, 29.

# 12

# Gravitational Collapse of Massive Stars, Supernovae, and SN 1987A in the Large Magellanic Cloud

## V. S. Imshennik
*Institute of Theoretical and Experimental Physics*

There are two directions in the modern hydrodynamic theory of the collapse of iron stellar cores of massive stars ($M_{ms} \gtrsim 10\ M_\odot$): for sufficiently small masses of the iron core $M_{Fe} \gtrsim 1.2\ M_\odot$, the theory of "cold" collapse, and for sufficiently large masses $M_{Fe} \lesssim 2.0\ M_\odot$, the theory of "hot" collapse.

A comparison presented in this study of these versions of the theory shows that the determining value in the hydrodynamics of collapse is the behavior of the specific entropy S calculated per single nucleon. This quantity in both versions of the theory changes comparatively little so that $S \cong S_o$, but the exact behavior is determined by the equations derived of entropy production. For $S_o \lesssim 1\ k_B$/nucleon the theory of cold collapse is correct and for $S_o \gtrsim (2-3)\ k_B$/nucleon the theory of hot collapse is correct ($k_B$ is Boltzmann's constant).

The mechanism of a supernova explosion has a special significance in the theory of collapse which has not been solved up to now in both versions of the theory. We present in this study an analysis of all existing attempts to solve this problem and indicate possible paths of their solution. At the same time we point out the unarguable indications in the SN 1987A burst of the presence of an effective explosion mechanism just in the center of the star, the progenitor Sk-69°202. In all probability in this case the explosion occurred in a quite massive iron core for which the theory of hot collapse is applicable.

It is completely possible that a one-dimensional supernova explosion mechanism does not exist and, in fact, the main role in the explosion is played by non one-dimensional effects of rotation in the magnetic field. A short discussion on the role of these effects is given.

## 1. INTRODUCTION. PHYSICAL AND ASTRONOMICAL CLASSIFICATION OF SUPERNOVAE INCLUDING SN1987A

Supernovae (SN) in the modern physical classification are divided into two types from the point of view of their physical mechanism or, more precisely, from the trigger mechanism. Attempts at a physical classification of SN were made long ago and persistently. (We consider as important here the studies of Arnett (1982), Shklovsky (1982), Ivanova et al. (1983), and the new work on this problem of Imshennik and Nadezhin (1989).) The first type of SN concerns those for which the trigger mechanism is assumed to be the hydrodynamic process of gravitational collapse of the central core of the star with a characteristic hydrodynamic time $\tau_{hd}$. We note that at the latest stages of a hydrostatic equilibrium evolution a star of quite large mass (on the main sequence) with $M_{ms} \gtrsim (8-10)$ $M_\odot$ consists of a central iron core (the chemical composition are elements of the iron peak), the mantle, having an onion structure from successive layers of different chemical composition (in the mantle layer are basically nucleides $^{12}C$, $^{16}O$, $^{20}Ne$, $^{24}Mg$, $^{28}Si$, $^{32}S$, $^{36}Ar$, and $^{40}Ca$), and finally, the initial hydrogen-helium shell with a small admixture (weight concentration of the admixture: $0 \lesssim Z \sim Z_\odot = 0.02$) of heavy elements (in the admixture composition: O, C, Ne, Fe, N, Si, etc. in order of decrease of the weight concentration according to Cameron (1982)). These ideas are based on the existing classical theory of the evolution of single (isolated from birth) stars. The first detailed study of these stars as presupernova ones was that of Fowler and Hoyle (1965). It is understood in this theory that some actual, but complex processes were not accounted for such as mass exchange from the stellar companions (and close double and multiple systems) and the interstellar medium, large-scale mixing of the material of neighbouring layers and the star as a whole, fairly strong rotations, and the magnetic field.

Nevertheless, the parameters of the central iron cores of stars in all studies of the hydrodynamic process of gravitational collapse, generally speaking, are adopted from the classical theory of evolution. In this connection we introduce the main parameter of the iron core, its mass $M_{Fe}$. According to recent data it varies in a quite narrow range: $M_{Fe}^{min} = 1.18$ $M_\odot$ for $M_{ms} = 13$ $M_\odot$ (Nomoto and Hashimoto 1988) and $M_{Fe}^{max} = 1.85$ $M_\odot$ for $M_{ms} = 25$ $M_\odot$ (Woosley and Weaver 1988). The quantity $M_{Fe}$ increases with an increase of stellar mass on the main sequence $M_{ms}$. There is some tendency for a concentration of $M_{Fe}$ in the neighbourhood of 1.4 $M_\odot$.

For the other type of supernovae with a limiting mass from above $M_{ms}$ of $(8-10)$ $M_\odot$ as the trigger mechanism a thermonuclear explosion according to the modern physical classification in the central carbon-oxygen core of the star is proposed which occurs in the center of the star taking the place of the iron core. In its remaining structure the star may be similar to the one described above. One qualitative difference should be noted. In the stellar core very strong relativistic degeneracy of electrons occurs so that, in particular, the thermal capacity of matter reduces to the thermal capacity of ions. This

## 12. GRAVITATIONAL COLLAPSE

creates exceedingly favorable conditions for an explosive (hydrodynamic) thermonuclear burning of a mixture of carbon with oxygen. If, in the course of the prior hydrostatic equilibrium evolution of the star its characteristic time of thermonuclear burning, $\tau_{nuc}$, was much larger than the characteristic hydrodynamic time, $\tau_{hd}$, then during the explosion $\tau_{nuc} \sim \tau_{hd}$. We note that in iron cores a degeneracy of electrons also occurs, but it is not as strong as in carbon-oxygen stellar cores.

The typically considered initial temperatures and densities (in the center with index "c") for studies of hydrodynamic processes in stellar cores are the following: $T_c \lesssim 10^9$ K and $\rho_c \cong (10^9-10^{10})$ g cm$^{-3}$ (in carbon-oxygen cores) (Imshennik and Nadezhin 1983); $T_c \cong 10^{10}$ K, $\rho_c \cong (10^9-10^{10})$ g cm$^{-3}$ (in iron cores) (Müller 1990; Woosley and Weaver 1986). The range of density variations is identical in both cases, in the first case it is due to different evolutionary features of carbon-oxygen cores such as processes of neutrino loss, initiation of convection, and processes of accretion in close binary systems. In this case the mass, $M_{CO}$, is close to the Chandrasekhar limit $M_{CO} \lesssim 1.44$ $M_\odot$. In the second case this range is possibly even somewhat broader and due to the variation of mass of iron cores, $M_{Fe}$, in the range indicated above. The central temperatures $T_c$ correspond to the corresponding stage of thermonuclear burning.

Together with the physical classification, a part of which concerning the mechanism of explosion we described, a well-known astrophysical classification of supernova explosions exist which in its simplest form reduces to two types: supernovae of type I and II (SN I and SN II) (Blinnokov, Lozinskaya, and Chugai 1988; Shklovsky 1985). This classification actually has a single spectral signature: the presence (SN II) or absence (SN I) of hydrogen in the supernova envelope. It is very important that in principle there exists no simple unique connection between the physical and astrophysical classification of supernovae. In fact, at the same time as the physical mechanism of the supernova explosion (in particular its trigger mechanism) acts close to the center of the star, all kinds of different burst phenomena develop in its external layers, beginning from the surface of the star itself. Finally, the central part of the star is visible through the expanding shell, but not immediately and not clearly.

However, according to modern ideas it is considered that SN II explosions are caused by the mechanism of gravitational collapse in iron cores of sufficiently massive stars and SN I explosions are caused by the mechanism of a thermonuclear explosion in almost bare carbon-oxygen cores of comparatively low mass stars (see, e.g., Müller (1990)). A whole combination of astronomical studies of supernovae, basically of extragalactic origin, serve as the cause for this, strictly speaking, hypothesis. The problematics of this mixture of two classifications of supernovae appears especially in attempts to classify in more detail definite subtypes existing in nature of supernova explosions (see, e.g., Blinnikov, Lozinskaya, and Chugai (1988)). In this study we limit the analysis to the mechanism of gravitational collapse in massive stars which, as before, lead to supernova explosions of type II (SN Ib?).

An important role in this theory is now played by the recent explosion SN 1987A in the Large Magellanic Cloud (Imshennik and Nadezhin 1989; Arnett et al. 1989). According to the astronomical classification it was a SN II as evidenced by the strong hydrogen lines of the P Cygni type starting from the first days of observation. According to the physical classification it actually belongs to the first type of explosion mechanism since the first of all other manifestations recorded at the Earth by underground neutrino observatories (at 7h 35m UT on February 23) were an unusual series of events which were reliably interpreted as the characteristic neutrino pulse of the gravitational collapse of a stellar core into a neutron star (or, very improbably, into a black hole). Thus, the explosion SN 1987A provided an important argument for the use of the conventional relation of classifications described above.

In addition, the long observation of SN 1987A provided a means using the hydrodynamic theory of explosions to reliably estimate the mass of the shell as (10–25) $M_\odot$ (Imshennik and Nadezhin 1989), i.e., to determine the massivity of this supernova. If one takes into account that previously in the process of its evolution (approximately 100 thousand years ago) several $M_\odot$ were thrown off, but a part of the initial mass ($\lesssim 2\ M_\odot$) turned into a neutron star, then the conclusion about massivity becomes even more valid. We add that the analysis carried out of the evolutionary path of the formation of the progenitor of SN 1987A, the blue supergiant Sk-69°202 also gave a mass in a similar range (15–25) $M_\odot$, but these data are still less reliable than the previous ones from hydrodynamic theory.

It is very likely that we have in SN 1987A one of the most massive phenomena in the range of actual SN II with gravitational collapse. In fact, the initial mass of the progenitor of this supernova can be estimated from previous data on the main sequence as (15–25) $M_\odot$. In particular, the mass of the iron core $M_{Fe}$ in Sk-69°202 immediately before the collapse could have been around $2 M_\odot$ (Woosley and Weaver 1986). An independent piece of evidence for the use of very massive stars experiencing SN II explosions as the result, in all probability, of gravitational collapse with a mass $M_{ms} \gtrsim 20\ M_\odot$ is the large excess of oxygen in the remnants of some supernova, e.g., in Cas A (Müller 1990).

In concluding this introduction we want to express sincere deep regret that Joseph Samuelovich Shklovsky was not able to see the SN 1987A explosion, for its light arrived at the Earth only two years after his death. In one of his last papers, Shklovsky (1984) answered an important question: why typical SN II are not observed in irregular galaxies? He presented the supposition that, due to the low metal abundances in these galaxies presupernova with extended atmospheres (red supergiants) would not be formed. Therefore, Shklovsky continued, in these galaxies explosions of dim SN II occur due to the compact initial structure of presupernova there. He cited here our study (Grassberg et al. 1971). Thus, Shklovsky, understanding deeply the physics of supernova explosions, actually was the prophet of the explosion SN 1987 A, an explosion of a dim SN II with a compact presupernova in the irregular galaxy of the Large Magellanic Cloud.

## 2. THEORY OF GRAVITATIONAL COLLAPSE OF STELLAR IRON CORES

At the present time the theory of gravitational collapse in massive stars is divided from the point of view of the physics of this hydrodynamic process into two limiting cases, the so-called cold (C) and hot (H) collapses. The collapse process in this case is determined basically by the initial specific entropy of the matter. The important role of this quantity was first indicated in the work of Bethe et al. (1979). The C case corresponds to low entropy $S_o \lesssim 1$ $k_B$/nucleon[†] which applies to the majority of numerical and analytic studies of the process of gravitational collapse (Nomoto and Hashimoto 1988; Woosley and Weaver 1988; Müller 1990). If this initial condition is completely valid in the theory of stellar evolution for not very massive iron cores in which $\rho_c \lesssim 10^{10}$ g cm$^{-3}$, then for more massive cores with $M_{Fe} \cong 2$ $M_\odot$ for which $\rho_c \lesssim 10^9$ g cm$^{-3}$, the supposition about the low initial entropy indicated above is, of course, inapplicable. An elementary examination in the approximation of a polytrope with index n = 3 for the case $M_{Fe} = 2$ $M_\odot$ gives the estimate $\rho_c = 4.45 \times 10^8$ g cm$^{-3}$ (Nadezhin 1977b). The density $\rho_c = 2 \times 10^9$ g cm$^{-3}$ is obtained in an evolutionary calculation of a presupernova star with $M_{ms} = 25$ $M_\odot$(Woosley and Weaver 1988)[‡] One then obtains for the entropy a correspondingly large quantity $S_o \cong (3-2)$ $k_B$/nucleon. For $S_o \gtrsim 2k_B$/nucleon the H case occurs and the process of gravitational collapse has a qualitatively different character from the C case (Nadezhin 1977a,b; see also Imshennik and Nadezhin 1983).

It should be especially emphasized that due to known inadequacies of stellar evolution theory (applied to the late stages of their evolution) in general the parametric means of specifying the initial state of an iron stellar core in which the quantity $S_o$ plays the role of a parameter can correspond both to the C case and to the H case. The actual collapse processes and supernova explosions sometimes, in all probability, relate to the H case, especially for SN 1987A.

Below we discuss both modern versions of the theory, emphasizing primarily the differences in their conclusions. For this purpose we will use the reviews of Müller (1990) and Imshennik and Nadezhin (1983) (see the references there to the original works with the exclusion of some indicated below in the text). It is worthy of note that the roughest conclusions of the theory including the character of the neutrino pulse basically coincide in the versions of cold and hot collapse (in C and H cases, respectively).

---

[†]The dimensionality of the quantity $S_o$ corresponds with the dimensionality of Boltzmann's constant $k_B$.

[‡]It should be noted that for an iron core of very large mass which was not considered excluded in general previously (Zeldovich and Novikov 1971), $M_{Fe} = 10$ $M_\odot$, and in the polytrope approximation taking into account the contribution of radiation in the equation of state one obtains $\rho_c \sim 10^7$ g cm$^{-3}$ and correspondingly $S_o \sim 6$ $k_B$/nucleon (Ivanova et al. 1969).

## 2.1. Physical Statement of the Problem and the Dynamics of Gravitational Collapse

The equation of state of matter (EOS) in the C version for $\rho \gtrsim 10^{12}$ g cm$^{-3}$ takes into account the nonideal nucleus-nucleon gas and for $\rho \sim 10^{14}$ g cm$^{-3}$ the phase transition to asymmetric nuclear matter ($Y_e < 0.5$). The EOS in this case contains a series of indeterminacies for the nucleus-nucleon component. In the H version up to nuclear densities of matter $\rho \gtrsim 10^{14}$ g cm$^{-3}$, thanks to the high characteristic temperatures, one uses the comparatively simple approximation of statistical equilibrium of the nucleus-nucleon ideal gas (NSE). Actually the NSE approximation is correct only for $T \gtrsim 5\times10^9$ K (necessary condition).

The calculation of weak interactions and the contribution of leptons to both versions of the theory differ little in general, but in the H version the approximation of kinetic $\beta$ equilibrium turns out to be applicable for $\beta$-interactions of nuclei and nucleons when the characteristic time of weak interactions $\tau_{weak} \ll \tau_{hd}$.[†] Physically this is explained by the comparatively large fraction of free nucleons due to the relatively high temperatures. We introduce an estimate of the time $\tau_{weak}$ for the case $T \cong 4\times10^{10}$ K and $\mu_e/k_BT \cong 3$ (Nadezhin 1977a): $\tau_{weak} \cong 3\times10^{-4}$ s in comparison with $\tau_{hd} \cong 10^{-1}$ s. In the low entropy C version: $\tau_{weak} \sim \tau_{hd}$. Therefore, if in the C version the change of the electron fraction $Y_e$ (number of electrons per unit nucleon) is small (from $Y_{eo} = 0.42$–$0.44$ to $Y_e = 0.35$ for $\rho_c \sim 10^{12}$ g cm$^{-3}$), then in the H version the parameter $Y_e$ has significant variations (from those initial values $Y_{eo}$ to $Y_e \sim 0.01$ for $\rho_c \sim 10^{12}$ g cm$^{-3}$). We note that in the presence of positrons with positron fraction $Y_{\tilde{e}}$ the connection of the quantities $Y_e$ and $Y_{\tilde{e}}$ with a ratio $\theta = n_n/n_p$ ($n_n$ and $n_p$ are the concentrations of neutrons and protons, both free and bound in nuclei) is such that

$$Y_e - Y_{\tilde{e}} = (1 + \theta)^{-1} \qquad (1)$$

The quantity $\theta$ actually characterizes the neutronization degree of matter.

Such a large difference of $Y_e$ values is necessitated in some discussions although one immediately notes the influence of the $Y_e$ parameter on the EOS is small because the main role in the H version is played by the contribution to it of free nucleons and not the lepton component and in the C version the variation of $Y_e$ is small. One can say that the formally strong variation of $Y_e$ in the H version is due to the approximation of kinetic $\beta$-equilibrium. In spite of the above indicated validity with the relation of characteristic times it should be kept in mind that even a small deviation from the state of kinetic $\beta$-equilibrium with the quantity $Y_e$ may lead to a change of rates of the process of

---

[†]We emphasize that the processes of weak interaction do not reach thermodynamic (statistical) equilibrium until an uninhibited neutrino exit occurs. The processes of neutronization of the matter by the kinetic $\beta$-equilibrium have a nonequilibrium character.

neutroniztion of matter more strongly than the larger degeneracy of electrons. One can say that for fixed $\rho$ and T the approximation of kinetic $\beta$-equilibrium gives the maximum degree of neutronization of matter, i.e., the smallest possible value of the $Y_e$ parameter. Therefore, it would probably be desirable to carry out calculations of actual numerical models of the H version without using the approximation of kinetic $\beta$-equilibrium.

In turn deep neutronization in the H version cooperates to transform all nuclei including neutron enriched ones into free nucleons due to the well-known process of cold neutronization of matter for $\rho \sim 10^{12}$ g cm$^{-3}$ and $T \lesssim 1.5 \times 10^{10}$ K.[†] Consequently, the processes of neutronization of matter and the breakdown of nuclei into free nucleons naturally facilitate each other. In this connection it also becomes understandable why in the C version a significant fraction of nuclei continue during the collapse to nuclear densities of matter.

The behavior of entropy in both C and H versions is quite similar. The basic feature is the comparatively small change of the entropy S and in any case $|\Delta S|$ ($\Delta S$ is the entropy change) is considerably smaller than its initial value $S_o$, $|\Delta S| \ll S_o$. The quantity $\Delta S$ may be negative due to neutrino energy losses, but a positive contribution to its change arises due to the nonequilibrium character of weak interactions so that the total change $\Delta S$ may even be positive. As a result of weak interactions the collapse process turns out to be almost isentropic. This fact is used especially often in the C version where a polytrope EOS has been fruitfully applied. However, we emphasize that the polytropic approximation, strictly speaking, is connected with an adiabatic process, i.e., with the absence of energy losses, and not with an isentropic one. In the H version this effect provides a quite high value of entropy at the time of neutrino capture, in spite of the intense neutrino energy losses by the modified Urca process (for kinetic $\beta$-equilibrium). One can write in a general form an equation of the entropy production (Imshennik and Chechetkin 1970; Müller 1990) in Lagranian coordinates (m,t):

$$T \frac{dS}{dt} = -q_{\nu\tilde{\nu}} + (\mu_p + \mu_e - \mu_n)\left(\frac{dY_e}{dt} - \frac{dY_{\tilde{e}}}{dt}\right), \qquad (2)$$

where the quantities S and $q_{\nu\tilde{\nu}}$ (neutrino energy losses) are calculated for a single nucleon. In the C version positrons are neglected ($Y_{\tilde{e}} = 0$) and $q_{\nu\tilde{\nu}}$ corresponds to processes of electron capture ($q_{\nu\tilde{\nu}} = -\langle\varepsilon_\nu\rangle_{emit} dY_e/dt$). In the H version $q_{\nu\tilde{\nu}} = a_\nu m_p T^6$ where $a_\nu = 7.8 \times 10^{-43}$ erg g$^{-1}$ s$^{-1}$ K$^{-6}$ for an asymptotic power-law in the kinetic $\beta$-

---

[†]The cold neutronization of matter is obtained in the NSE approximation just for the density of matter indicated (Imshennik and Nadezhin 1965). For low temperatures (T $\ll 5 \times 10^9$ K) this effect of incomplete cold neutronization is well known (Landau and Lifshitz 1976). We note that in the work cited the approximation $\mu_\nu = 0$ was used which, as can be shown, is almost equivalent to the approximation of kinetic $\beta$-equilibrium.

equilibrium. Equation (2) is correct also in the presence of heavy nuclei, but, of course, in the NSE approximation.

For a density $\rho_c \sim 10^{12}$ g cm$^{-3}$ (more precisely, for $\rho_c = (0.3-5) \times 10^{12}$ g cm$^{-3}$ and $T_c = (4-8) \times 10^{10}$ K) the more important phenomenon of neutrino capture occurs, in other words, opaqueness begins for neutrino emission. In the H and C versions the trapping mechanisms differ somewhat although the threshold density is in the range indicated above. In the H version the opaqueness corresponds primarily to charged fluxes of weak interactions (absorptions $\nu_e$ on neutrons and $\tilde{\nu}_e$ on protons) and in the C version to neutral fluxes (coherent scattering of $\nu_e$ on nuclei). After neutrino capture the spectra $\nu_e$ and $\tilde{\nu}_e$ (and then $\nu_\mu$ and $\tilde{\nu}_\mu$, and $\nu_\tau$ and $\tilde{\nu}_\tau$) become almost Fermi-Dirac ones with chemical potentials $\mu_{\tilde{\nu}_i} = -\mu_{\nu_i}$. In the interior the star has the appearance of a neutrino core with a surface (neutrinosphere) which radiates a Fermi-Dirac neutrino spectrum (strictly speaking, a series of neutrinospheres for different sorts of neutrinos). In the C and H versions there are significant differences in the quantitative description of neutrino transfer. The H version uses the approximation of neutrino thermal conductivity inside the neutrino core (its boundary is somewhat higher than the neutrinosphere) and in C versions other, to our way of thinking less adequate, approximations are used for large optical depths of the collapsing stellar core.

The behavior of entropy in the neutrino opaque region, i.e., under the neutrinosphere is determined by a description of neutrino transfer. In the approximation of a neutrino thermal conductivity one can write a quite simple equation for the entropy production analogous to Eq. (2):

$$T \frac{dS}{dt} = -4\pi m_p \left[ \frac{\partial}{\partial m}(r^2 H) + \mu_\nu \frac{\partial}{\partial m}(r^2 F) \right], \quad (3)$$

where H and F are the energy flux and lepton charge flux, respectively, expressed through the gradients of temperature T and chemical neutrino potential $\mu_\nu$ (Imshennik and Nadezhin 1972; Imshennik and Nadezhin 1983). The second term in brackets partially compensates the decrease in entropy due to energy losses determined by the first term as shown by numerical calculations.

In studies with the C version it has been noted repeatedly that the description of neutrino transfer plays a critical role for both described explosion mechanisms (prompt and delayed) (see below). Therefore, it would probably be useful to carry out calculations of some numerical models of collapsing stellar cores in the C version using the approximation of neutrino thermal conductivity. On the other hand, in the H version it would be desirable to employ a more adequate description of the beginning of neutrino capture using kinetic equations of neutrino transfer until optical depths are still not very large and also to introduce a direct description of muon and tau-neutrinos in the numerical models. We emphasize that all these remarks relate to the problem of the explosion of the shell of the collapsing star which is a quite refined effect (see below).

## 2.2. Gravitational Collapse Before Arrestation of Compression

Together with a large number of model calculations of gravitational collapse of iron stellar cores, some useful self-similar solutions were obtained (Nadezhin 1968; Goldreich and Weber 1980; Yahil and Lattimer 1982; Yahil 1983). The first of the solutions noted relates to the H version of the theory and takes into account neutrino energy losses for which an asymptotic power-law is used $q_{v\tilde{v}} = a_v \rho^s T^k$ (in actual physical conditions, s = 0, k = 6, see above). In this solution entropy is not conserved. Other self-similar solutions concern the C version and start from the condition of adiabatic collapse. In this case the EOS reduces to the polytrope relation $p = K\rho^\gamma$ while in the solution of Nadezhin (1968) the EOS has the form: $p = b_n \rho T$ and $E = (\gamma-1)^{-1} p \rho^{-1}$ ($\gamma$ and K are constants).

A comparison of the self-similar solutions with numerical models provides a means of interpreting some important qualitative properties of collapse. Above all, one obtains a description of the inhomogeneity of the collapse process when some part of the core close to the center is compressed in a homologous manner and the remainder of the core outside it lags behind considerably. The differentiation of a homologous internal core is obtained well in the framework of self-similar solutions if $\gamma \neq 4/3$. The typical profile of the compression velocity v is: close to the center $v \propto r$, reaching at some point $r = r_m$, the maximum negative value $v_m$ and then for $r > r_m$ the modulus of the velocity decreases to zero. It is just the point $r = r_m$ that determines the boundary of the internal core during collapse. It is characteristic that somewhat nearer to the center a so-called sound point is located where $v = c_s$ while outside it there is a zone of supersonic infall ($c_s$ is the sound speed). The interior core in the end forms the embryo of the hot neutron star. However, this occurs only in the framework of numerical models where the onset of neutrino opaqueness and (or) stiff NEOS of nucleus-nucleon matter is taken into account. Of course, in self-similar solutions there is no stopping of collapse and an unbounded compression occurs (in the center after the singularity $\rho_c \to \infty$ a point mass arises), but, being intermediate asymptotics of collapse, these solutions are qualitatively correct up to the stoppage in the framework of numerical models.

The self-similar solutions found are classical solutions of the first type (Sedov 1957; Zeldovich and Raizer 1963), where the self-similar index $\alpha$ is found from dimensionality relations: the self-similar variable $\xi = Qrt^{-\alpha}$ in which the constants Q in both of the above-mentioned solutions are, of course, different. Nevertheless, these solutions have a larger similarity among themselves than differences which can be seen from Table 12.1, where $v_{ff}$ is the free-fall velocity, the indices o, m, and a correspond to the center, minimum velocity, and asymptotic velocity ($\xi \to \infty$), respectively. The similarity of solution characteristics occurs, in spite of the fact that they did not transform into each other: in the solution of Nadezhin (1968) $\alpha > 2-\gamma$ while in other solutions $\alpha = 2-\gamma$. The impossibility of a transition from one solution to another reflects the behavior of the quantity $K' = b_n T \rho^{1-\gamma}$ which is determined by the solution of Nadezhin (1968). Using the

Table 12.1. Self-similar Solution Characteristics†

| α | γ | $-v/v_{ff}$ o | $-v/v_{ff}$ m | $-v/v_{ff}$ a | $-v/c_s$ m | $-v/c_s$ a | Remarks |
|---|---|---|---|---|---|---|---|
| 2/3 | 4/3 | 0.08 | 1 | — | ∞ | — | G.W., 1980 |
| 0.70 | 1.30 | 0.17 | 0.54 | 0.66 | 2.1 | 2.9 | Y.L., 1982 |
| 0.80 | 1.20 | 0.30 | 0.56 | 0.62 | 1.9 | 2.3 | and Y., 1983 |
| 0.90 | 5/3 | 0.41 | 0.62 | 0.64 | 1.6 | 1.8 | Nadezhin (1968) |
| 0.90 | 1.14 | 0.43 | 0.65 | 0.66 | 2.1 | 2.3 | I. et al., 1969 |
| 0.90 | 1.100001 | 0.45 | 0.55 | 0.68 | 2.3 | 2.5 | |

†The data in columns for the point of minimum velocity m was extracted from graphs of the original study of Nadezhin (1968) so that their accuracy, probably, is small and, in the best case, suffices for the two signs presented in the table of values.

quantity K' the EOS can be written in the form $p = K'\rho^\gamma$, but here K' is already not the constant coefficient K as it was in solutions of other authors. The quantities K' and K are clearly proportional to the entropy of matter,

$$K' = b_n \frac{T}{\rho^{\gamma-1}} = \frac{\alpha^{4-2\gamma}A^2}{(4\pi G)^{1-\gamma}} \cdot \frac{\tau(\xi)}{g^{\gamma-1}(\xi)} t^{2(\alpha+\gamma)-4} \propto t^{2(\alpha+\gamma)-4}, \qquad (4)$$

in which one can see that $K' \to 0$ for $t \to 0$, when $2(\alpha+\gamma)-4 > 0$ or $\alpha > 2-\gamma$, i.e., in the whole region of existence of the Nadezhin (1968) solution. In Eq. (4) all designations are from the study of Nadezhin (1968).

Both forms of solutions do not exist for $\alpha < 2/3$, for $\alpha > 0.8$ the total energy diverges for $\xi \to \infty$, and for $2/3 < \alpha \le 0.8$ the total energy for $\xi \to \infty$ is zero. In the solutions with $\alpha = 0.9$, consequently, the energies diverge for $\xi \to \infty$. We note that the divergence of energy values is not a hindrance for comparison of self-similar solutions with numerical models (see, e.g., the solution about a short impulse: Zeldovich (1956) and Zeldovich and Raizer (1963)) as well as the divergence of mass occurring in all solutions of Table 12.1.

The data of Table 12.1 indicate the general properties of collapse: 1) a significant difference of flow velocities from velocities of free fall; 2) a supersonic character of flow at the point m and asymptotically; 3) due to energy losses a ratio $(-v/v_{ff})$ in the center (o) considerably higher than this ratio for $\alpha \le 0.80$; and 4) the ratio $(-v/v_{ff})$ grow in both solutions with a decrease of γ, but the ratio $(-v/c_s)$ behaves differently for each one.

From the various applications of self-similar solutions we here delineate one of the most important, the determination of the mass of the internal core collapsing in an almost homologous manner, $M_{IC}$. In the solution of Goldreich and Weber (1980), $M_{IC} = (KG^{-1})^{3/2}m(\xi_m)$, where $m(\xi_m) = 4.76$ and for $Y_e = 0.464$, $K = 0.447 \times 10^{15}$ (for degenerate electrons in $^{56}$Fe) we obtain $M_{IC} = 1.31$ M$_\odot$. In the solution of Yahil and Lattimer (1982) for $\gamma = 1.30$ (more appropriate for comparison with numerical models of the C version) $M_{IC} = K^{3/2}G^{-29/20}(-t)^{1/10}m(\xi_m)$ already has a time dependence t. We determine the time $t = t_f$ connecting it with the density $\rho_c = 10^{12}$g cm$^{-3}$ corresponding to neutrino capture. We obtain $-t_f = 5.12 \times 10^{-3}$ s. For the same value of K and $m(\xi_m) = 5.98$ we obtain $M_{IC} = 0.57$ M$_\odot$. Finally, in the solution of Nadezhin (1968) with $\gamma = 5/3$ the mass $M_{IC}$, also depending on $t_f$, has the form $M_{IC} = \alpha^2 A^3 G^{-1} t_f^{0.7} m(\xi_m)$. For the same value of $Y_e$ (for nondegenerate electrons in $^{56}$Fe), we have $A = 6.514 \times 10^8$, $\alpha = 0.9$, $m(\xi_m) = 46.0$, and $t_f = 2.16 \times 10^{-3}$ s. Finally we obtain $M_{IC} = 1.06$ M$_\odot$. Practically also in numerical models the values obtained of mass $M_{IC}$ are approximately half the mass of corresponding iron cores while for the case $\gamma = 4/3$ the quantity $M_{IC}$ clearly overestimates in the C version.

The mass estimates $M_{IC}$ given above were carried out for $Y_e = 0.464$. However, the parameter $Y_e$ decreases due to neutrino emission of matter and in the C version $K \propto Y_e^{4/3}$. In the H version nucleon pressure occurs in place of electron pressure. In the framework of self-similar solutions there is no sense in considering these effects because it is impossible to take into account the time dependence of coefficients $K(t)$ and $b_n(t)$. One can see qualitatively in general that these effects decrease the mass $M_{IC}$.

## 3. THE SUPERNOVA EXPLOSION MECHANISM PROBLEM

It is well known that for a typical supernova explosion one requires not as much as 1% release during gravitational collapse of the iron stellar core energy which is, e.g., $1.3 \times 10^{53}$ erg for $M_{Fe} = 1$ M$_\odot$. In the case of an arbitrary mass $M_{Fe}$ the gravitational binding energy of the neutron star whose formation ends the gravitational collapse of the iron core is given approximately by the equation of Lattimer and Yahil (1989):

$$\mathcal{E}_G = 1.5 \times 10^{53} (M_g/M_\odot)^2, \quad M_{Fe} = M_g + 0.084(M_{Fe}^2/M_\odot) \tag{5}$$

In the case of SN 1987A the kinetic energy of the expanding shell was practically measured, $\mathcal{E}_{kin} \cong 1.5 \times 10^{51}$ erg (with an accuracy better than 30%). In the hydrodynamic theory of gravitational collapse the problem of the supernova explosion mechanism has not been solved up to now, i.e., the mechanism of deriving the indicated small fraction of energy. At the same time the theory describes excellently the processes of dividing of the basic part of the gravitational binding energy in the form of a pulse of neutrino emission and of the formation of a neutron star (Nadezhin 1978; Blinnikov et al. 1988). The recording of neutrino signals from SN 1987A verified that the neutrino pulse actually removes almost the whole gravitational binding energy (theoretically $\geq 99\%$).

## 3.1. Attempts at Solving the Problem in the Theory of Cold (C) Collapse

We discuss the problems of the mechanism of supernova explosions in the C and H versions of the theory. We begin with the necessary condition of an explosion, the stoppage of gravitational collapse of the stellar core. In the C version the stoppage is due more or less to the sharp increase of the adiabatic index for an approach of the matter density to nuclear density. Depending on the assumptions about the properties of nuclear matter the stoppage occurs for small (hard NEOS) or large (soft NEOS) densities, overshooting the corresponding stationary neutron star position of equilibrium. It is then clear that the phenomenon of elastic rebound (bounce) develops in the homologous collapse by the internal stellar core. At its boundary, more precisely, at the sonic point, for very small time (~1 ms) a strong shock wave forms at whose front the matter of the exterior core thermalizes layer by layer. In this case, when the speed of matter behind the shock front is positive, the shock wave may become an expanding shock and an explosion occurs in the envelope of the star. This mechanism of explosion is called prompt explosion. The basis of this mechanism in the C version of the theory is that the stoppage of collapse occurs approximately in an adiabatic manner because energy losses are relatively small after neutrino capture. The corresponding EOS (NEOS) is as before approximately a polytrope relation. On the basis of numerical calculations the initial kinetic energy of the shock wave is estimated as $\varepsilon_{sw}^{(i)} \cong \varepsilon_{kin}^{IC} \cong (4-10) \times 10^{51}$ erg, where $\varepsilon_{kin}^{IC}$ is the kinetic energy of the internal core at the time of the last "good" homogeny before stoppage. However, this energy is rapidly spent on the disintegration of iron in layers of the external core and neutrino energy losses after exit of the shock wave beyond the neutrinosphere. First of all, no addition of energy from the layers of the external core to $\varepsilon_{sw}^{(i)}$ occurs because their kinetic energy passes to the internal energy of the rebounding stellar core. Secondly, only in sufficiently small mass iron cores $M_{Fe}$ the disintegration of iron may be implemented due to the initial energy of the shock wave $\varepsilon_{sw}^{(i)}$ so that the prompt explosion mechanism occurs for $M_{Fe} < 1.35\ M_\odot$ or for $M_{ms} < 12\ M_\odot$. In the remaining cases the shock wave turns into an accretion shock, stalling close beyond the collapsing stellar core.

For a physical classification of shock waves in the stoppage of gravitational collapse it is useful to return to the relation of continuity of mass flux through the wave front, moving with velocity $\mathcal{D}$:

$$\mathcal{D} = -\frac{\rho_0}{\rho-\rho_0} v_{ac} + \frac{\rho}{\rho-\rho_0} v, \qquad (6)$$

where $\rho_0$ and $v_{ac}$ are the density and matter velocity in front of the shock front and $\rho$ and $v$ are the corresponding values behind the shock front. For $v < 0$ the second term in Eq. (6) makes the velocity $\mathcal{D}$ zero ($v_{ac} < 0$) and for $v = 0$ the velocity $\mathcal{D} > 0$, but with a

## 12. GRAVITATIONAL COLLAPSE

significantly (by several times) smaller value of $|v_{ac}|$. This is the case of an accretion shock. On the contrary, sufficiently large positive velocities v in Eq. (6) lead to a growth of $\mathcal{D}$ so that one reaches $\mathcal{D} \gtrsim |v_{ac}|$ and an expanding shock arises. Thus, it can be considered that the velocity condition $v = 0$ divides the cases of accretion and expanding shock waves.

In the C version the problem of the supernova explosion of more massive stars ($M_{ms} > 12\ M_\odot$) was considered starting in 1985 in the framework of a delayed explosion mechanism. The essence of this mechanism consists in developing the hypothesis of Colgate and White (1966) about the effect of a neutrino deposition as the mechanism of explosion. At that time this hypothesis was put into doubt by refinement of physics in the studies of Arnett (1966, 1967) and Ivanova et al. (1969) (see for more details the discussion in Zeldovich and Novikov (1971)). A revival of the hypothesis of neutrino deposition occurred as the result of some new numerical calculations, beginning with the study of Wilson et al. (1985). After transformation of the expanding shock into an accretion shock the flux of neutrino radiation from the neutrinosphere continuously (for hundreds of ms) heated the matter behind the shock front. According to the calculations, altogether only 0.1% of this flux is absorbed in this region, but in the end this absorption causes expansion of the matter and the accreting shock again becomes an expanding shock ("revives"). Unfortunately, this explosion is comparatively weak, with an energy $\mathcal{E}_{kin} \lesssim 4 \times 10^{50}$ erg. Clearly, the temperature and radius of the neutrinosphere play a critical role in this mechanism as well as the complex process of neutrino interaction with the matter of the semitransparent region considered. In turn these characteristics are very sensitive to the description of neutrino energy transfer and to the detailed EOS properties. Probably, for these reasons the numerical calculations of the delayed explosion mechanism were called into question (Arnett 1986), but the small energy in this mechanism, apparently, is the unavoidable inadequacy (from the point of view of supernova explosions). The heated region of matter begins to expand when only an excess of thermal energy reaches the value of the binding energy of the envelope layer's positioned higher and this value $\sim \kappa \times 10^{50}$ erg, where $\kappa \cong 1-5$ (Hillebrandt, 1987). These considerations indicate that the delayed explosion mechanism in actuality probably does not work. For a complete picture it is necessary to recall here some modifications of the mechanism discussed proposed in Arnett (1986). Here it was pointed out that there was a negative entropy gradient in the region between the shock front and the neutrinosphere due to heating of the stalling shock (during its transformation to an accretion shock stalling). This region is convectively unstable and the convective energy flux in the case of development of this instability may strengthen the flow of energy to the shock front in order to "revive" it. However, it should be recognized that in essence this mechanism of explosion already has a non-one-dimensional character so that its validity lies outside the framework of the one-dimensional theory of gravitational collapse considered here (Arnett 1987).

## 3.2. Attempts at Solving the Problem in the Theory of Hot (H) Collapse

How is the situation with the problem of the supernova explosion in the H version of the theory? Some quantitative data below will be taken from a numerical model with $M_{Fe} = 2M_\odot$ of Nadezhin (1977a,b). Stoppage of collapse in this case occurs in connection with onset of neutrino opaqueness because in this case the rate of growth of energy losses sharply decreases. Therefore, already for $\rho_c \sim 10^{13}$ g cm$^{-3}$ the bounce phenomenon develops. The adiabatic index $\gamma \cong 5/3$ because the basic contribution in the EOS is made by free neutrons, and not by the electron gas ($\mu_e/k_B T \sim 3$). While neutrino losses grew in conditions of transparency, the collapse continued unimpeded in agreement with the self-similar solution although it was $\gamma > 4/3$. The strong shock wave arising in stoppage of collapse immediately took on an accretionary character, expending its initial energy in neutrino radiation. This energy $\mathcal{E}_{kin}^{(i)} \cong 3\times 10^{51}$ erg is converted into internal energy in back of the shock front during a time ~0.02 s. The remainder of the kinetic energy of the whole collapsing stellar core is $\mathcal{E}_{kin} \cong 10^{50}$ erg at a time of 0.1 s after stoppage of collapse (the stoppage time is taken as the beginning of the calculation time). Of course, there were no energy losses due to disintegration of iron here because the concentration of iron was practically zero.

Probably, the shock wave stalling was effectively accelerated by the fact that the region behind the shock front was always located in conditions of neutrino thermal conductivity. Actually, the boundary of the opaque region was situated initially inside and then immediately behind the "spreading" artificial viscosity of the shock front. It is worth noting that the whole iron stellar core after onset of neutrino opaqueness was divided into two regions: an internal one with neutrino thermal conductivity and an external one with semitransparency for neutrino radiation. The boundary of these regions, as noted above, was determined according to criterion $\tau_\nu \cong 0.01$–$0.03$, where $\tau_\nu$ is the effective optical depth outside for $\nu_e$ and $\tilde{\nu}_e$.

One can conclude that the mechanism of collapse stoppage in the H version was different than in the C version, occurring earlier, e.g., in the parameter $\rho_c$. Nevertheless, as the result of stoppage the exact same configuration arose: a neutrino opaque internal core with an accretion shock at its surface. In the H version the mass of the internal core was ~0.8 $M_\odot$ (at a time ~0.04s) and the neutrinosphere was situated at $\gtrsim 0.1$ $M_\odot$ behind the shock front.

Careful numerical and theoretical study of all effects enabling the development of an expanding shock did not lead (Nadezhin 1977a,b) to an explosion of the external core in this numerical calculation. Thermonuclear burning of oxygen (initially in the core there was 0.18 $M_\odot$ of oxygen which gave altogether ~$3\times 10^{50}$ erg) occurred too slowly for layer-by-layer accretion so that it did not influence the collapse. The effect of the explosion due to other thermonuclear reactions in the mantle and exterior stellar envelope (outside the core) were also practically excluded (Ivanova et al. 1969). The deposition of

## 12. GRAVITATIONAL COLLAPSE

neutrino radiation in the semitransparent region was practically completely compensated by volume energy losses. We note that in principle if there were no compensation the neutrino energy deposition could cause an effect: after ~100 ms for a neutrino luminosity $L_{\nu\tilde{\nu}} \cong 2\times10^{53}$ erg/s for an optical thickness of the semitransparent region $\tau_\nu \cong 0.01$ an energy ~$2\times10^{50}$ erg would have been absorbed. Consideration of the additional effect of deposition due to coherent scattering of neutrinos in iron nuclei also showed that the cross section of coherent scattering was insufficient (approximately by ten times) for transferring significant momentum to the matter of the external core. Finally, an analysis was carried out of the special role of muon neutrinos. Volume losses of energy due to $\nu_\mu$ and $\tilde{\nu}_\mu$ were deliberately included in a sharp manner and were continued for some time after formation of an opaque core for $\nu_e$ and $\tilde{\nu}_e$. The calculations showed the rapid stalling of the strong shock wave arising as the result of this inclusion from the secondary bounce. Thus in the H version of the theory it was also not possible to clarify the mechanism of a supernova explosion during collapse of a massive iron core with $M_{Fe} = 2\ M_\odot$.

In terms of the theory from the C version no hints on the prompt explosion mechanism of the calculation of the H version were detected. This role could be played by the effect noted above of a later initiation of opaqueness from muon (tau) neutrinos. The absence of a basic extinction of the shock wave bounce, the effect of iron disintegration, supports the use of this mechanism in the H case. However, neutrino radiation even in conditions of an earlier onset of the phenomenon of collapse stoppage and bounce turned out to be the means of extinguishing the expanding shock and turning it into an accretion shock. It should be noted here that the development of the bounce phenomenon coincided in the H version with a transition from a single system of equations of the problem to another one taking into account the opaqueness of the region of the collapsing core. Of course, this "jump" in the calculation method is fraught with several inadequacies capable, generally speaking, of masking the prompt explosion mechanism. Concerning the delayed explosion mechanism which is more characteristic for such massive iron cores, the calculations of the H version also show its absence. The effects of neutrino deposition (absorption and coherent scattering) did not lead to an explosion with an energy $\mathcal{E}_{kin} \gtrsim 10^{50}$ erg. By the way it is worth noting that the region of action of this mechanism occurred in the semitransparent region of neutrino transfer where it would have been better to solve exact equations of neutrino radiation transfer.

In contrast to the C version the numerical model with $M_{Fe} = 2\ M_\odot$ in the H version was calculated to very late stages of evolution of the hot proto-neutron star when the pulse of neutrino radiation had practically completely finished. It is possible in relation to the explosion problem at these late stages that some hope remains on the pulsation of a hot proto-neutron star because, apparently, a pulsational instability (Zentsova and Nadezhin 1975) is characteristic for this star. It is true that in a numerical model these pulsations are hardly noticeable.

## 4. SOME LESSONS OF THE SN 1987A EXPLOSION IN RELATION TO THE MECHANISM OF SUPERNOVA EXPLOSIONS

At the present time nothing is known about the relic star in the remnant of SN 1987A and it is difficult to make a complete analysis of the problem of the explosion mechanism of type II supernovae on the basis of this explosion. However, a whole series of conclusions are extremely useful in this connection.

The main conclusion which we partially discussed above (see the beginning of Section 3), of course, is that the explosion was preceded by the gravitational collapse of the iron core of the presupernova of SN 1987A, the blue supergiant Sk-69°202. It almost surely preceded the explosion, i.e., the explosive energy release inside this star. In any case there are unarguable indications that the explosion took place in the central region of this massive star and occurred approximately at the time when the neutrino pulse was formed, i.e., when gravitational collapse occurred. Thus, the another conclusion emerging from studies of the SN 1987A explosion consist in these indications which we will consider below based primarily on the reviews of Imshennik and Nadezhin (1989), Arnett et al. (1989), Schramm and Truran (1990), and Müller (1990).

The hydrodynamic calculations using as initial conditions the internal structure of the star Sk-69°202 give a propagation time of the shock wave to the stellar surface of (0.5–2.5) h as a function of all possible characteristics of initial structure and the allowable dispersion of the total explosion energy $(1.5–3) \times 10^{51}$ erg. Taking into account that the time of the first optical recording of 10 h 40 m UT was separated by a small time interval (about 1 h) from the time of the emergence of the shock wave to the surface, these times are in complete agreement with the supposition that the instantaneous energy release occurred at a time very close to the time of detection of the neutrino pulse at 7 h 35 m UT and practically in the stellar center. Unfortunately, the strong pulse of hard radiation (UV- and X ranges) due to the shock wave emergence was not detected, but its "soft tail" in the visible V band was detected only at the time of the first optical recording. However, indirect signatures of this short ($\sim 10^2$ s) pulse can actually be observed. One of them above all is the rapid decay in the far UV range after the first week of the explosion. Thus, it was first shown that the supernova explosion begins with the passage of a strong expanding shock wave through the whole thickness of the progenitor, from its very center where the shock wave originates close to the time of collapse.

An important fact for verifying the latter assertion was the discovery in SN 1987A of a radioactive energy source (in addition to the instantaneous energy release) in the form $\sim 0.07$ $M_\odot$ $^{56}$Co. All near maximum ($\sim 100$ d) and post-maximum (up to 1000 d) stages of the explosion in the optical and infrared ranges in which the basic energy of the total radiation of SN 1987A are included were determined by this source. This source also caused hard X-ray emission discovered in the first half of August, 1987, as well as the characteristic γ-lines of $^{56}$Co discovered at the end of this year. These data in fact became the main basis for the immediate application of the theory of explosive nucleosynthesis in SN 1987A (Woosley et al. 1988; Thielemann et al. 1990). In particular, in the latter work it was shown that to obtain in the shell of the supernova 0.07 M$_\odot$ $^{56}$Co it is necessary to suppose that a core of 1.6 M$_\odot$ collapsed. In explosive nucleosynthesis the nucleide $^{56}$Ni actually forms only for sufficiently high temperatures T $\gtrsim$ 5×10$^9$ K in conditions of applicability of the NSE approximation. In back of the shock front these temperatures can occur within the radius r$_{sw}$ which is given by the approximate equation (Thielemann et al. 1990):

$$r_{sw} \cong \left(\frac{3\mathcal{E}}{4\pi aT^4}\right)^{1/3}, \tag{7}$$

where it is assumed that all the explosion energy $\mathcal{E}$ behind the shock front inside the radius r$_{sw}$ is in the form of radiative energy with a density aT$^4$ (radiation dominated shock wave) which have is no dependence on the matter density $\rho$. For the known explosion energy $\mathcal{E}$ = 10$^{51}$ erg for T = 5×10$^9$ K we obtain from Eq. (7) r$_{sw}$ $\cong$ 3700 km, and thus is, in turn, the radius of a mass $\gtrsim$ 1.7 M$_\odot$. Thus, the nucleide $^{56}$Ni is formed in a layer immediately surrounding the collapsing stellar core. We note that some part of $^{56}$Ni may be already contained in the outer layers of the iron core before the explosion, but it is insufficient there due to the significant neutronization of matter in the course of evolution.

In the paper cited the total mass of the star was 20 M$_\odot$ and the coincidence of these results with other calculations was only qualitative. It is completely clear that mass fractions of different nucleides of the iron peak are very sensitive functions of temperature and density around the boundary of the stellar disconnexion, the boundary of the collapsing core, and the ejected shell. The briefly described results of calculations of explosive nucleosynthesis nevertheless convincingly show that necessary conditions of $^{56}$Ni nucleide synthesis may only be created at the center of the star.

In addition, the main problem of the theory of nucleosynthesis is immediately connected with the $^{56}$Ni synthesis which we emphasize. The nucleide $^{56}$Ni has the largest binding energy among $\alpha$-particle nucleides. Inside the iron core of the star, its fraction becomes small as neutronization occurs to Y$_{eo}$ $\cong$ 0.42–0.44 (see above). In the conditions of explosive nucleosynthesis in the surrounding layers of the mantle of the iron core with Y$_e$ $\cong$ 0.5 for some increase of temperature critical for establishing NSE, T $\gtrsim$ 5×10$^9$ K, the $^{56}$Ni nucleide is synthesized in the largest quantities. Thus, not only the place of the explosion (center of the star), but also the main point of explosive nucleosynthesis (the predominating family of $\alpha$-particle nucleides among all elements of the iron peak) is verified by a study of the SN 1987A explosion.

An application of hydrodynamic theory to a study of the late stages of the SN 1987A explosion could give valuable information on the interaction of the shell with the collapsing remnant. At the end of May 1987 the velocities in the photosphere reached ~2000 km/s and the fraction of untransparent mass decreased to ~1.6 $M_\odot$. Unfortunately an interpretation of these observations is strongly hampered by destruction of the approximation of radiative heat conduction received in the majority of hydrodynamic calculations. It is first of all evidenced by the nebula character of spectra at this stage. However, there is still another complicating circumstance. The fact is that the structure of these internal low velocity layers is strongly changed by the significant mixing of matter which, most of all, occurred immediately after passage of the shock wave. The effect of the mixing should also be taken into account in hydrodynamic models of the shell before going to an interpretation of late stages of the observations of the SN 1987A explosion. If the difficulties indicated can be overcome, then a correct reproduction in the numerical models of the theory of the expansion dynamics of the deepest layers of the shell of this explosion would give valuable new information. It is just in the layers situated between the collapsing remnant and the external massive and high speed shell that the unrevealed mechanism of the supernova explosion would have worked. By a correct reproduction, of course, one must understand the agreement of theory and the observed SN 1987A spectra.

## 5. CONCLUSION. THE POSSIBILITY OF NON-ONE-DIMENSIONAL SUPERNOVA EXPLOSION MECHANISMS

It has already been many years that in the framework of a one-dimensional spherically symmetric statement of the problem of gravitational collapse of iron stellar cores that there has been no success in clarifying any kind of effective mechanism of supernova explosions. Below, as in all our studies, the conversation concerns, of course, type II supernovae according to the modern rough identification of the physical and astronomical classification of supernovae. Could one make the skeptical conclusion about the hopelessness of further attempts of all kinds to solve the problem by this means? It seems to us that it is still early to lay down arms. In the previous discussion a whole series of weak points were noted both in the version of the theory of C collapse (small mass cores) and in the version of H collapse (most massive cores). Unfortunately these weaknesses relate to the most difficult physical problems: 1) the equation of state of matter especially close to nuclear densities and 2) the transfer of neutrinos especially for an intermediate optical depth of unity. These questions are subject to clearing up on the basis of the newest accomplishments of theoretical physics. In addition, the discussion given above shows in our opinion that there is some disparity, at least methodological and numerical, between the C and H versions of the theory. For example, would it turn out that the prompt explosion mechanism is deliberately simplified if iron is almost absent in the external core of the collapsing star and the matter consists practically of free nucleons. At the same time neutrino emission, possibly, is not able to extinguish the expanding shock arising as a result of the bounce of the internal core. It is hardly necessary a priori that an almost complete compensation of the deposition of neutrino energy and volume neutrino energy losses in the semitransparent region should occur where, as a rule, the expanding shock is converted into an accretion shock. Concerning the delayed explosion mechanism, it is completely clear that systematic numerical modeling is necessary for the long stage of accretion of the external core onto the internal one and the subsequent cooling of the hot protoneutron star. The latter remark relates basically to calculations of collapse of massive iron cores, i.e., to the region of applicability of the H version.

Independently of the state of art in collapse theory, a solution of the supernova explosion problem in general will certainly depend on new observational data of neutrino events from collapsing stars. In the review of Schramm and Truran (1990) quite convincing arguments are presented for using the well-known frequency of collapses in the galaxy (Zeldovich and Novikov 1971) estimated as (1/10) yr$^{-1}$. The SN 1987A explosion in the Large Magellanic Cloud shows that many explosions both in the Galaxy and in neighboring galaxies could be missed due not only to covering from gas and dust in the galactic disk, but also due to a decrease of brightness of supernovae of the SN 1987A type (or Cas A?). Then neutrino occurrences in the Galaxy detect these closed and dim supernova explosions or, it is completely possible, even silent collapses of stellar cores in general without the phenomena of supernova explosions. An analysis of these observations prompts how often in nature collapses are accompanied by explosions and how large is the correlation of these grandiose natural phenomena. If this correlation is small, then there would be a significant basis to solve the problem of supernova explosions outside of the contemporary statement of the problem. Probably then, in particular, the explosion mechanism would be more reliably connected with effects of rotation and the magnetic field in the collapsing stellar core.

Together with the one-dimensional statement of the problem of collapse, a large number of theoretical studies have already been made on the collapse of rotating stellar cores. A review of these works was given by Müller (1990). We also note the studies of Dyachenko et al. (1968) and Imshennik and Nadezhin (1977, 1991). In many supernova explosions as well as in the SN 1987A explosion deviations were observed from a spherically symmetric explosion. In turn, these deviations could be due to the effect of rotation in the central part of the star where the explosion began, i.e., the stellar core and its near vicinity although other deviations could be due to inhomogeneities of the distribution of the circumstellar medium. Hydrodynamic theory should show which peculiarities of gravitational collapse come from the effects of rotation given in the initial state of the collapsing stellar core. It was shown in the works cited that for sufficiently strong initial rotation (period $\tau_{rot}^{(i)} \sim 1$ s) a stoppage of collapse occurs for significantly smaller stellar core parameters (for example, $\rho_c^{(rot)} \sim 0.1\, \rho_c$) with very strong rotation in the hydrodynamically equilibrium protoneutron star (period $\tau_{rot}^{(f)} \gtrsim 1$ ms). Unfortunately the expanding shock in general is again transformed into an accretion shock and the supernova explosion in the calculations performed also does not develop.

In the two-dimensional problem of collapse taking into account rotation, of course, it is very hard to maintain a completely physical statement of the problem of gravitational collapse reached in the one-dimensional statement of the problem in the discussion above. Therefore, final conclusions are worth clearing up and verifying. Of course, in a non-one-dimensional statement of the problem there are many different still unstudied possibilities. For example, it is sufficient to recall large-scale convection behind the accretion shock front considered in the framework of the delayed explosion mechanism of the C version of the theory. Non-one-dimensional models of stellar collapse, however, should not move to a second plan of searching for a supernova explosion mechanism in the old framework of the one-dimensional statement of the problem of the collapse of stellar cores if observations do not show in the end that the explosion of a star of supernova explosion scale is a quite rare phenomena during the collapse of its core.[†] Then, the role of additional effects, first of all effects of rotation and the magnetic field could become determining in the supernova explosion mechanism.

We would like to express our sincere thanks to S. I. Blinnikov and D. K. Nadezhin for a fruitful discussion of this study, transmission of information, and valuable advice.

## References

Arnett, W. D. (1966) *Can. J. Phys.* **44**, 2553; (1967) *Can. J. Phys.* **45**, 1621.

Arnett, W. D. (1982) in Supernovae: A Survey of Current Research, M. J. Ress and R. J. Stoneham (eds.), Reidel, Dordrecht, p. 221.

Arnett, W. D. (1986) in The Origin and Evolution of Neutron Stars, D. J. Helfand and J.-H. Huang (eds.), Reidel, Dordrecht, p. 273.

Arnett, W. D. (1987) *Astrophys. J.* **319**, 136.

Arnett, W. D., Bahcall, J. N., Kirshner, R. P., and Woosley, S. E. (1989) *Ann. Rev. Astron. Astrophys.* **27**, 629.

Bethe, H. A., Brown, G. E., Applegate, J. H., and Lattimer, J. M. (1979) *Nucl. Phys.* **A324**, 487.

Blinnikov, S. I., Lozinskaya, T. A., and Chugai, N. N. (1988) *Sov. Sci. Rev., Sec. E, Astronophys. Space Phys.* **6**, 195.

Blinnikov, S. I., Imshennik, V. S., and Nadezhin, D. K. (1988) *Astrophys. Space Sci.* **150**, 273.

Cameron, A. G. W. (1982) in Essays in Nuclear Astrophysics, C. A. Barnes, D. D. Clayton, and D. N. Schramm (eds.), Cambridge University Press, Cambridge.

Colgate, S. A., and White, R. H. (1966) *Astrophys. J.* **143**, 626.

---

[†]Could it be that a supernova explosion due to gravitational collapse is a random accompanying event? This situation, not completely excluded by the modern state of our knowledge, nevertheless is inconsistent from the philosophical point of view.

Dyachenko, V. F., Zeldovich, Ya. B., Imshennik, V. S., and Paleichik, V. V. (1968) *Astrophysika* **4**, 2, 159.

Fowler, W., and Hoyle, F. (1965) Neutrino Processes and Pair Formation in Massive Stars and Supernovae, The University of Chicago Press, Chicago.

Goldreich, P., and Weber, S. V. (1980) *Astrophys. J.* **238**, 991.

Grassberg, E. K., Imshennik, V. S., and Nadezhin, D. K. (1971) *Astrophys. Space Sci.* **10**, 28.

Hillebrandt, W. (1987) in NATO ASI on High Energy Phenomena Around Collapsed Stars, F. Pacini (ed.), Reidel, Dordrecht, p. 73.

Imshennik, V. S., and Nadezhin, D. K. (1965) *Astron. Zh.* **42**, 1154.

Imshennik, V. S., and Chechetkin, V. M. (1970) *Astron. Zh.* **47**, 929.

Imshennik, V. S., and Nadezhin, D. K. (1972) *Zh. Eksp. Teor. Fiz.* **63**, 1548.

Imshennik, V. S., and Nadezhin, D. K. (1977) *Pis'ma Astron. Zh.* **3**, 8, 355.

Imshennik, V. S., and Nadezhin, D. K. (1983) *Sov. Sci. Rev. Sec. E. Astrophys. Space Phys.* **2**, 75.

Imshennik, V. S., and Nadezhin, D. K. (1989) *Sov. Sci. Rev. Sec. E. Astrophys. Space Phys.* **8**, 1.

Imshennik, V. S., and Nadezhin, D. K. (1989) in Contemporary Problems of Physics and Stellar Evolution, A. G. Masevich (ed.), Nauka, Moscow, p. 224.

Imshennik, V. S., and Nadezhin, D. K. (1991) Supernova 1987A and the Formation of Rotating Neutron Stars, in Physics of Neutron Stars, in press.

Ivanova, L. N., Imshennik, V. S., and Nadezhin, D. K. (1969) *Nauchnye Inf. Astron. Sovet AN SSSR* **13**, 3.

Ivanova, L. N., Imshennik, V. S., and Chechetkin, V. M. (1983) Hydrodynamics in Deflagration Models of Supernova Explosions, ITEP Prepr. **109**.

Landau, L. D., and Lifshitz, E. M. (1976) Statistical Physics, Nauka, Moscow.

Lattimer, J. M., and Yahil, A. (1989) *Astrophys. J.* **340**, 426.

Müller, E. (1990) Supernovae: Observations, Theory, Models, and Nucleosynthesis, MPA Prepr. **514**.

Nadezhin, D. K. (1968) *Astron. Zh.* **45**, 1166.

Nadezhin, D. K. (1977a) *Astrophys. Space Sci.* **49**, 399; (1977b) *Astrophys. Space Sci.* **51**, 283.

Nadezhin, D. K. (1978) *Astrophys. Space Sci.* **53**, 131.

Nomoto, K., and Hashimoto, M. (1988) *Phys. Rep.* **163**, 13.

Shklovsky, I. S. (1982) *Pis'ma Astron. Zh.* **8**, 347.

Shklovsky, I. S. (1984) *Pis'ma Astron. Zh.* **10**, 723.

Shklovsky, I. S. (1985) Stars: Their Birth, Life, and Death, Nauka, Moscow.

Schramm, D. N., and Truran, J. M. (1990) *Phys. Rep.* **189**, 89.

Sedov, L. I. (1957) Similarity and Dimensionality Methods in Mechanics, Gostekhizdat, Moscow.

Thielemann, F.-K., Hashimoto, M., and Nomoto, K. (1990) *Astrophys. J.* **349**, 222.

Wilson, J. R. (1985) in Numerical Astrophysics, J. M. Centrella, J. M. LeBlanc, and R. L. Bowers (eds.), Jones and Bartlett, Boston, p. 422.
Woosley, S. E., and Weaver, T. A. (1986) *Ann. Rev. Astron. Astrophys.* **24**, 205.
Woosley, S. E., Pinto, P. A., and Weaver, T. A. (1988) *Proc. Astron. Soc. Austr.* **7**, 355.
Woosley, S. E., and Weaver, T. A. (1988) *Phys. Rep.* **163**, 79.
Yahil, A., and Lattimer, J. M. (1982) in Supernovae: A Survey of Current Research, M. J. Ress and R. J. Stoneham (eds.), Reidel, Dordrecht, p. 53.
Yahil, A. (1983) *Astrophys. J.* **265**, 1047.
Zeldovich, Ya. B. (1956), *Akust. Zh.* **2**, 28.
Zeldovich, Ya. B., and Raizer, Yu. P. (1963) Physics of Shock Waves and High-Temperature Hydrodynamic Phenomena, GIF-ML, Moscow.
Zeldovich, Ya. B., and Novikov, I. D. (1971) Theory of Gravitation and Stellar Evolution, Nauka, Moscow.
Zentsova, A. S., and Nadezhin, D. K. (1975) *Astron. Zh.* **52**, 234.

# 13

# Radio Emission from Supernovae

## V. I. Slysh
*Astro Space Center, Lebedev Physical Institute, Moscow*

The radio emission from supernovae is observed in the first days or years after explosion. It has a power-law spectrum with a low-frequency cut-off, whose position moves with time to lower frequencies. The size of the radio emitting region increases with time at a rate consistent with the rate of supernova envelope expansion. This kind of spectrum evolution was described by the free-free absorption in the circumstellar matter accumulated by the stellar wind of the pre-supernova or by synchrotron self-absorption. It is found that the synchrotron self-absorption model is consistent with observed parameters of type II supernovae including the supernova SN 1987a in the Large Magellanic Cloud. For type Ib supernovae SN 1983n and SN 1984l the synchrotron self-absorption model predicts a radio envelope expansion velocity of about 60,000 km s$^{-1}$, an order of magnitude larger than the optically determined expansion velocity.

The synchrotron emission of supernovae is produced by relativistic electrons which are most likely accelerated by the shock associated with the expanding envelope. Three different mechanisms of shock acceleration are discussed. The first-order Fermi, or diffusive shock acceleration can provide observed flux densities of all detected supernovae, but requires very high mass-loss rates and magnetic fields. Of the two quasi-perpendicular mechanisms acceleration by lower-hybrid waves seems not to be efficient enough to produce relativistic electrons, while shock drift acceleration can accelerate a sufficient number of electrons to relativistic energies. This conclusion was derived from scaling of the Earth's bow shock electron energy density to supernova shock parameters. The shock drift acceleration can supply a sufficient number of relativistic electrons for equipartion with the magnetic field.

## 1. INTRODUCTION: SUPERNOVA-SNR CONNECTION: A TRIBUTE TO J. S. SHKLOVSKY

Compared to supernova remnants (SNR), the radio emission which accompanies explosions of supernovae at early phases (several years after the outburst) was discovered quite recently (Gottesman *et al.* 1972). This is a completely new phenomenon and its properties cannot be found by a simple extrapolation back from SNRs. The secular variation of the SNR radio flux density so ingeniously predicted by J. S. Shklovsky (1960) and successfully confirmed by subsequent observations of Cassiopea A is caused by the SNR envelope expansion. If the process were reversed and Cassiopea A were compressed to a size which the supernova 1979c had at the age of 2.3 years, that is to $8 \times 10^{16}$ cm, its radio luminosity would be 3 to 7 orders of magnitude higher than that of the supernova 1979c depending on whether the energy of relativistic electrons is decreasing due to adiabatic expansion or the decrease is compensated by some further acceleration. This means that the radio emission of a SNR can not be regarded as the emission of the highly expanded envelope of the supernova. There is a greater distinction between SNRs and supernovae: during the process of evolution from the supernova phase to the phase of SNR the rate of the decrease of radio luminosity changes to larger values.

Another distinction between supernovae and SNRs may be caused by neutron star activity in type II supernovae. At the supernova phase the envelope is dense enough to completely block any emission from the neutron star. As the envelope is expanded it becomes transparent not only to radiation of the rotating neutron star (a pulsar), but also to relativistic particles accelerated in its magnetosphere. A supply to the radio emitting region of the SNR of relativistic particles becomes possible. Thus a manifestation of the neutron star activity is more probable at a later time during the SNR phase.

Young SNRs as a rule have a rather regular circular shape which means the existence of a high spherical symmetry (not counting the Crab nebula and similar SNRs powered by pulsars). The spherical symmetry of SNRs is a result of the independence of the expansion from initial conditions in self-similar (Sedov) explosions. It was J. S. Shklovsky who had applied the self-similar solution to SNRs (1962). The situation is quite different during the supernova phase when the envelope is not yet decelerated and is in free flight. The free flight may be asymmetric for many reasons, for example, due to asymmetry of the explosion when the envelope is moving in one direction and the neutron star in another direction (Shklovsky 1980). The first VLBI radio image of a supernova shows a lack of symmetry in SN 1986j (Bartel 1990). Thus there are several important differences between radio supernovae and radio SNRs.

In one of his last papers Shklovsky (1985) discussed radio emission from the type Ib supernova SN 1983n and came to the conclusion that the radio envelope of the supernova was expanding with a velocity of about 60,000 km s$^{-1}$, much faster than the expansion of the supernova envelope measured from optical spectra. This conclusion was further supported both for SN 1983n and for another type Ib supernova SN 1984l. But direct VLBI measurements of the expansion velocity of type Ib supernovae are not yet available.

## 2. OBSERVATIONS

As was mentioned in the Introduction the first observations of the radio emission from a supernova were carried out by Gottesman *et al.* (1972) who determined very roughly the radio flux of the supernova 1970g. Only rather recently much more detailed observations became possible. Here some results of the observations are briefly reviewed.

### 1979c

The radio emission of this supernova in the galaxy NGC 4321 (M100) was investigated in great detail due to its exceptional brightness. Figure 13.1 shows 2 radio images of the galaxy M100 obtained with the VLA one year apart. On the left one sees the central part of the galaxy while on the right picture taken one year later one can see also a bright point source in the lower left corner. Light curves at two frequencies are shown in Figure 13.2. The light curves show some typical features. First, the radio emission does not appear simultaneously with the explosion, but with some delay, the lower the frequency the larger the delay. Second, the rise of the radio emission is much faster than its decay. Third, maximum flux density was reached earlier at high frequencies and later at lower frequencies. No maximum was observed at the highest frequency used 15 GHz, perhaps because it occurred before the first observation was made.

Spectra of the radio emission at different times all have maxima with a sharp cut-off at low frequencies and apparently a power-law fall-off at high frequencies with an exponent $\alpha = -0.75$ (Figure 13.3). The position of the maximum has shifted from 5 GHz on day 450 after the explosion to 1 GHz on day 1415.

The angular size of the SN 1979c radio envelope was measured with VLBI at several epochs. Figure 13.4 shows visibility functions at 5 GHz, while Figure 13.5 shows variation of the angular size with time calculated from the visibility functions (Bartel 1990). One can see from the Figure 13.5 that in the 3.5 years between the first and the last measurements the angular size has increased substantially. If the variation of the angular size with time is described by a power-law $\theta \propto t^m$, then observations would correspond to $m = 1.03 \pm 0.15$ which is consistent with uniform expansion ($m = 1$). The angular velocity of the uniform expansion shown on Figure 13.5 by a straight line is $\theta = 0.11$ mas yr$^{-1}$. If the expansion velocity of the radio envelope is the same as that measured from optical lines, $v = 11,000$ km s$^{-1}$, then an estimate of the distance to the supernova can be made $D = 22$ Mpc. This result was used by Bartel *et al.* (1985) for an independent determination of the Hubble constant.

### 1980k

Similar light curves are shown on Figure 13.6 for the supernova 1980k in the galaxy NGC 6946. The only substantial difference is a much faster evolution: as is evident on

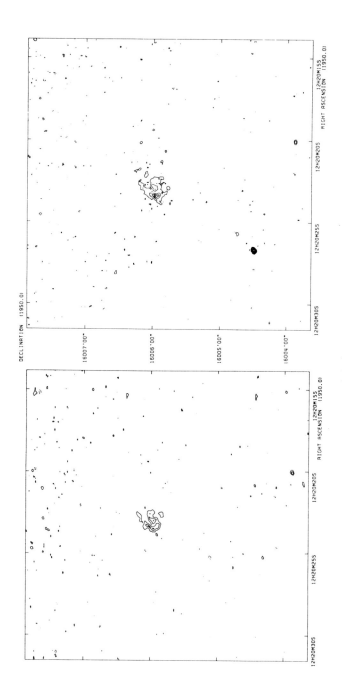

Figure 13.1. Contour map showing SN 1979c before (top) and after (bottom) explosion in the galaxy M100 (NGC 4321) (Weiler and Sramek 1988).

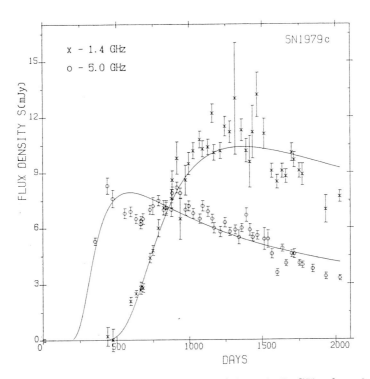

**Figure 13.2.** Radio light curves at 1.4 and 5 GHz for the supernova SN 1979c (after Weiler et al. 1986).

Figure 13.6 the time scale for the supernova 1980k is a factor of 3 less than for the supernova 1979c; the flux density of this supernova is also a factor of 3 lower. VLBI measurements at 13 cm 930 days after explosion have determined only an upper limit on the angular size of 1 mas (Bartel 1990), which is consistent with optical data.

## 1983n

The radio spectrum of the supernova 1983n in the galaxy M83 is evolving still faster (Figure 13.7). The half-power width of the light curve at 6 cm is only 28 days while for 1980k it was 400 days and for 1979c 1500 days.

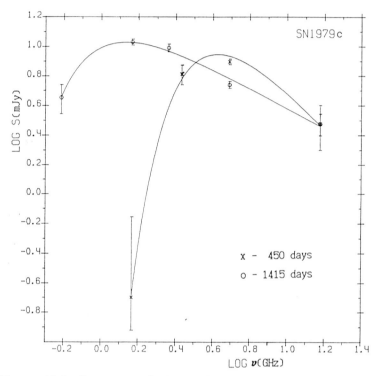

**Figure 13.3.** Spectrum of the radio emission from SN 1979c 450 and 1415 days after explosion.

## 1986j

This is thus far the brightest radio supernova known: at a wavelength of 6 cm a flux density of 120 mJy was reached. The light curves measured at several wavelengths show a rather slow evolution (Figure 13.8) (Weiler *et al.* 1990). The radio emission at 6 cm initially rose rapidly, passed a maximum about 1400 days after the explosion, and started a slow decline. The light curve at a wavelength of 20 cm showed a similar behavior with a distinction that the maximum was reached 2000 days after the explosion. At shorter wavelengths the data are less complete: the first observations at wavelengths of 2 and

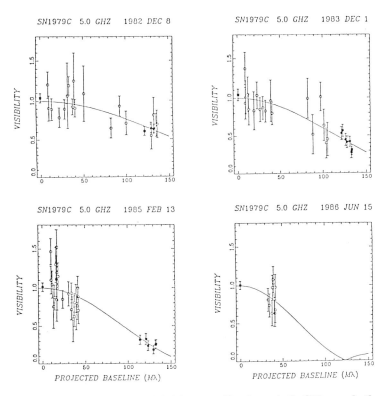

Figure 13.4. Measured visibility amplitudes at 5 GHz and the fit of a uniform spherical model (Bartel 1990).

1.35 cm were made after the maximum and only a slow decline was observed. Two measurements at a wavelength of 1.2 mm were made showing a rather fast decline between days 1500 and 1800 after the explosion. The value of the flux density at 1.2 mm however is not consistent with the results of the lower frequency observations. A possible maximum of the light curve around day 2050 was observed at the longest wavelength of 92 cm, but the shape of the light curve and the level of the flux density is not consistent with more detailed and more accurate data at wavelengths of 6 and 20 cm. It is possible that free-free absorption in the ionized gas of the galactic disc of the edge-on galaxy NGC 891 is important or absorption in the near-by HII-region. Bartel (1990)

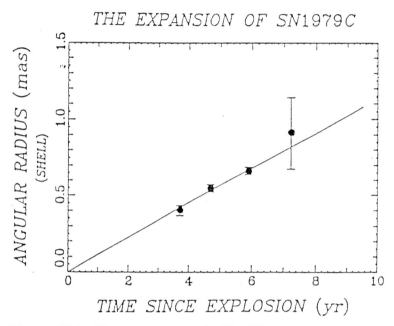

**Figure 13.5.** The expansion of SN 1979c. The solid line represents uniform expansion (m = 1) which is consistent with the weighted least-squares solution m = 1.03 ± 0.15 (Bartel 1990).

has constructed a VLBI image of this supernova (Figure 13.9) which shows an irregularly shaped envelope 1.6 mas in diameter. The supernova was 2200 days old at this moment. At a distance of 5 Mpc assumed for the parent galaxy NGC 891 this corresponds to a uniform expansion velocity of 3000 km s$^{-1}$ in reasonable agreement with an optical estimate of 3500 km s$^{-1}$ (Kirshner and Blair 1980).

## 1987a

The radio emission from the supernova in the Large Magellanic Cloud was discovered only 2 days after the explosion with the timing of the explosion being very accurately determined by the neutrino burst (Turtle *et al.* 1987). The light curves at frequencies of 0.843 and 1.4 GHz are similar to the light curves of other supernovae with a fast rise of flux and slow decline after the maximum (Figure 13.10). A delay of the maximum at

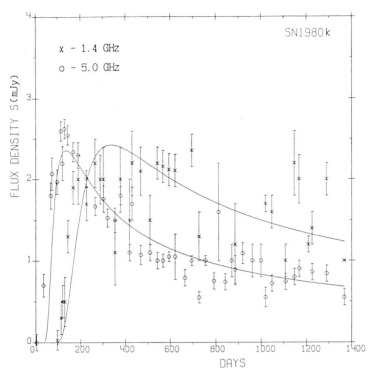

**Figure 13.6.** Radio light curves at 1.4 and 5 GHz for the supernova SN 1980k in NGC 6946 (after Weiler et al. 1986).

low frequencies relative to that at high frequencies is evident. At 2.3 and 8.4 GHz observations started apparently after the maximum. As compared to other supernovae the evolution of the radio spectrum of the supernova 1987a was a factor of 10 (1983n) or even a factor of 500 (1979c) faster and the radio luminosity was 3 to 4 orders of magnitude lower. VLBI observations at 2.3 GHz undertaken 5 days after the explosion showed that the source was completely resolved on the base line Australia- South Africa which means an angular diameter larger than 2.5 mas; under the assumption of an uniform expansion this corresponds to an expansion velocity higher than 19,000 km s$^{-1}$ (Jauncey et al. 1988); this is in agreement with the expansion velocity determined from the H$_\alpha$ absorption line.

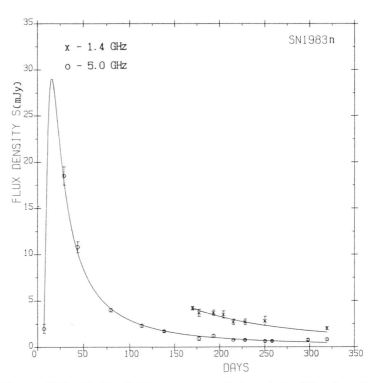

Figure 13.7. Radio light curves at 1.4 and 5 GHz for the supernova SN 1983n in M83 (after Weiler et al. 1986).

## 3. INTERPRETATION OF RADIO SPECTRA EVOLUTION

### 3.1. Interaction with the Stellar Wind of the Pre-supernova

As we mentioned earlier one of the most characteristic features of supernova radio emission is the presence of a peak both in the light curves and in the radio spectra. The natural hypothesis about the cause of the presence of the peaks would be the effect of an absorption. Chevalier (1982) suggested that the absorption was caused by the circumstellar ionized gas surrounding the supernova envelope. The free-free absorption

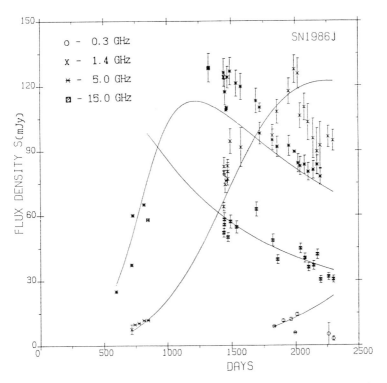

**Figure 13.8.** Radio light curves at 0.3, 1.4, 5, and 15 GHz for the supernova SN 1986j in NGC 891 (after Weiler, Panagia, and Sramek 1990).

produces a cut-off of the spectra at low frequencies. The density of the gas is higher closer to the center so the absorption was stronger at earlier phases when the envelope had a smaller radius; as the envelope is expanding the gas density becomes lower and the absorption smaller. This process explains the initial flux rise on the curves. By applying this model Weiler *et al.* (1986) were able to fit light curves of several supernovae and to determine some parameters.

In the model it was assumed that the free-free absorption optical depth $\tau$ was formed on the path starting from the edge of the radio envelope R to infinity

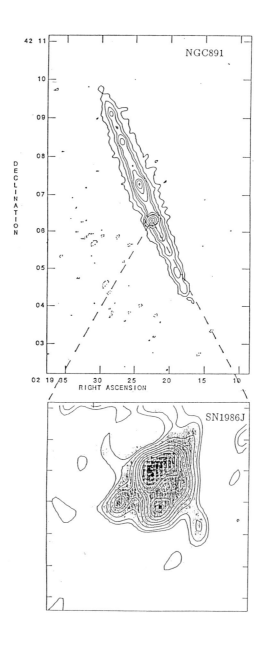

13. RADIO EMISSION FROM SUPERNOVAE 201

Figure 13.9 (facing page). VLBI radio image of SN 1986j on day 2200 after explosion (Bartel 1990).

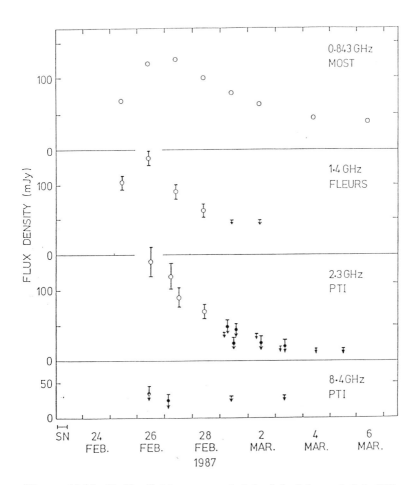

Figure 13.10. Radio light curves at 0.8, 1.4, 2.3, and 8.4 GHz for supernova 1987a in the Large Magellanic Cloud (Turtle et al. 1987).

$$\tau = \int_R^\infty n_e n_i k_{f-f} \, dr, \qquad (1)$$

where $n_e$ and $n_i$ are the densities of electrons and ions,

$$k_{f-f} = 3.62 \times 10^{-27} \left(\frac{\nu}{5 \text{ GHz}}\right)^{-2.1} \left(\frac{T}{10^4 \text{ K}}\right)^{-1.35} \qquad (2)$$

is the absorption coefficient, $\nu$ is the frequency, and $T$ is the temperature. It is assumed that the circumstellar matter was accumulated by mass loss in a stellar wind before the supernova explosion; the density of the stellar wind falls off as $r^{-2}$:

$$n_e = \frac{\dot{M}}{4\pi r^2 v \mu m_H}, \qquad (3)$$

where $\dot{M}$ is the mass loss rate (g s$^{-1}$), v is the stellar wind velocity (cm s$^{-1}$), $m_H$ is the hydrogen atom mass, and

$$\mu m_H = \frac{\Sigma X_j m_j}{\Sigma X_j Z_j}, \qquad (4)$$

where $X_j$, $Z_j$, and $m_j$ are the fraction by number, charge, and atomic mass, respectively, of the jth ion; the ion density is

$$n_i = \bar{Z} n_e, \qquad (5)$$

here

$$\bar{Z} = \frac{\Sigma X_j Z_j^2}{\Sigma X_j Z_j}. \qquad (5')$$

Substituting $n_e$ and $n_i$ in equation (1) one obtains

$$\tau = \frac{\dot{M}^2 \bar{Z} k_{f-f}}{3(4\pi)^2 R^3 \mu^2 m_H^2 v^2}. \qquad (6)$$

Thus optical depth in the process of expansion decreases as $R^{-3}$. In the Chevalier model it was assumed that the envelope is expanding according to a power law,

$$R \propto t^m, \qquad (7)$$

where $m = (n-3)/(n-2)$. This law of expansion follows from a consideration of the interaction between the supernova envelope with a density distribution $\rho = r^{-n}$ and the stellar wind gas with a density distribution $\rho \propto r^{-2}$ (Chevalier 1982). Thus the time dependence of the optical depth may be found:

$$\tau \propto R^{-3} \propto t^\delta, \qquad (8)$$

where $\delta = -3m = -3(n-3)/(n-2)$. Chevalier (1982) also proposed that the synchrotron emission of the envelope is generated by relativistic electrons which are in equipartition with the magnetic field and their energy density is proportional to the thermal energy behind the shock front. Under these assumptions it is possible to find a dependence of the radio emission flux density on radius and ultimately on time:

$$S \propto K_e B_\perp^{(1+\gamma)/2} \nu^{(1-\gamma)/2} R^3, \qquad (9)$$

where $K_e$ is a coefficient proportional to the density of relativistic electrons in the power law energy E distribution function

$$N(E) = K_e E^{-\gamma}, \qquad (10)$$

with an exponent $\gamma$. According to the assumption that the energy density of relativistic electrons is proportional to the thermal energy density behind the shock front one has

$$K_e \propto n_e m_i v_s^2, \qquad (11)$$

where $v_s$ is the shock front velocity, which according to equation (7) is equal to

$$v_s = dR/dt \propto t^{m-1}. \qquad (12)$$

In the stellar wind $n_e \propto R^{-2}$ so taking into account equation (7) and the assumption of equipartition one obtains

$$K_e = B_\perp^2 \propto t^{-2} \qquad (13)$$

Substitution in equation (9) gives

$$S \propto t^\beta \nu^\alpha, \qquad (14)$$

where $\alpha = (1-\gamma)/2$, $\beta = 3m+\alpha-3$.

The light curves from observations were approximated by the following expressions (Weiler 1986)

$$S(\text{mJy}) = K_1 \left(\frac{\nu}{5 \text{ GHz}}\right)^\alpha \left(\frac{t-t_0}{1 \text{ day}}\right)^\beta e^{-\tau}, \qquad (15)$$

where

$$\tau = K_2 \left(\frac{\nu}{5 \text{ GHz}}\right)^{-2.1} \left(\frac{t-t_0}{1 \text{ day}}\right)^\delta, \qquad (16)$$

and parameters $\alpha$, $\beta$, and $\gamma$ have the same meaning as in equations (8) and (14) describing the model of Chevalier. A rather good fit of equations (15) and (16) to the observed light curves for supernovae was obtained. Table 13.1 gives values of the parameters of the fit. Comparing them with the model, one can determine some properties of the envelope and stellar wind. Thus the parameter $\delta$ according to equation (8) gives $m = -\delta/3$, a parameter which describes the rate of deceleration of the envelope. One can see that in all cases $m$ exceeds 0.8 and is close to unity; this means that the envelope expands almost without deceleration: for $m > 0.81$ the parameter $n$ which describes the density distribution in the envelope is always larger than 7; this means that the density in the envelope falls off very sharply.

If the envelope expands without deceleration ($m = 1$ and $n = \infty$), then equation (14) transforms to

$$S \propto t^\alpha \nu^\alpha, \qquad (17)$$

that is $\alpha = \beta$. As is evident from Table 13.1, $\beta$ is always quite close to $\alpha$ as predicted by the Chevalier model (with the exceptions of pts. 3 to 5, where the model was put into the fitting procedure).

## 3.2. Synchrotron Self-absorption

Although the model with free-free absorption in the gas of the stellar wind of the pre-supernova was quite successful in describing the evolution of radio spectra, it requires a rather large stellar mass-loss rate. For supernova 1979c the rate should be $6 \times 10^{-5}$ solar masses per year, and for supernova 1986j—as large as $2.4 \times 10^{-4}$ solar masses per year (Weiler et al. 1986, 1990). It was suggested that the peak in the spectra of supernova radio emission was not caused by the absorption by thermal, but rather relativistic electrons which are responsible for the synchrotron emission (Slysh 1990). Synchrotron

Table 13.1. Fitting parameters

| SN Name | $K_1$ | $\alpha$ | $\beta$ | $\delta$† | $K_2$ | m | Ref |
|---|---|---|---|---|---|---|---|
| 1979c | 810±100 | −0.76±0.05 | −0.69±0.08 | −2.81±0.06 | $(1.4\pm0.1)\times10^7$ | 0.94 | 1 |
| 1980k | 69±11 | −0.50±0.10 | −0.64±0.07 | −2.82±0.27 | $(2.4\pm0.3)\times10^5$ | 0.94 | 1 |
| 1983n | 4400±600 | −1.03±0.06 | −1.59±0.08 | ≠2.44±0.05 | 530±30 | 0.81 | 1 |
| 1984j | $275^{+927}_{-212}$ | −1.01±0.17 | −1.48±0.33 | −2.54±0.36 | $690^{+2775}_{-554}$ | 0.85 | 2 |
| 1986j | $\left(6.7^{+2.5}_{-2.9}\right)\times10^5$ | $-0.67^{+0.12}_{-0.04}$ | $-1.18^{+0.06}_{-0.04}$ | $-2.49^{+0.30}_{-0.20}$ | $(3\pm3)\times10^6$ | 0.83 | 3 |
| 1987a | 411±51 | −0.46±0.22 | −1.37±0.33 | −2.64±0.09 | 0.3±0.3 | 0.88 | 4 |

References: (1) Weiler *et al.* (1986); (2) Panagia, Sramek, and Weller (1986); (3) Weiler, Panagia, and Sramek (1990), additional absorption in the envelope; (4) present work.
†For SN 1983n and 1984l it was assumed that $\beta = \alpha-3-\delta$ (Chevalier model).

self-absorption does not require any absorbing agent except the emitting relativistic electrons themselves. To estimate the importance of synchrotron self-absorption in the formation of the peak in the spectra one may compare the angular size of a radio supernova with the angular size calculated by synchrotron theory (Slysh 1990).

The angular size of a radio source with a low-frequency cut-off in its spectrum can be determined by its flux density S at the frequency ν of the maximum of the spectrum and by the value of the magnetic field B (Slysh 1963). Using convenient units one can write an equation for the angular size θ as follows

$$\theta(\text{mas}) = 5\, S^{1/2}(\text{Jy})\, \nu^{-5/4}\, (\text{GHz})\, (B/10^{-2}G)^{1/4}. \qquad (18)$$

The magnetic field in equation (18) is the only unknown parameter although the angular size dependence on it is quite weak. It is a widely accepted practice to estimate magnetic field in radio sources assuming equipartition between the energy density of the magnetic field and relativistic electrons. To make such an estimate one has to know the distance; it is always known more or less precisely for supernovae. Scott and Readhead (1977) give the following expression for the angular size of a radio source with synchrotron self-absorption under the condition of equipartition

$$\theta'' = F(\alpha)\, (1-(1+z)^{-1/2})^{-1/17}\, (1+z)^{(15-2\alpha)/34}\, S^{8/17}\, \nu^{-(2\alpha+35)/34}, \qquad (19)$$

where $F(\alpha) = 3.25$ for $\alpha = -1$, S is in Jansky, ν is in MHz, and z is the redshift. The expression is given assuming an Einstein–de Sitter model of the Universe with $q_o = 1/2$ and $H = 50$ km s$^{-1}$ Mpc$^{-1}$. For near-by galaxies in which supernovae were observed it is

Table 13.2. Comparison of self-absorption, expansion and VLBI angular sizes of radio supernovae

| SN Name | Frequency of the Maximum (GHz) | Maximum Flux (mJy) | Age (days) | $\theta_{ABS}$ (mas) | Expansion Velocity (km s$^{-1}$) | Distance (Mpc) | $\theta_{ENV}$ (mas) | $\theta_{VLBI}$ (mas) |
|---|---|---|---|---|---|---|---|---|
| 1979c | 1.465 | $7.1^{+4.1}_{-2.6}$ | 874±422 | $0.39^{+0.10}_{-0.07}$ | 9200[a] | 17[a] | 0.56±0.27 | 0.70[e] |
| 1980k | 1.465 | $1.6^{+1.1}_{-0.7}$ | 202±103 | $0.20^{+0.06}_{-0.04}$ | 5700[a] | 7[a] | 0.20±0.10 | <2[e] |
| 1983n | 1.465 | $18^{+7}_{-5}$ | 38±3 | $0.64^{+0.12}_{-0.09}$ | 17000[a] | 7[a] | 0.11±0.009 | — |
| 1984l | 1.465 | $1.7^{+15}_{-1.5}$ | $36^{+45}_{-20}$ | $0.20^{+0.37}_{-0.13}$ | 17000[a] | 24[b] | $0.03^{+0.04}_{\neq 0.02}$ | — |
| 1986j | $3.3^{+0.9}_{-0.7}$ | 135±5 | 1460±300 | $0.74^{+0.19}_{-0.15}$ | 2000[c] | 7.6[c] | 0.46±0.09 | 0.62±0.11[e] |
| 1987a | $1.415^{+0.36}_{-0.29}$ | 190±30 | 3.1±1 | $2.7^{+0.8}_{-0.6}$ | 30000[d] | 0.05 | 2.3±0.7 | >2.5[f] |

[a]Weiler et al. (1986); [b]Panagia, Sramek, and Weiler (1986); [c]Rupen et al. (1987); [d]blue limit of the H$_\alpha$ absorption, Hanuschik and Dachs (1987); [e]Bartel (1990); [f]Jauncey et al. (1988).

more convenient instead of the redshift which can be distorted by peculiar motions to use distance D (Mpc). Transforming equation (19) in this way and putting $\alpha = -1$ one has

$$\theta(\text{mas}) = 6.92 \, S^{8/17} \, (\text{Jy}) \, D^{-1/17} \, (\text{Mpc}) \, v^{-33/34} \, (\text{GHz}), \quad (20)$$

Comparing equations (18) and (20) one can see that the angular size is the same for S = 1 Jy at $v = 1$ GHz and D = 1 Mpc with the magnetic field B = 0.037 G which is evidently the equipartition value of the field in a radio source having the above parameters.

Using equation (20) one can estimate how important synchrotron self-absorption is in radio supernovae. To do this compare the angular size calculated using equation (20) with that of a supernova assuming that the radio emitting region expands with the velocity of the envelope. VLBI results agree with this assumption in the case of 1979c (Bartel 1990). Now we do not invoke free-free absorption and attribute the observed cut-off in the spectra of radio emission of supernovae to synchrotron self-absorption. The flux density at the maximum of the spectrum and the time elapsed after explosion during which the maximum has moved to frequency 1.485 GHz was calculated using approximation Eqs. (15, 16) and coefficients from Table 13.1 for supernovae 1979c, 1980k, 1983n, and 1984l. For supernova 1986j the position of the maximum of the spectrum was determined directly from the data of Rupen et al. (1987) for September 22, 1986 (at an age of about 4 years). The maximum in the spectrum of radio emission from the supernova 1987a occurred at frequency 1.415 GHz with a flux density 190 mJy, 3.2

days after the explosion (Turtle *et al.* 1987; Storey and Manchester 1987). The results of calculations are given in Table 13.2. Errors of the angular size $\theta_{ABS}$ determined from the synchrotron self-absorption depend mostly on the errors of the flux density S, since the frequency of the maximum was fixed, and the error of distance is not important because it enters in Eq. (18) with the very small power 1/17. The flux errors were estimated from the errors of the coefficients of Table 13.1 (for supernovae 1979c, 1980k, 1983n, and 1984l). For supernovae 1986j and 1987a the errors of the frequency of the maximum and of the maximum flux density were estimated from experimental data given by Rupen *et al.* (1987), Turtle *et al.* (1987), and Storey and Manchester (1987). Using the approximation Eq. (16) for absorption the age of supernovae was calculated at the moment when the optical depth was equal to unity at a frequency of 1.465 GHz for the first four supernovae from Table 13.2. The age of supernova 1986j was taken to be 4 years. The angular size of a supernova was computed under the assumption of a constant expansion velocity also given in Table 13.2. In the last row of Table 13.2 angular sizes measured by VLBI are given. A comparison of these sizes with those computed from the expansion velocity and age shows good agreement in two cases (1979c and 1986j) and a consistency with the upper (1980k) or lower (1987a) limits. Unfortunately no VLBI measurements are available for supernovae 1983n and 1984l.

The angular size determined by self-absorption $\theta_{ABS}$ also coincides within error limits with the angular size computed from expansion velocity and age in all cases except 1983n and 1984l. A correlation between the two types of angular size is clearly seen on Figure 13.11 where a dashed line corresponds to the equality of angular sizes determined from self-absorption and expansion. Since the latter is the physical size of radio supernovae as follows from the VLBI measurements the coincidence between it and that computed from synchrotron self-absorption implies a prevailing role of synchrotron self-absorption in forming the low-frequency cut-off in the spectrum of radio emission. If synchrotron self-absorption were not important and the cut-off was caused by other reason, e.g., free-free absorption in the ionized gas around the radio source or intermixed with it, the position of such a radio source on the plot of Figure 13.11 would be to the right and below the dashed line. In other words the angular size determined by synchrotron self-absorption will be less than the physical size determined from the expansion velocity. One can see on Figure 13.11 that there is no object among the known radio supernovae located in this portion of the diagram. Therefore, synchrotron self-absorption is the only cause of the cut-off in the spectra and there is no reason to consider any additional absorbing agent such as ionized gas.

### 3.3. Supernovae of Type 1b

Two supernovae on Figure 13.11 1983n and 1984l fall in the region which is not permitted for synchrotron emission to the left and above the dashed line. Their physical

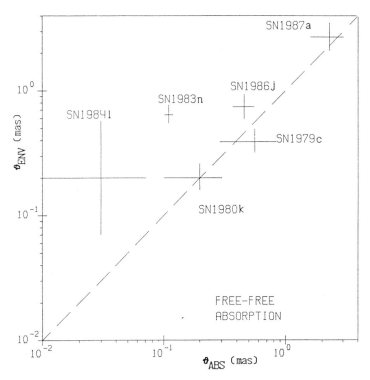

Figure 13.11. Comparison of angular sizes of radio supernovae determined from synchrotron self-absorption ($\vartheta_{ABS}$) and from the envelope expansion ($\vartheta_{ENV}$). The dashed line corresponds to the equality $\vartheta_{ABS} = \vartheta_{ENV}$. The region of free-free absorption is below and to the right of the dashed line.

size as determined by the expansion velocity happens to be a factor of 5 to 7 lower than size which is permitted by synchrotron self-absorption. In other words their brightness temperature exceeds the maximum possible for synchrotron emission by a factor of 20 to 60. For supernova 1983n this was first noticed by Shklovsky (1985). Analyzing possible causes of the discrepancy between angular sizes determined by synchrotron self-absorption and by the expansion Shklovsky came to the conclusion that the former is the right one. If this is the case, then the expansion velocity of the radio envelope should be higher than the velocity measured from optical spectra and must exceed 60,000 km s$^{-1}$. Our estimate of the radio envelope expansion velocity which is consistent with synchrotron self-absorption is about 100,000 km s$^{-1}$ for both supernovae 1983n and 1981l which is a factor of 6 higher than the expansion velocity of the optical envelope. If the radio emitting envelope is related to a shock wave travelling ahead of the supernova envelope then the shock wave should travel with acceleration which is possible in a medium with a sharply decreasing density. As was mentioned earlier from VLBI measurements it follows that the radio envelope expands with about the same velocity as the optical envelope. It is true that such measurements were not done for supernovae 1983n and 1984l; also the two supernovae belong to type Ib while the rest of the radio supernovae belong to type II. Optical properties of the two types of supernovae are different as well as properties of their radio emission: a shorter time scale and a steeper radio spectrum for type Ib supernovae. Therefore, the difference of other properties of their radio envelope including different expansion velocities is quite possible. One can not determine the parameter $m$ from Table 13.1 for type Ib supernovae independent of other parameters, the parameter which tells whether acceleration or deceleration of the expansion is taking place (in the model of Chevalier which of course assumes free-free, not synchrotron absorption). VLBI measurements are needed to determine the expansion velocity of the radio envelope.

If the expansion velocity of the radio envelope will prove to be close to the expansion velocity of the optical envelope as is the case of type II supernovae, then the reason for the discrepancy between the angular size determined by synchrotron self-absorption and physical angular size must be connected with assumptions made in the calculations. First, the radio emission of type Ib supernovae may not be synchrotron in origin, but some other non-thermal mechanism, for example, plasma waves (Bisnovatyi-Kogan, Illarionov, and Slysh 1990; Benz and Spicer 1990). In that case due to the coherent nature of the radio emission from plasma waves its brightness temperature may greatly exceed that of synchrotron emission as is observed in solar type II bursts. However, in contrast to the solar corona in the circumstellar medium it is practically impossible to generate emission at the plasma frequency or at the harmonic frequency without a very significant free-free absorption of this emission. If the synchrotron mechanism of radio emission is operating in type Ib supernovae, then the assumption of the equipartition of the magnetic energy density and relativistic particle energy density should be questioned, the assumption used in calculations with the Eq. (20). The brightness temperature can be increased by increasing the relativistic particle energy and by correspondingly decreasing the magnetic field to keep the same frequency of emission. For the supernova 1983n case field should be about $2 \times 10^{-4}$ G. However, other problems arise with such a weak magnetic field. First, in such a field the relativistic particles will not be confined any longer. Second, the magnitude of the field probably is too small to accelerate particles at the shock front both in quasi-parallel and in quasi-perpendicular cases (see below). The main argument, however, against the weak field and accordingly against the high energy of relativistic particles is their short life time caused by the inverse Compton effect in the radiation field of the optical emission of the supernova. The radio emission maximum at 6 cm of the supernova 1983n occurred practically simultaneously with the optical maximum (Weiler *et al.* 1986). At that time the bolometric luminosity was $L = 3 \times 10^{42}$ erg s$^{-1}$ and the envelope radius was $R = 2 \times 10^{15}$ cm; therefore the radiation energy density was

$$u = \frac{L}{4\pi R^2 c} = 2 \text{ erg cm}^{-3}, \quad (21)$$

With this radiation field energy density the life time of relativistic electrons with energy $E = 2.4 \times 10^9$ eV responsible for radio emission at a wavelength of 6 cm in a magnetic field $B = 2 \times 10^{-4}$ G is

$$t_{compt} = \frac{3 \times 10^7}{uE/mc^2} = 3100 \text{ s} = 52 \text{ min.}, \quad (22)$$

which is much less than the age of the supernova at that moment: 13 days. If, however, the magnetic field was in equipartition with relativistic particles, then its magnitude should be of the order of 0.3 G, and at 6 cm relativistic electrons with an energy $6 \times 10^7$ eV were radiating; the envelope radius in this case was substantially greater (as determined from synchrotron self-absorption) $R = 1.2 \times 10^{16}$ cm; at this distance $u = 0.055$ erg cm$^{-3}$, and the life time $t_{compt} = 53$ days, greater than the age of the supernova. Therefore, inverse Compton losses preclude the existence of high energy relativistic electrons radiating in a weak magnetic field, but do allow an equipartition value of the relativistic electron energy density and that of the magnetic field with a larger envelope radius compared to the optical envelope. All these arguments confirm the original suggestion by Shklovsky about the large expansion velocity of the type Ib supernova radio envelope.

The nature of the absorbing agent which determines the shape of the spectrum of radio supernovae and its evolution remains uncertain. On the one hand, the light curves were fitted quite well by the approximation formulae consistent with the Chevalier model in which the absorbing agent is the circumstellar gas of the presupernova stellar wind. On the other hand, the intensity of the supernova radio emission is so high that synchrotron self-absorption has to appear in their spectra; the shape of the spectrum and its time evolution will be qualitatively the same as in the previous case. However, some minor differences may exist between free-free and synchrotron absorption that can be revealed by numerical modelling of the light curves; no such modeling was done yet for synchrotron self-absorption.

There is one more experimental test for synchrotron self-absorption connected to the polarization of the emission. The synchrotron emission is known to be linearly polarized in a uniform magnetic field, the degree of polarization being

$$\Pi_1 = \frac{\gamma+1}{\gamma+7/3}. \qquad (23)$$

When the optical depth increases the degree of polarization decreases to

$$\Pi_2 = \frac{1}{2\gamma+13/3} \qquad (24)$$

and the position angle changes by 90° (Ginzburg, Sazonov, and Syrovatsky 1968). Thus when measuring a light curve at a fixed frequency a rise of the degree of polarization should take place by a factor of $\Pi_1/\Pi_2$ (which is normally 6–8) and a change of the positional angle by 90° when a transition from the self-absorption phase to the optically thin emission phase takes place. No such change of the polarization properties is expected for free-free absorption except Faraday rotation. Unfortunately, the degree of polarization of the supernova radio emission is quite small (Rupen *et al.* 1987) and such measurements are difficult.

## 4. ACCELERATION OF RELATIVISTIC ELECTRONS

### 4.1 The Mechanism and the Place of Acceleration

The most likely place where acceleration to relativistic energies of particles responsible for radio emission can occur is the shock wave travelling ahead of the expanding supernova envelope. It was also suggested that pulsar acceleration by the stellar remnant in the center of the supernova may be responsible for relativistic particles (Shklovsky 1981; Pacini and Salvati 1973). Although this mechanism seems to be very attractive, it can not provide a supply of relativistic particles for the radio emission at early phases since the particles are prevented from coming out through a very thick supernova envelope. The acceleration by neutrinos released when a neutron star is forming was suggested by Bisnovatyi-Kogan, Illarionov, and Slysh (1990). They showed that the interaction of the neutrinos with the matter of the supernova envelope may lead to acceleration of particles (mostly positrons) to an energy of about 15 MeV. The principal reaction is:

$$\tilde{\nu}_e + p \rightarrow n + e^+, \qquad (25)$$

which has a cross-section $\sigma = 2 \times 10^{-41}$ cm$^2$. The total number of fast positrons released by the collapsing star is about $2 \times 10^{41}$. This is far less than is needed to produce radio emission even from the intrinsically weakest supernova 1987a. Additional acceleration in the shock region was required by the authors.

In the model of Chevalier (1982) no details of the acceleration were specified. It was suggested that some modification of the statistical acceleration mechanism proposed by Gull (1973) for SNRs is responsible for the production of relativistic particles in radio supernovae. The mechanism is second-order Fermi acceleration on turbulent irregularities arising due to a Rayleigh-Taylor instability. For such a fast phenomenon as a supernova explosion, the statistical second-order Fermi acceleration has too low a rate of energy increase. First-order Fermi acceleration like the shock wave acceleration proposed by Krymsky (1977) is more promising in this respect. The time scale for first-order Fermi acceleration is $(v_s/v_A)^2 \cong 10^4$ less than the time scale of the statistical second-order Fermi acceleration ($v_s$ is the shock velocity, about 10,000-20,000 km s$^{-1}$, $v_A$ is the Alfven velocity of about 200 km s$^{-1}$) (Berezhko and Krymsky 1988). Therefore only first-order acceleration will be considered here.

## 4.2 First-order Fermi Acceleration

We will discuss the earliest stage of supernova evolution when the envelope is expanding freely, practically without any deceleration in the surrounding medium. The expanding envelope acts as a piston on the surrounding matter resulting in a shock wave travelling ahead of it. The surrounding medium is assumed to be the gas of the stellar wind which was emitted by the pre-supernova before the explosion; the gas contains a magnetic field frozen in it. The gas density in the stellar wind varies with the distance from the star r as $r^{-2}$; in that case the shock wave travels ahead of the piston with a velocity 1.19 times the velocity of the piston for strong shocks (Parker 1961). In the region between the shock front and the piston (envelope) all gas is located that was swept up by the shock wave during expansion. Its density at the shock front is 4 times the density of the undisturbed stellar wind gas adjacent to the shock front and increases towards the envelope. The gas velocity just behind the shock front is 3/4 of the shock velocity $v_s$ and the temperature is $kT = (3/32)m_p v_s^2$, where $m_p$ is the proton mass (Parker 1961); the temperature decreases towards the envelope. With a shock front velocity of 10,000–20,000 km s$^{-1}$ typical for supernovae, the temperature behind the shock may reach (1–4) × 10$^9$ K or 100–400 keV.

The most energetic particles can traverse the shock front without appreciable curving of their trajectory and will increase their energy by $\Delta E = (v_s/c)E$ (for relativistic particles) after each crossing the front (Krymsky 1977). On both sides of the shock front there are regions of intense turbulence from which the particles will be reflected and cross the front many times. As a result a systematic acceleration of the particles takes place and a power law energy distribution is established with an exponent very close to what is observed. The time scale of the acceleration can be estimated as follows. The energy of the particles

## 13. RADIO EMISSION FROM SUPERNOVAE

is increasing according to

$$\ln(E/E_o) = (4/3) i (u_1 - u_2)/c, \quad (26)$$

where $E_o$ is the initial energy of the particle, i is the number of shock front crossings by the particle, and $u_1$ and $u_2$ are velocities of the upstream gas and of the gas flowing from the shock front downstream, respectively, in the coordinate frame co-moving with the shock front (Bell 1978a). For strong shocks $(u_1 - u_2) = (3/4) v_s$. The number of crossings needed to acquire an energy E is found from Eq. (26),

$$i = (c/v_s) \ln(E/E_o). \quad (26')$$

If the initial energy of electrons $E_o \sim m_e c^2 = 511$ keV and the shock velocity is $v_s = 20{,}000$ km s$^{-1}$, then the energy $E = 200\, m_e c^2$ (a typical energy for emission at 6 cm) will be acquired by an electron after 80 crossings. The thickness of the region behind the shock front where most of the accelerated particles are concentrated is about $0.07 R_s$ ($R_s$ is the radius of the shock front) (see below). At each crossing a particle traverses this region twice (back and forth). Therefore, the total acceleration time is

$$t_{acc} = 2 \times 0.07\, R_s\, i/v \quad (27)$$

The particle velocity is $v \sim c$, so using Eq. (26') one has

$$t_{acc} = 0.14 t_{exp} \ln(E/E_o), \quad (27')$$

where $t_{exp} = R_s/v_s$ is the envelope expansion time. For $E = 200\, m_e c^2$, $t_{acc} = 0.74\, t_{exp}$ which means that at every stage of the expansion the particles will have enough time to be accelerated up to an energy of $200\, m_e c^2$. Using Eq. (27'), one can estimate the maximum energy which the particles may reach during expansion: $E/E_o = 1265$ or about 600 MeV.

For this type of acceleration Bell (1978b) calculated the volume synchrotron emissivity of particles accelerated by the shock:

$$\varepsilon(\nu) = 2.94 \times 10^{-34}(1.435 \times 10^5)^{0.75-\alpha}\xi(2\alpha+1) \times$$

$$\left(\frac{\varphi_e}{10^{-3}}\right)\left(\frac{\psi_e}{4}\right)^{2\alpha}\left(\frac{\alpha}{0.75}\right)\frac{n_e}{\text{cm}^{-3}} \times \left(\frac{\nu}{\text{GHz}}\right)^{-\alpha}\left(\frac{v_s}{10^4 \text{km s}^{-1}}\right)^{2\alpha} \times$$

$$\left(\frac{B}{10^{-4}\text{G}}\right)^{\alpha+1}\left[1 + \left(\frac{\psi_e}{4}\right)^{-1}\left(\frac{v_s}{7000 \text{ km s}^{-1}}\right)^{-2}\right]^{\alpha} \text{Wm}^{-3}\text{Hz}^{-1}, \qquad (28)$$

(the exponent of $v_s$ must be $2\alpha$, not $4\alpha$ as quoted by Bell), where $\alpha$ is the spectral index, $\nu$ is the frequency, B is the magnetic field, $\xi(2\alpha+1)$ is a function of order unity weakly depending on $\alpha$, $n_e$ is the plasma electron density behind the shock front, and $\varphi_e$ is the fraction of electrons with an energy exceeding $m_p v_s^2/2$ by $\psi_e$ times. It is the so-called injection energy needed to start the acceleration mechanism, that is to provide crossing of the shock front by particles. The mechanism of supply of such particles was not specified; this could be some type of plasma instability. Using experimental data from the Earth's bow shock, Bell (1978b) made an estimate of the fraction of the injected electrons $\psi_e = 10^{-3}$ and of their energy $E_o = 4 \times (m_p v_s^2/2)$ or $\varphi_e = 4$.

The piston velocity in young type II supernovae may be determined using the radial velocity of Balmer absorption lines. The effective volume of the radio emitting region can be roughly calculated using the Krymsky and Petukhov (1980) distribution of the accelerated particle density which is increasing approaching the shock front (behind it) as $r^{12}$ where r is the radius. Upstream, ahead of the shock front, the particle density falls off exponentially and the contribution from this region may be neglected. This distribution was obtained for the model of a point explosion while here we are dealing with a piston; nevertheless, we will use the distribution for rough estimates. It is easy to show that the effective volume of that distribution terminated at $R_s$ is

$$V_{\text{eff}} = 4\pi R_s^3/15, \qquad (29)$$

and the effective thickness is $0.07 R_s$.

Now we are in position to determine physical properties of the stellar wind and magnetic field needed to generate radio emission from young supernovae. According to the model assuming that generation takes place in the region just behind the shock front and the shock is strong, the density in this region must be 4 times the ambient density of the undisturbed gas of the stellar wind. The electron density upstream of the shock front can be calculated assuming free-free absorption following the Chevalier (1982) and Weiler et al. (1986) models of absorption by the stellar wind gas. The temperature is taken as $10^4$ K. The results of such calculations are given in Table 13.3. The velocity of

Table 13.3. Magnetic field and presupernova stellar wind as determined from the first-order Fermi acceleration

| SN Name | Age (days) | Radius (cm) | 5 GHz Volums Emissivity (W m$^{-3}$ Hz$^{-1}$) | $n_e \left(\dfrac{B}{10^{-4}}\right)^{1+\alpha}$ | $n_{eo}$ (cm$^{-3}$) | $B_o$ (Gauss) | $v_A$ (km s$^{-1}$) | $\dot{M}$ ($M_\odot$ yr$^{-1}$) |
|---|---|---|---|---|---|---|---|---|
| 1979c | 349 | 3.3x10$^{16}$ | 1.5x10$^{-23}$ | 1.5x10$^{11}$ | 1.6x10$^{5}$ | 0.028 | 157 | 4.3x10$^{-3}$ |
| 1980k | 81 | 4.8x10$^{15}$ | 2.5x10$^{-22}$ | 1.6x10$^{11}$ | 4.2x10$^{5}$ | 0.053 | 183 | 2.7x10$^{-4}$ |
| 1983n | 13 | 2.3x10$^{15}$ | 4.2x10$^{-22}$ | 4.3x10$^{15}$ | 6.0x10$^{5}$ | 0.90 | 2600 | 1.3x10$^{-3}$ |
| 1986j | 399 | 8.2x10$^{15}$ | 8.1x10$^{-21}$ | 5.36x10$^{13}$ | 3.2x10$^{5}$ | 0.92 | 3600 | 1.2x10$^{-2}$ |
| 1987a | 0.63 | 1.9x10$^{14}$ | 3.6x10$^{-20}$ | 4.7x10$^{12}$ | 2.1x10$^{6}$ | 0.22 | 340 | 4.2x10$^{-6}$ |

the shock front was taken to be 1.19 of the envelope velocity as given by the optical spectra. Using Eq. (28) the quantity $n_e B^{1+\alpha}$ was calculated.. Taking into account a factor of 4 compression in the region of generation behind the shock front of the gas density and magnetic field and using an electron density $n_{eo}$ determined from free-free absorption ($n_e = 4\, n_{eo}$), one can find the magnitude of the magnetic field $B_o$ in the undisturbed region ($B = 4\, B_o$) upstream of the shock front. There are also values of the Alfven velocity in the stellar wind given in Table 13.3, $v_A = B_o/(4\pi n_o m_p)^{1/2}$. The velocity of the stellar wind has to be much greater than the Alfven velocity; otherwise, the gas motion will be controlled by the magnetic field. The solar wind velocity is, for example, an order of magnitude higher than the Alfven velocity: that is why the coronal and interplanetary magnetic field is pulled out by the solar wind in the radial direction and its magnitude is inversely proportional to the distance. The magnetic field in the stellar wind should be of a similar configuration. If we make the assumption that the stellar wind velocity is 5 times the Alfven velocity, then from the values of the Alfven velocity given in Table 13.3, a stellar wind velocity exceeding 1000 km s$^{-1}$ is deduced which is typical for the stellar wind of hot OB stars. The last row of Table 13.3 gives stellar wind rates calculated with this assumption. For the supernova 1987a it is known almost certainly that the pre-supernova was a hot blue supergiant of type B3Ia which could have a stellar wind with the velocity 550 km s$^{-1}$ and a rate of 2 x 10$^{-6}$ solar mass per year (Chevalier and Fransson 1987) in reasonable agreement with Table 13.3. Therefore in the case of the supernova 1987a in the Large Magellanic Cloud the stellar wind properties of the pre-supernova may satisfy requirements for first-order Fermi acceleration at the shock front to provide the observed power of radio emission.

Other type II supernovae 1980k and especially 1979c had radio luminosities much higher than supernova 1987a mostly because of the much larger emitting volume. The parameters of the stellar wind were almost the same as for 1987a, but at much greater distances from the stars. As a result the stellar wind rate has to be greater by 2 to 3 orders of magnitude and exceeds $10^{-3}$ solar mass per year with a velocity of 700 to 1000 km s$^{-1}$. This means that the pre-supernovae had to be hot supergiants of spectral type 0 having very intense stellar winds. This conclusion is different from the results of Chevalier and Fransson (1987) who proposed a slow dusty wind from red supergiants for the supernovae 1979c and 1980k. Weiler *et al.* (1986) also assumed a slow stellar wind with a velocity 10 km s$^{-1}$ and for this reason their estimate of the stellar wind rate is much lower than ours. One can reduce the electron density by giving up the assumption of a constant gas temperature $10^4$ K. Because of radiation and adiabatic cooling the stellar wind temperature may drop to a much lower level. In the case of supernovae 1979c and 1980k radio emission at the frequency 1.4 GHz has appeared 870 and 200 days after the explosion, respectively, when the intensity of the ultraviolet ionizing radiation providing heating had dropped to a very low level. The time scale of the radiation cooling of the stellar wind with density $5 \times 10^4$ cm$^{-3}$ and temperature $10^4$ K is about 150 days. Therefore, the temperature may drop to several thousand K. We have performed calculations for a temperature of 1000 K. Although the electron density needed decreased almost a factor of 5, the Alfven velocity increased by a greater amount and the stellar wind rate became somewhat higher. Thus the decrease of the temperature does not reduce the rate of the stellar wind.

### 4.3. Shock Drift Acceleration

Diffusive or first-order Fermi acceleration is generally invoked for shocks which propagate parallel to the magnetic field direction. At oblique or quasi-perpendicular shocks the shock drift acceleration is more effective (Pesses, Decker, and Armstrong 1982). Such shocks contain a strong electric field in the shock front due to the motion of the plasma across the magnetic field

$$\mathcal{E} = -\frac{v_s}{c} B \sin \vartheta_{Bn}, \qquad (30)$$

where $\vartheta_{Bn}$ is the angle between the shock normal and the magnetic field. The electric field is directed perpendicular both to the shock normal and to the magnetic field so its vector lays in the plane of the shock. Energetic particles encounter the shock and drift along it in the direction of the electric field due to the gradient of the magnetic field at the shock with a drift velocity:

$$v_d = \frac{c \overline{\nabla} \times \overline{B}}{4\pi n e}. \qquad (31)$$

and can be accelerated by the electric field, if they remain in the shock front long enough, or can return to the shock frequently enough. If a particle can travel a distance $\Lambda$ without changing direction, then the maximum energy gained is

$$E = e \frac{v_s}{c} B \Lambda. \qquad (32)$$

If the ambient particle density and energy density of waves is low enough, $\Lambda$ is of the order of the characteristic scale of the shock. Thus, for the Earth's bow shock $\Lambda \sim 10{,}000$ km, and with $B = 3 \times 10^{-4}$ G and $v_s = 300$ km s$^{-1}$, $E = 90$ keV in agreement with observations. For supernovae $\Lambda = 10^{15}$–$10^{16}$ cm, $B = 0.1$ G, $v_s$, = 10,000–30,000 km s$^{-1}$, and $E = 10^{15}$–$3 \times 10^{16}$ eV far in excess of the particle energy necessary to produce radio emission. This flux of energetic particles will be unstable and will generate plasma waves which can stabilize the particle flux at a certain level. Unfortunately, there is no way at this moment to exactly calculate the energy distribution and density of accelerated particles and one can only rely on experimental data from measurements in the bow shock.

### 4.4. Acceleration by Lower-Hybrid Waves

This is a mechanism also efficient in quasi-perpendicular shocks. For solar bursts of type II associated with coronal shocks it was proposed (Lampe and Papadopoulos 1977) that the drift current $\nabla \times \mathbf{B}$ generates lower-hybrid waves which can stochastically accelerate electrons up to relativistic velocity. The acceleration is possible if the velocity of the waves is greater than the electron thermal velocity and it may be as great as the speed of light. This is possible for waves with a wave vector highly inclined to the magnetic field direction. Another possibility to excite the lower-hybrid waves is by ions reflected from the shock through the two-stream instability (Galeev 1984). The energy which can be acquired by electrons with a thermal distribution of temperature $T_e$ is (Lesch, Appl, and Camenzand 1989)

$$E = 0.6 \text{ GeV} \left(\frac{B}{\text{mG}}\right)^{-1/2} \left(\frac{T_e}{10^8}\right)^{-1/4} \left(\frac{n_e}{10^{-4}}\right)^{1/2}. \qquad (33)$$

For supernovae with $n_e = 10^5$ cm$^{-3}$, $T_e = 10^4$ K, and $B = 0.1$ G, $E = 200$ eV which is too small. On the other hand one can get higher energies using the Galeev (1985) estimate of the maximum energy of accelerated electrons:

$$E_{max} = m_e c^2 \left(\frac{m_i v_b^3}{m_e c^3}\right)^2 \frac{\omega_{pe}}{\omega_{ce}}, \qquad (34)$$

where $v_b$ is the reflected ion velocity approximately equal to the shock velocity, and $\omega_{pe}$ and $\omega_{ce}$ are the electron plasma and cyclotron frequencies. With $v_b = 30{,}000$ km s$^{-1}$ and the above parameters for supernovae $E_{max}$ = 17 MeV. This is still about an order of magnitude less than is needed to generate radio emission. Moreover, the relaxation parameter z describing the excitation of waves by the reflected ions (Galeev 1984) is so small in this case that the energy balance between ions and waves is not established and Eq. (33) does not hold. We conclude that acceleration by lower-hybrid waves is not able to supply electrons of sufficient energy for radio supernovae.

## 4.5. Experimental Evidence for Shock Acceleration

The evidence for shock related energetic electrons comes from measurements made in planetary bow shocks, interplanetary travelling shocks, and coronal shocks associated with type II solar bursts. For the latter a so called herringbone structure may serve as a manifestation of energetic electron acceleration. The herringbone structure is a series of short radio bursts similar to type III bursts which emanate from the main type II burst generated by the coronal shock (Cairns and Robinson 1987). From the frequency drift of the herringbone bursts an electron velocity of 0.05 c–0.5 c was deduced. In interplanetary travelling shock electrons accelerated up to 2 MeV were observed by spacecraft particle counters (Sarris and Krimigis 1985). Finally, near the Earth's bow shock energetic electrons with energies up to hundreds of electron-volts were observed in great detail (e.g., Anderson 1974). So there is ample experimental evidence that shocks are able to accelerate energetic electrons. Numerical simulations also show the possibility of accelerating electrons to relativistic energies by shock waves (Ohsawa and Saki 1988). To know whether supernova shocks can produce electrons with higher energies it is necessary to extrapolate from existing experimental data or to use a good theoretical model. As compared to interplanetary or planetary shocks the supernova shocks have much higher velocities (up to 0.1c) and Mach numbers. From the observed evidence that the herringbone structure in type II solar bursts is present only in the strongest bursts (Cane and White 1987) and that the intensity of type II bursts is proportional to the shock velocity as a power of 2.4–3.1 (Lengyelle-Frey and Stone 1989), one can expect that supernova shocks may produce more energetic electrons of higher energy. All available experimental evidence indicates that quasi-perpendicular shocks are more efficient than quasi-parallel shocks in acceleration of electrons. For interplanetary shocks this was shown by Pyle *et al.* (1984): to accelerate electrons above 2 keV a minimum shock velocity of 20 km s$^{-1}$ is sufficient for quasi-perpendicular shocks while for quasi-parallel shocks the velocity should exceed 160 km s$^{-1}$. The quantity which is more relevant is $v_{sB} = v_s/\cos\theta_{Bn}$, where $\theta_{Bn}$ is the angle between the shock normal and the magnetic field, and $v_s$ is the shock velocity. According to Pyle *et al.* (1984), energetic particles were observed for shocks with $v_{sB} \geq 400$ km s$^{-1}$, that is for large $v_s$ and small $\cos\theta_{Bn}$ or quasi-perpendicular shocks. In the Earth's bow shock energetic electrons are commonly found near perpendicular portions of the shock where the interplanetary magnetic field lines are tangent to the bow shock, but are rarely present near quasi-parallel portions (Gosling et al. 1989). So all the evidence seems to show that quasi-perpendicular acceleration is responsible for the acceleration of electrons. Quasi-parallel or diffusive shock acceleration evidently is not important. As was discussed in the previous section two acceleration mechanisms were proposed for quasi-perpendicular shocks: shock drift and lower-hybrid waves. The latter requires that a high lower-hybrid wave turbulence be present upstream and at the shock. This is not observed. The energy density of accelerated electrons is 5-6 orders of magnitude higher than the energy density of waves. Neugubauer, Russell, and Olson (1971) have measured the energy density of electrons in the energy range 100–200 eV as 1 to 5 keV cm$^{-3}$ or from $1.6 \times 10^{-9}$ to $8 \times 10^{-9}$ erg cm$^{-3}$ while the intensity of the waves near a frequency of 10 Hz where the lower-hybrid frequencies are located can be estimated as about $200 \times 10^{-5}$ $\gamma$ Hz$^{-1/2}$ or $4 \times 10^{-15}$ erg cm$^{-3}$. Similar values for the electromagnetic turbulence in the frequency interval from 20 Hz to 4 kHz of $2.4 \times 10^{-15}$ erg cm$^{-3}$ were given by Rodriguez and Gurnett (1976). Thus the waves are only a minor energy component at the shock and cannot be the source of energetic electrons. Besides, as was shown in the previous section the lower-hybrid wave mechanism is not able to supply electrons of sufficiently high energy. So one is left with shock drift as a promising acceleration mechanism. Measurements of suprathermal electrons at the Earth's bow shock by Gosling, Tomsen, and Bame (1989) may provide some details of the acceleration process. It was found that at energies below 20 keV the suprathermal electrons are most intense immediately downstream from the shock and decrease in intensity deeper into the magnetosheath. The energetic electron spectrum extends smoothly out of the shocked solar wind spectrum as a power law in energy with an exponent in the range from –3 to –4, their angular distribution is generally isotropic immediately downstream of the shock ramp, but with increasing penetration into the magnetosheath an anisotropy perpendicular to the magnetic field develops. Upstream the suprathermal electrons are observed escaping along the magnetic field. These observations are interpreted as an acceleration of suprathermal electrons out of the solar wind thermal population as the solar wind convects across the shock. In this case the number of accelerated electrons should be proportional to the upstream density of the wind. In the case of supernovae the upstream density $n_{eo} \sim 10^5$–$10^6$ cm$^{-3}$ (see Table 13.3), or 4 to 5 orders of magnitude higher than the solar wind density. The energy of electrons as estimated in section 4.3 can be as high as $10^{15}$–$10^{16}$ eV or a factor of $10^{10}$–$10^{11}$ higher than in the Earth's bow shock if the shock drift mechanism is operating. So the energy density of relativistic electrons in supernova shocks may be a factor of $10^{14}$–$10^{16}$ higher than in the Earth's bow shock or $10^{17}$–$10^{19}$ eV cm$^{-3}$ = $1.6 \times 10^5$–$1.6 \times 10^7$ erg cm$^{-3}$. This is unrealistic since the energy density can not exceed the kinetic energy density of the shock which is only 0.1–1 erg cm$^{-3}$ and must be close to the magnetic field energy density $3 \times 10^{-5}$–$3 \times 10^{-2}$ erg cm$^{-3}$ as in the Earth's bow shock. The limiting energy density of energetic electrons seems to be close to equipartition with the magnetic field energy density.

## 5. SUMMARY

The radio emission which accompanies optical emission just after supernova explosions may be a common phenomenon. This emission is produced by relativistic electrons accelerated by the expanding envelope of supernovae. The exact mechanism of the acceleration at present can not be specified although there are evidences that the interaction of the supernova envelope with the circumstellar matter plays an important role. The second possibility, acceleration by a pulsar, seems to be unrealistic. There are observational indications of the existence of the circumstellar matter around supernova stars. Its origin could be stellar wind emitted by the star before explosion. This matter may be responsible for the low-frequency absorption in the spectrum of radio emission although the synchrotron self-absorption could as well give rise to the low-frequency cut-off. A shock wave travelling ahead of the envelope is the most likely place where the acceleration occurs. The Earth's bow shock resulting from the interaction of the solar wind with the magnetosphere is known to accelerate electrons to high energies. By analogy one can speculate that the same process on a much larger scale is taking place at supernova shocks. Three possible acceleration mechanisms were considered here. The first-order Fermi, or diffusive shock acceleration was applied to the supernova acceleration problem following Bell's (1979b) prescription. A set of reasonable parameters was deduced for known radio emitting supernovae, but the magnetic field proved to be rather high. The mass loss rates of stellar winds were also found to be too high. Also, diffusive acceleration is evidently not responsible for acceleration at the Earth's bow shock. Thus, two mechanisms which can operate at quasi-perpendicular shocks were also considered. The acceleration by lower-hybrid waves generated at the shock was found to be ineffective for supernovae, but shock drift acceleration is a more profitable mechanism. At the Earth's bow shock it can produce energetic electrons with an energy density close to the magnetic field energy density. By extrapolating to the supernova shock parameters it was possible to get relativistic electrons in quantities sufficient for equipartition with the magnetic field. It seems likely that the radio emission from supernovae can be produced by electrons accelerated by the shock drift mechanism. Two types of radio supernovae can be identified: one in which the expansion velocity is the same both for optical and radio envelopes, and the second where the radio envelope is expanding much faster than the optical envelope. The latter is a type Ib supernova, while the former is a type II supernova. No radio emission was found from type Ia supernovae.

## REFERENCES

Anderson, K. A. (1974) *J. Geophysics* **40**, 701.
Bartel, N. (1990) In *Supernovae*, S. E. Woosley (ed.), Springer Verlag, New York, in press.

Bartel, N., Rogers, A. E. E., Shapiro, I. I., et al. (1985) Nature **318**, 25.
Bell, A. R. (1978a) Mon. Not. Roy. Astron. Soc. **182**, 147.
Bell, A. R. (1978b) Mon. Not. Roy. Astron. Soc. **182**, 443.
Benz, A. O., and Spicer, D. S. (1990) Astron. Ap. **228**, L13.
Berezhko, E. G., and Krymsky, G. F. (1988) Usp. Fiz. Nauk **154**, 49.
Bisnovatyii-Kogan, G. S., Illarionov, A. F., and Slysh, V. I. (1990) Proc. Joint Varenna-Abastumani-ESA-Nagoya-Potsdam Workshop on Plasma Astrophysics (ESA SP-311), p. 289.
Cairns, I. H., and Robinson, R. D. (1987) Solar Phys. **111**, 365.
Cane, V., and White, S. M (1989) Solar Phys. **120**, 137.
Chevalier, R. A. (1982) Astrophys. J. **259**, 302.
Chevalier, R. A., and Fransson, C. (1987) Nature **328**, 44.
Galeev, A. A. (1984) Zh. Eksp. Teor. Fiz. **86**, 1655.
Galeev, A. A. (1985) in Advances in Space Plasma Physics, B. Buti (ed.) World Scientific, Singapore.
Ginzburg, V. L., Sazonov, V. N., and Syrovatsky, S. I. (1968) Usp. Fiz. Nauk **24**, 63.
Gosling, J. T., Thomsen, M. F., and Bame, S. J. (1989) J. Geophys. Res. **94**, 10011.
Gottesman, S. T., Broderick, J. J., Brown, R. L. and Balick, B. (1972) Astrophys. J. **174**, 383.
Gull, S. F. (1973) MNRAS **161**, 47.
Hanuschik, R. W., and Dachs, J. (1987) Astron. Ap. **182**, L29.
Jauncey, D. L., Kemball, A., Bartel, N., et al. (1988) Nature **334**, 412.
Kirshner, R. P., and Blair, W. P. (1980) Astrophys. J. **236**, 135.
Krymsky, G. F. (1977) Sov. Phy. Doklady **22**, 327.
Krymsky, G. F., and Petukhov, S. I. (1980) Pis'ma Astron. Zh. **6**, 227.
Lampe, M. and Papadopoulos, K. (1977) Astrophys. J. **212**, 886.
Lengyell-Frey, D., and Stone, R. G. (1989) J. Geophys. Res. **94**, 159.
Lesch, H., Appl, S. and Camenzand, M. (1989) Astron. Ap. **225**, 341.
Neugebauer, M., Russel, C. T., and Olson, J. V. (1971) J. Geophys. Res. **76**, 4366.
Ohsawa, Y., and Sakai, J. (1988) Astrophys. J. **332**, 439.
Pacini, F., and Salvati, M. (1973) Astrophys. J. **186**, 249.
Panagia, N., Sramek, R. A., and Weiler, K. W. (1986) Astrophys. J. **300**, L55.
Parker, E. N. (1961) Astrophys. J. **133**, 1014.
Pesses, M. S., Decker, R. B., and Armstrong, T. P. (1982) Space Sci. Review **32**, 185.
Pyle, K. R., Simpson, J. A., Barnes, A., and Mihalov, J. D. (1984) Astrophys. J .**282**, L107.
Rodriguez, P., and Gurnett, D. (1976) J. Geophys. Res. **81**, 2871.
Rupen, M. P., van Gorkom, J. H., Knapp, G. R., et al. (1987) Astron. J. **94**, 61.
Sarris, E. T., and Krimigis, S. M. (1985) Astrophys. J. **298**, 676.
Scott, M. A., and Readhead, A. C. S. (1977) MNRAS **180**, 539.
Shklovsky, J. S. (1960) Astron. Zh. **37**, 256.

Shklovsky, J. S. (1962) *Astron. Zh.* **39**, 209.
Shklovsky, J. S. (1980) *Astron. Zh.* **57**, 673.
Shklovsky, J. S. (1981) *Sov. Astron. Lett.* **7**, 263.
Shklovsky, J. S. (1985) *Pis'ma Astron. Zh.* **11**, 261.
Slysh, V. I. (1963) *Nature* **199**, 682.
Slysh, V. I. (1990) *Pis'ma Astron. Zh.* **16**, 790.
Storey, M. C. and Manchester, R. N. (1987) *Nature* **329**, 421.
Turtle, A. J., Campbell-Wilson, D., Bunton, J. D., *et al.* (1987) *Nature* **327**, 38.
Weiler, K. W., Sramek, R. A., Panagia, N. *et al.* (1986) *Astrophys. J.* **301**, 790.
Weiler, K. W., and Sramek, R. A. (1988) *Ann. Rev. Astron. Ap.* **26**, 295.
Weiler, K. W., Panagia, N.,Sramek, R. A. (1990) *Astrophys. J.* **364**, 611.

# 14

# Stellar Winds and Supernovae in the Interstellar Medium

## T. A. Lozinskaya
*Shternberg Astronomical Institute, Moscow*

## INTRODUCTION: HISTORY AND PERSPECTIVES

The problem of the interaction of supernovae and stellar winds with the interstellar medium was the center of the crossing of the scientific interest of J. S. Shklovsky and S. B. Pikel'ner as no other problem.

J. S. Shklovsky was among those who laid the basis of the science of supernovae. He continued this science with strength and interest to his last days. One of his last works in 1984 "Why Supernovae of Type II Do Not Occur in Irregular Galaxies" turned out to be very relevant when SN 1987A exploded in the Large Magellanic Cloud. This peculiar SN II actually turned out to be unlike explosions of this type, in principle as J. S. Shklovsky indicated: the star before the explosion was not a typical red supergiant. We can only regret that J. S.Shklovsky did not live to see this supernova. One can only imagine how enlivened J. S.Shklovsky would have been and how many multiple and brilliant ideas he would have generated.

The book "Supernova Stars and the Problems Connected with Them" by J. S. Shklovsky was published twice with an intervening time of ten years and twice presented the grand total of modern ideas on supernovae, stated new questions, and opened new perspectives.

The interaction of the stellar wind with the interstellar gas was first considered by S. B. Pikel'ner in 1968. He proposed a model of a wind-blown bubble which turned out to be the basis of all further theoretical studies and it actually has not undergone significant changes to this day.

We would also like to mention three studies of S. B. Pikel'ner in the region of supernovae, each of which in its time was the answer to a riddle which the supernova phenomena constantly puts before us.

In 1921 the expansion of Crab Nebula filaments was detected and in 1938 Baade showed that they expand with acceleration which was completely nonunderstandable in the framework of the classical emission of the then-known classical nebulae. S. B.

Pikel'ner gave the explanation. Polarization observations of the Crab were carried out in 1954–1955 with his participation and the distribution of the polarization vector in the nebula was first obtained. On the basis of these data S. B. Pikel'ner (1961) showed that the system of filaments confines a cloud of relativistic plasma due to the tension of magnetic field lines, and the pressure of relativistic particles and the magnetic field is the cause of the acceleration of filaments. Having determined this pressure with synchrotron radiation, he was the first to estimate the mass of the filaments, i.e., the mass of material ejected in the supernova explosion. Succeeding estimates of the mass according to the optical radiation of filaments did not change this value significantly.

We give another pioneering observational study: having obtained the spectrum of the Cygnus Loop supernova remnant, S. B. Pikel'ner proved that this spectrum cannot be explained by the standard radiation of a plasma at any temperature. In 1954 he made the first calculation of the emission spectrum of the cooling gas behind a shock wave and showed that for an adequate explanation of the radiation of filaments interaction of shock waves has to be included in consideration, caused, for example, by deflection due to inhomogeneities of the interstellar gas.

We note that the mechanism of filament radiation of the Cygnus Loop has been discussed intensely up to the present. At the present time there is a controversy over what are filaments: ropes or layers seen edge-on. Recently unique photographs of the second external layer of weak filaments of the Cygnus Loop were obtained (Hester 1987) and the argument heated up to a new level.

Finally in 1975 when the first X-ray spectra of supernova remnants appeared, Pikel'ner and Bychkov (1975) explained the discrepancy between the velocity of the shock wave necessary for heating the plasma to a temperature corresponding to the observed spectrum and the velocity of filament expansion by association of X-ray emission with the gas behind the shock wave front in the rarefied interstellar medium and the optical emission of the filaments with the rapid radiative cooling of gas behind the shock wave in dense clouds.

We should stop here. It is impossible to emphasize how much of what is done in this interesting region of astrophysics is based on the fundamental pioneering studies of J. S.Shklovsky and S. B. Pikel'ner.

To predict the development of studies in the regions of supernovae, stellar winds, and their interaction with the interstellar medium is simultaneously difficult and easy.

It is difficult because the program of studies connected with this problem continues to expand and promises us not only answers to long-standing questions, but also new riddles. It isn't necessary to go far for examples.

In the last several years the classification of supernovae has changed which reflects a new degree in our understanding of the final stages of stellar evolution and the crowning supernova explosion. A series of new questions immediately arises. One of the most trivial of these is: can one compare the three types of supernova explosions (SN Ia,

SN Ib, and SN II) with the three presently known types of young supernova remnants? There is still no exhaustive answer to this question.

The flight of the special IR observatory IRAS "discovered" a new region, quite promising for studies of supernova remnants. This was the third revolutionary jump in the study of their nature. (The first was connected with the beginning of the era of radio observations and the second with the opening of the X-ray region, and namely with the flight of the Einstein Observatory.) In this connection the most pressing problem seems to be the further development of the theory of radiation of dust. The complexity is that the same shock wave which provides the heating of dust in supernova remnants by collisions with the hot plasma behind the front changes the chemical composition of dust particles and their distribution in sizes.

Finally, the 1987A supernova explosion shed new light on the whole science of supernovae. Our ideas about stellar explosions have already changed and will be changed in the coming years and decades on how young supernova remnants develop. Moreover, this bright and close explosion opened new perspectives for studying the interstellar and intergalactic gas. It is sufficient to recall that observations of the SN 1987A light echo gave the unprecedented possibility of three-dimensional tomography of the interstellar medium.

One of the predictions for the closest decade seems quite obvious. The most promising new direction of study is the "combination" of two problems: supernovae and stellar winds. It is now completely obvious that an adequate analysis of supernovae and supernova remnants of any age is impossible without taking into account the action of radiation and the stellar wind of the stellar precursor on the surrounding interstellar medium.

We know quite well (hopefully we know) how supernova explosions and stellar winds individually influence the interstellar medium. In both cases we know the energetics of the phenomenon, the size of the perturbed region, the lifetime of the perturbation, and the frequencies of star formation and supernova explosions. Knowing these things it is easy to be convinced that the physical state of the interstellar medium in galaxies similar to ours is regulated by supernovae, and near OB associations by stellar winds and supernovae. This means that the nature of any supernova remnant is determined not only by its individual explosion and the action of its precursor on the ambient gas, but by the whole complex of the interactions of the stellar and gaseous population supergiant of the parent molecular cloud.

An analysis of the nature of supernova remnants taking into account the combined action of the progenitor star and nearby OB stars on the surrounding gas is an incredibly difficult problem. At the present time only separate blocks of this problem have been developed. We consider briefly the physical state of the interstellar gas in the neighborhood of a massive star, a supernova precursor: single (§1) and belonging to a young cluster or OB association (§2) and touch some aspects of the nature and evolution of supernova remnants expanding in this "previously prepared" medium (§3). To combine the individual knowledge from these blocks is a project for the future because presently we have only formulated the basic problems and noted paths for their solution.

## 1. INTERSTELLAR MEDIUM IN THE VICINITY OF THE PRE-SUPERNOVA

The evolution of massive stars is accompanied by the loss of material in the form of a stellar wind; the characteristic rate of outflow is about $\dot{M} = 10^{-6}-10^{-7}$ $M_\odot$/year for 05 V–09 stars and reaches $\dot{M} = 10^{-5}-10^{-4}$ $M_\odot$/year in WR and $O_f$ stars; the wind terminal velocity is $V_w = 1000-3000$ km/s. A classical scheme is shown in Fig. 14.1 of a wind-blown bubble in a homogeneous interstellar medium. One can distinguish three stages in the evolution of this multilayered envelope (Pikel'ner 1968; Avedisova 1971; Weaver et al. 1977; Dyson 1981): 1) a short adiabatic stage (radiation losses are small in comparison with the influx of energy of the wind throughout the bubble), 2) continuous "energy-driven" stage (the swept-up gas in layer "c" radiatively cools and collapses into a dense thin shell, but the hot wind in layer "b" still expands adiabatically), and 3) "momentum-driven" stage (radiation cooling and collapse of the hot wind in layer "b").

The transition into the energy-driven stage occurs rapidly during a characteristic time of several thousand years. The collapse of the hot isobaric wind in layer "b," generally speaking, may not occur during the lifetime of the star if the influx of gas evaporated from layer "c" significantly regulates the rate of cooling of the hot plasma in layer "b." Thus, a geometrically thin dense comparatively cold swept shell should be observed for

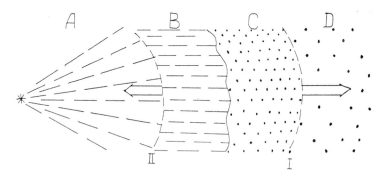

Figure 14.1. The scheme of a bubble blown out by a wind in a homogeneous interstellar medium: "a") freely expanding wind; "b") a wind heated by a reverse shock wave II; "c") interstellar gas swept-up and heated by a shock wave I.

## 14. STELLAR WINDS AND SUPERNOVAE

a large part of the life of the O star around it (ionized completely or surrounded by a layer of neutral gas) filled by a hot (T(b) = $10^6$–$10^7$ K) plasma of low density (n(b) = ~$10^{-2}$ cm$^{-3}$). The expansion of the shell at this stage is described by the equations (Weaver et al. 1977):

$$R = 28 \left(\frac{L_{36}}{n_o}\right)^{0.2} t_6^{0.6} \text{ [pc]}, \quad v = 16 \left(\frac{L_{36}}{n_o}\right)^{0.2} t_6^{-0.4} \text{ [km/s]}$$

where $L_{36}$ is the power of the wind in units of $10^{36}$ erg/s, $n_o$ is the number density of the unperturbed gas, and $t_6 \equiv t/10^6$ yr.

A geometrically thick ring-like HII region may be formed around the star of an early spectral class even without a wind as the result of the reactive effect of photoionization evaporation of small-scale clouds (McKee et al. 1984). In this case a cavity is formed around the star filled predominantly by the evaporated gas of clouds at the same time that dense clumps are concentrated on the periphery. During the lifetime of O4 V–O9 V stars the radius of the cavity reaches about $50<n>^{-0.3}$ pc and the rate of expansion of the peripheral layer of clouds is about $7(S_{48}/<n>)^{0.25}$ km/s, where $<n>$ is the average density of gas of clouds spread over the whole volume and $S_{48}$ is the flux of ionizing radiation in units of $10^{48}$ photons/s (McKee et al. 1984).

If the star passes through the WR stage or the blue supergiant one, then in this cavity a strong stellar wind is turned on and depending on the relation between the mechanical energy of the wind and the ionizing radiation, the wind-blown shell remains inside the evaporated gas, reaches the peripheral cloud layer, or partially encloses it.

Undoubtedly, this idealized scheme is overly simplified. In fact the spherically symmetric picture is distorted by the large-scale inhomogeneity of the interstellar medium and the motion of the star; the shell swept up by the wind is broken up due to thermal instability and small-scale fluctuations of density; the rate of mass loss, the wind velocity, and the flux of ionizing radiation change in the process of stellar evolution.

An ejection of a shell in the process of stellar evolution may also be added to this quasi-stationary structure of the near-stellar medium regulated by ionizing radiation and stellar wind.

These are the purely theoretical paths of influence of a massive star, a supernova precursor, on the structure and kinematics of the surrounding interstellar gas. Observations indicate that all the possibilities enumerated actually occur. The observations of nebulae connected with $O_f$ and WR stars are the most interesting in this connection because, on the one hand, they are characterized by the strongest stellar winds, and on the other hand, these massive rapidly evolving stars are possible supernova precursors. In complete agreement with the mechanisms mentioned above of the action of the star on the interstellar medium we delineate three types of ring nebulae connected with WR and $O_f$ stars: wind-blown bubbles, stellar ejecta, and ring HII regions. Their basic characteristics can be summarized as follows (Chu 1981; Chu et al. 1983; Lozinskaya 1982, 1991).

1. Wind-blown bubble: finely filamented geometrically thin shells of radius $R = 1–5$ pc, $\Delta R/R = 0.1–0.2$, expansion velocity 30–100 km/s, and mass 10–100 $M_\odot$, examples of which are NGC 2359, NGC 7635, and the shell that is second from the center in the system NGC 6164–5.
2. Stellar ejecta: clumpy shells of radius 1–3 pc, expansion velocity 30–300 km/s, and mass 1–5 $M_\odot$, examples of which are the central bright "S" form nebula in the system NGC 6164–5, M1-67, and RCW 58.
3. Diffuse massive geometrically thick evolved HII regions with a ring-like morphology: sizes of order 10–100 pc, $\Delta R/R = 0.3–0.5$, expansion velocity less than 10–20 km/s, and mass $10^2–10^3$ $M_\odot$, examples of which are the nebulae Sh 119 and Sh 132.

Without exception all ring-like nebulae are sources of thermal radio emission. In the series of objects related to the first two types anomalous chemical compositions are detected (increased content of heavy elements) supporting a stellar origin of the shell material (Kwitter 1984; Chu 1990 and references therein). The shells swept up by the wind and ejected by the star are often characterized by an elongated form which, possibly, evidences an asymmetry (bipolarity) of material loss.

Today such a division of ring nebulae into three types is somewhat outdated because observations evidence a complex action of ionizing radiation, the stellar wind, and the ejection of a shell. The majority of wind-blown bubbles and ejecta are localized inside diffuse or ring-like HII regions, inside caverns in HI or CO distributions, and IRAS observations detect extended dust shells connected with many WR stars (see Lozinskaya (1991); Cappa de Nicolau and Niemela (1984); Cappa de Nicolau et al. (1986, 1987); Van Buren and McCray (1988) and references therein).

The most striking example is the system of four shells of NGC 6164–5 around the $O_f$ star HD 148937; the bright clumpy ejecta surrounded by a thin filamentary shell blown out by the wind are localized in the center of an extended diffuse ring-like HII region with an external dust shell (Bruhweiller et al. 1981; Dufour, Parker, and Henize 1988; Leitherer and Chavarria 1987). This structure shows that the ejection of the shell and the stellar wind originate from the same star and both processes did not differ strongly in time.

Many ring nebula, previously considered as isolated, are in fact localized inside extended shells and supershells connected with OB associations. In particular, the most studied nebula NGC 6888, the prototype of a wind-blown bubble, is found in a cavity inside an extended shell connected with the association Cyg OB1 (see §2). This localization provides a basis for supposing the NGC 6888 is a stellar ejecta swept up by the stellar wind because the density inside the cavity is small for the formation of the bright nebula NGC 6888 from the interstellar gas. A stellar origin is also evidenced by the anomalous chemical composition of bright filaments. At the same time the X-ray emission of NGC 6888 (Bochkarev 1987) shows that the shell ejected by the star and swept up by wind according to the classical model (Fig. 14.1) is filled by the hot gas of the shocked stellar wind.

Thus the observations of ring nebulae around WR and $O_f$ stars in a variety of forms clearly demonstrate that the interstellar gas in the vicinity of massive stars of early spectral classes is characterized by a multi-layer shell structure. Do the observed shells exist until the supernova explosion? An answer to this question is not trivial. On the one hand, according to theory, the strong stellar wind in a homogeneous interstellar medium should sweep up the shells whose lifetime is comparable with the life of the wind. On the other hand, the characteristic kinematic age of shells connected with WR and $O_f$ stars is tens of thousands of years which is significantly less than the lifetime of intensive outflow. The fact that in the distance-completed sample of stars and nebulae about 80% of $O_f$ and WR stars are connected with HII regions and less than 40% of these are characterized by a ring structure (Lozinskaya 1982, 1983) also evidences the short lifetime of shells swept up by the wind.

In this connection it is very interesting to study the so-called oxygen sequence of Wolf-Rayet stars. The WO stars are the most advanced massive stars, probably only several thousand years before their SN explosion. Indeed, our observations of WR 102 (spectral type WO 1) indicate an absence of He in the star's envelope and one can suppose the absence of it in the interior as well; the star's position on the HR diagram also supports our interpretation as a CO core of a massive star M(initial) $\cong$ 60 $M_\odot$ (Dopita et al. 1990; Dopita and Lozinskaya 1990).

The WO stage is very short (the lifetime of the CO core of such a massive star consists of only about 0.1–1% of the He core burning time (Maeder and Maynet 1989)) and is characterized by an extremely fast "superwind" ($V_w \cong$ 4500 to 7400 km/s (Barlow and Hummer 1982; Dopita et al. 1990; Torres et al. 1986)) and the ambient gas is influenced by the previous "normal" outflow of the progenitor WR star.

We have shown that the four known WO stars in the galaxy and in the Magellanic Clouds appear to be associated with optical and/or IR shell-like structures, although the short-lived WO superwind does not dominate in the shell's formation (Lozinskaya 1990). The WO2 oxygen star WR 142 in the Galaxy and the WO+O4 star Sand 1 in the Small Magellanic Cloud both belong to young open clusters and they are located within large shells most probably created by the clusters and OB associations.

The G2.4+1.4 nebula around WR 102 (see Fig. 14.2) is the only object of the four which was shown to be a classical wind-blown bubble located on the edge of a dense

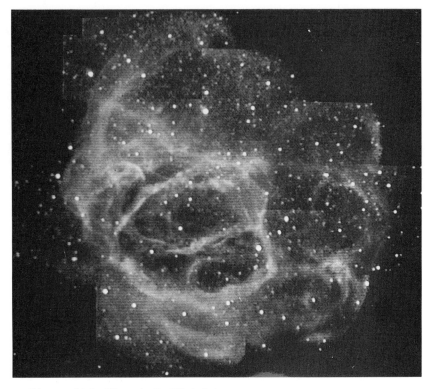

Figure 14.2. The shell G2.4+1.4 produced by the wind around the most evolved star WO WR 102 (R = 5–6 pc, $V_{exp}$ = 42 km/s) (Dopita et al. 1990; Dopita and Lozinskaya 1990).

cloud (Dopita et al. 1990; Dopita and Lozinskaya 1990). The shell's dynamical age $t \simeq 10^5$ yr is much longer than the WO superwind duration which indicates that the progenitor WR star created a ring nebula before the WO stage. Since the wind-blown bubble G2.4+1.4 is old and slowly evolving, one can expect the shell to survive until the supernova explosion.

Thus, observations of nebulae connected with WR and $O_f$ stars show how the action of massive SN Ib and SN II precursors may change the structure of the interstellar medium to distances of several pc to several tens of pc.

It is important to emphasize that the multi-shell quasi-symmetric structure of the interstellar medium is also formed around low-mass SN Ia precursors which are former central stars of planetary nebulae. As the result of multiple ejections or under the action of the wind of a central star the two-shell structure of planetary nebulae is formed.

According to the data of Chu et al. (1987) and Jewitt et al. (1986) more than half of planetary nebulae are characterized either by a two-shell structure, or a weak external halo, or a weak external shell around a bright central nebula.

## 2. SUPERNOVA PRECURSORS IN CLUSTERS AND OB ASSOCIATIONS

At least half of massive stars of early spectral classes belong to clusters and associations. This means that the majority of supernovae, in any case SN II and SN Ib, explode in OB associations.

The average frequency of supernova explosions is about 1 SN per $(1-5)\times10^5$ yr in one association (Cowie et al. 1979), i.e., during the characteristic lifetime of the association $t = 5\times10^7$ yr the total kinetic energy given by supernovae is about $10^{53}$ erg, taking an average value of the explosion kinetic energy $E = 5\times10^{50}$ erg (Lozinskaya 1991). The total power of the stellar wind is about $10^{38}$ erg/s or $\sim3\times10^{52}$ erg over the whole lifetime of the average OB association and an order of magnitude larger for rich associations of the type of the giant stellar complexes in the Cygnus and Carina (Abbot 1982; Van Buren 1985; Van der Hucht et al. 1987). Under the action of radiation, stellar wind, and supernova explosion shells and supershells around OB associations are formed with sizes from ~50 pc to ~1 kpc, expanding with velocities from several km/s to several tens of km/s.

Because a large part of the mass of the giant mother molecular cloud is conserved at the time of cluster formation, the formation of stars of the next generation continues in it, triggered by shock waves induced by the expanding shells. The stars of the second generation should be more massive than central stars of the first generation, they are localized in the peripheral parts of the super shell, and are comparatively short lived ($t \lesssim 10^7$ yr). This active phase of evolution of the molecular complex continues for about $10^7-10^8$ yr which exceeds the time of stellar condensation from the pre-stellar cloud, and the time of total dissipation of supernova remnants and cavities swept up by the winds of individual stars. Therefore, one observes genetically connected groups of clusters and associations, giant molecular clouds and HII regions, giant expanding shells formed by stellar winds and supernovae, new regions of star formation on their boundary, and small-scale shells that are the remnants of supernovae and cavities blown out by the winds of separate stars.

The observation of irregular galaxies of the Large Magellanic Cloud type where the development of supershells is facilitated by low metallicity, homogeneity of the interstellar medium, and the absence of spiral density waves actually supports the existence of a secondary cascade of star formation induced by the gravitational instability of supershells (Dopita 1987; Feitzinger 1987; Smith et al. 1987; McCray 1988) and the references in these works.

Several giant ring formations are observed in the Large Magellanic Cloud containing

OB associations, young clusters, HII regions, and supernova remnants which bound large caverns in the distribution of neutral hydrogen. The stellar aggregate Shapley III with a size of more than 600 pc and the ring chain from two dozen young OB associations around the Loop IV is the most impressive (Dopita et al. 1985). The clear ring regions of induced star formation were revealed by UV observations of the Large Magellanic Cloud (Smith et al. 1987). In the Galaxy the formation of similar gigantic formations is impeded by spiral density waves as well as by break-out of super shells due to instability in polar regions when their size becomes comparable with the thickness of the gas disk. Shell complexes of less grandiose scale in which the formation of massive stars continues are observed in our stellar system around rich OB associations. We recall as an example the super shell in Orion-Eridan (Reynolds and Ogden 1979; Goudis 1982) and the hierarcical system enclosing shells of different size in the region of Cyg OB1 (Lozinskaya and Sitnik 1988).

The shell complex in the direction Cyg OB1 is of special interest because it includes tens of WR and $O_f$ stars including the oxygen sequence star WR 142. The general scheme of the system of shells in Cygnus is presented in Fig. 14.3: the extended general

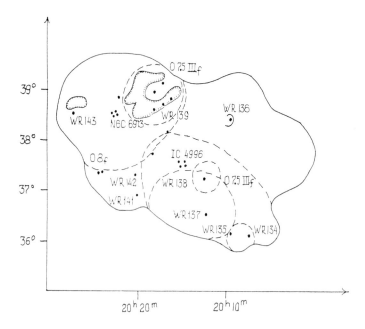

Figure 14.3. The system of shells in Cygnus: the common shell around Cyg OB1 and two of its components around the clusters NGC 6913 and IC 4996, and small-scale HII shells and cavities in the distribution HI and CO around WR and $O_f$ stars (Lozinskaya and Sitnik 1988; Lozinskaya and Repin 1990).

shell around Cyg OB1, two of its components around clusters NGC 6913 and IC 4996, small-scale HII shells and caverns in the distribution of HI and CO, and individual shells formed by the winds of WR and $O_f$ stars. The northwestern part of the general shell is clearly delineated by the optical emission of ionized gas and the eastern part by the IR radiation of heated dust in 60 and 100 micron bands of IRAS and radio continuum emission (Wendker 1984; Lozinskaya and Sitnik 1988; Lozinskaya and Repin 1990). Here two regions of active star formation ON2 and Sh 106 are located. Although the existence itself of the hierarchical cloud system in Cygnus as a single physically connected formation in space still requires proof, the region is a good model of a "prepared" interstellar medium around massive pre-supernovae which belong to clusters and associations. The majority of massive WR stars are concentrated in dense regions of the general shell and sweep out individual cavities. The most advanced oxygen sequence star WR 142, a probable precursor of the next supernova in this region, is situated inside an extended shell I at the boundary of a dense compact cloud.

## 3. SUPERNOVA REMNANTS

Modern all-wavelength methods of studying supernova remnants reveal a whole complex of phenomena which accompany the explosion of stars, enclose the region of the interstellar medium for tens of parsecs, and are observed for $10^4$–$10^5$ yr. We emphasize that this complex is quite complicated even if one does not take into account the action of the precursor on the ambient gas.

The primary scheme of young supernova remnants is shown in Fig. 14.4. In the framework of this scheme emission of the following components is observed in young supernova remnants.

If a pulsar is formed during the explosion, then the central cloud of relativistic plasma with the magnetic field generated by the pulsar is a source of synchrotron radiation in the radio, X-ray, IR, and optical ranges (the so-called plerion or Crab-like SNR). The pulsar may or may not be observed depending on orientation and in some plerions the compact X-ray source is observed connected with radiation of the hot neutron star surface.

The hot rarified plasma of circumstaller gas swept up and heated by the shock wave I and the gas ejected by the supernova and heated by the reverse shock wave II correspond to the thermal X-ray emission. Dense clumps of matter ejected by the supernova and dense condensations of the circumstellar gas (possibly thrown out by the precursor) are heated by shock waves travelling in the ejecta and in the circumstellar gas, respectively, and radiate in the optical and X-ray ranges. A new component was recently identified: IR radiation of the dust heated by inelastic collisions with the hot gas of the remnant.

The synchrotron radiation in shells around plerions and in remnants without a central pulsar at early evolutionary stages is due to relativistic particles accelerated at the shock front and the magnetic field amplified by a random-walk in the convective layer at the

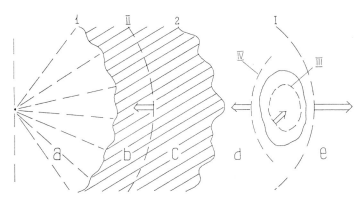

**Figure 14.4.** The conventional scheme of a young supernova remnant with a pulsar: a) relativistic plasma with the magnetic field injected by the pulsar; b) the shell ejected in the explosion; c) ejecta heated by the reverse shock wave II; d) circumstellar gas raked and heated by shock wave I; unperturbed circumstellar gas; 1 and 2) contact discontinuity between the pulsar plasma and ejecta, and between the ejecta and swept-up gas.

contact surface between the ejecta and swept-up gas. At late stages of evolution synchrotron radiation is most probably connected with amplification of the magnetic field frozen into the interstellar gas which is radiatively cooled and compressed in the dense layer behind the shock wave front, initially in dense clouds and later behind shock I, see Fig. 14.4.

These are the basic components of emission of young remnants in the quasi-homogeneous interstellar medium with small-scale density fluctuations. In the whole volume this scheme is not observed in a single one of the supernova remnants. It is complicated "by bricks" each of which has been studied in detail in one of these objects and is verified by a whole set of observations of the young remnants with a theoretical basis (Shklovsky 1976; Lozinskaya 1991; Blinnikov et al. 1988). We are not able to dwell on the fascinating process of composing a general scheme from observations of individual objects. We identify only the hopeful observational evidence that the expanding shell of the supernova does not interact with the standard interstellar medium, but with a medium on which the radiation and outflow of the precursor has acted.

The basic parameters of the remnants of the six historical supernovae exploding in our galaxy in the last thousand years are presented in Table 14.1.

SN 1006 and SN 1572 (Tycho) can be quite reliably identified with SN Ia explosions; it is probable that SN 1604 (Kepler) belongs to this type. They are characterized by a

Table 14.1. Historical Supernovae in the Galaxy

| Supernova | 1006 | 1572 | 1604 | 1054 | 1181 | 1670±3 |
|---|---|---|---|---|---|---|
| Remnant | PKS 1459–41 | SNR Tycho | SNR Kepler | Crab Nebula | 3C 58 | Cassiopeia A |
| SN type | SN Ia | SN Ia | SN Ia? | SN II | SN II | SN Ib? |
| r, kpc | 1.6 | 3 | 3.3 | 2 | 3–4.5 | 2.8 |
| D, pc | 14 | 6.6 | 2.6 | 3×4 | 11×7 | 3 |
| v = R/t, km/s | 7000 | 8000 | 3500 | 2000 | 5600 | — |
| $v_{meas}$, km/s | 6500 | 3600 | — | 2300 | 1000 | 6000 |
| $M_{ej}$, $M_\odot$ | 1–2 | 2 | 1–2 | ~1 | — | 2.4 |
| z, pc | 300 | 73 | 380 | 200 | 140 | 100 |
| Stellar remnant | — | — | — | pulsar | neutron star | — |
| $T_{ns}$, $10^6$ K | ≤0.7–0.8 | ≤1.2–1.5 | ≤1.2–1.4 | 2.0–2.5 | 2–2.4 | ≤1.5–1.7 |

The table lists: the year of the outburst, the most conventional name of the remnant, the type of supernova, the most probable estimate of the distance, the linear dimensions of the remnant, the expansion velocity: by the dimensions and the age, assuming free expansion of the envelope and by measurements of the proper motions or radial velocities of the filaments, the height above the galactic plane, and information on the compact stellar remnant. All the references are given in Lozinskaya (1991) and the neutron star's temperature is taken from the data of Nomoto and Tsurata (1983) in terms of the distance used here.

clearly expressed shell structure without the slightest signatures of radiation or a compact stellar remnant. In complete agreement with the scheme of Fig. 14.4 X-ray emission is detected in them of a diffuse SN-ejected shell containing dense clumps along with the emission of the swept-up circumstellar gas. The X-ray emission spectrum evidences an elevated heavy element content in the ejecta. Absorption lines of Fe II (at a velocity ±5000 km/s), Si II, IV (5200 km/s), S II (6000 km/s), and O I (6500 km/s) (Fesen et al. 1988a) were detected in the UV spectrum of the blue subdwarf on which the SN 1006 shell is projected which evidences the layered stratification in the exploding star and the election of separate dense clumps.

The mass of ejecta in a SN Ia (the data in Table 14.1 take into account the redistribution of density in the ejecta after the passage of the reverse shock wave (Hamilton et al. 1985) and the chemical composition of the ejecta) may agree with the explosion of a white dwarf.

The optical emission of the SN 1006 and Tycho remnants are represented by weak thin filaments with a purely hydrogen spectrum probably due to the shock excitation of neutral hydrogen atoms intersecting the shock wave front I, i.e., they display radiation of the ambient gas, and not the supernova ejecta (Kirshner and Chevalier 1978; Bychkov and Lebedev 1979; Chevalier et al. 1980).

What is the origin of this gas? There is only an indirect argument that a supernova of type Ia expands into a gas ejected by the precursor: the radio structure of the SNR Tycho and namely the thin external radio rim show poorly in a numerical model of the remnant if one does not introduce into the calculation a planetary nebula ejected before the explosion (Dickel and Jones 1985).

The absence of clear signatures of the influence of the precursor on the ambient gas in these two classical SN Ia remnants is completely understandable: the size of the SN 1006 and Tycho shells already significantly exceeds the size of the region disturbed by the low-mass precursor and namely the characteristic size of planetary nebulae.

The bright compact optical filaments and condensations in the SNR Kepler are apparently the ejecta of the precursor. Their kinematic age is $10^4$ yr and the high density (similar to $10^3$ cm$^{-3}$) evidences that they have not been significantly slowed down in the interstellar medium. A stellar origin of the filaments is also verified by their anomalous chemical composition (Dennefeld 1982). Taking into account the large translational velocity of filaments (similar 350 km/s) and the significant galactic height, Bandiera (1987) came to the conclusion that the pre-supernova was a massive run-away star, intensely losing matter, which, apparently, puts into doubt the identification of SN Kepler as a SN Ia type, but the question has not been resolved.

The young remnants of the SN II explosions of the Crab and 3C 58 in the Galaxy and 0540-69.3 in the Large Magellanic Cloud have close ages and all three were accompanied by the formation of a pulsar although they turn out to be quite varied in a detailed comparison. Due to limited space we are not able to consider each in detail, although, undoubtedly, it is just their comparison which should provoke a whole series of questions concerning the nature and evolution of supernova remnants with pulsars.

Today the remnant 0540-69.3 most completely satisfies the main scheme of a remnant with a central pulsar. Here one observes radio, X-ray, and optical synchrotron emission of the cloud of the pulsar plasma; bright high velocity optical filaments with an increased oxygen abundance represent supernova ejecta; the thermal X-ray emission may be associated with the ejected and swept-up gas of the shell; the weak extended halo around the bright nebula with traces of optical, radio, and X-ray radiation probably represent the interstellar gas disturbed by the stellar wind of the precursor (Chanan et al. 1984; Reynolds 1985; Midleditch et al. 1987; Kirshner et al. 1989) and references therein.

Synchrotron radiation of the cloud of pulsar plasma is observed in the Crab Nebula in the whole range from $10^8$ to $10^{23}$ Hz along with radiation of the system of optical filaments ejected in the explosion. The ejecta of SN 1054 is characterized by an anomalously low expansion velocity of filaments (see Table 14.1); the corresponding kinetic energy of the ejecta turns out to be an order of magnitude lower than the standard ones for SN I and SN II. All attempts to find the weak far external shell predicted by Chevalier (1977) and Shklovsky (1978) or the high velocity motions for the speed normal for supernovae turned out to be unsuccessful. It is possible that this is connected with the fact that the wind of the precursor swept-up the circumstellar gas. If SN 1054 exploded in a rarefied cavity, the radiation of gas behind the shock front I (see the scheme of Fig. 14.4) could not be observed at this stage of evolution (in the X-ray and optical due to the low emission measure of the gas behind the front; in the radio due to the fact that the instability generating a turbulent layer on the contact surface develops only when the mass of the swept up gas becomes comparable with the mass of ejecta). It seems that the question is beginning to be clarified. Observations in the 21 cm line and IRAS data (Romani et al. 1990) actually showed an extended cavity with a size of about 160 pc, which was apparently swept up by the precursor of SN 1054 and neighboring O stars. The authors found an initial density of the unperturbed gas in the region of about 0.7 cm$^{-3}$ and the density inside the cavity swept-up by the wind should be several times lower. This means that the mass of swept-up gas behind the shock front I (~0.1–0.2 M$_\odot$) is actually significantly less than the mass of the shell ejected in the explosion (~1 M$_\odot$) so that the emission of this gas can be completely below the threshold of detectability taking into account the interfering emission of the bright ejected shell and the pulsar plasma cloud.

Cassiopeia A and the so-called oxygen-rich supernova remnants related to it differ from remnants of both SN Ia and SN II. They can be associated with SN Ib explosions for which a large quantity of oxygen is ejected, but the question is still not clear. The most probable precursor of Cassiopeia A is a WR star. In Cassiopeia A the interaction of the supernova shell with the material ejected by the precursor has the most clear observational manifestation. The internal bright X-ray shell ($T_e \cong 10^7$ K) and the weak external plateau ($T_e \cong (5-7) \times 10^7$ K) are diffuse components of the ejecta and the swept-up circumstellar gas heated by direct and reverse shock waves. Dense clumps of supernova ejecta form a system of bright fast-moving optical knots moving out from the center with an average velocity of 6,000 km/s. The fast-moving knots radiate only in the lines O, S, and Ar while the stratification of radiation evidences the ejection of the weakly mixed fragments of the exploded core of the massive star. On the contrary, the system of optical stationary flocculae is rich in hydrogen, but contain enhanced He/H and N/H which evidences their stellar origin. The expansion velocity of 100–400 km/s corresponds to a kinematic age of $10^4$ yr, i.e., the stationary flocculae were ejected by the star long before the explosion. Recently at the boundary of the bright nebula fast-moving flocculae were detected, expanding with a maximum velocity 7600–8600 km/s, but radiating only in the lines H$\alpha$ and NII (Fesen et al. 1987). These most external layers of the exploded star show the presence of hydrogen and a strong nitrogen excess which evidence the explosion of a star of spectral class WN7–WN9. In this case the stationary flocculae can be naturally compared with the ring nebula around WR type stellar ejecta (see §1) in which an enhanced nitrogen content is often observed.

All O-rich SNRs observed with sufficient angular resolution have a toroidal geometry and are enriched by heavy elements. In Cassiopeia A the toroidal geometry was detected by the radial velocities of X-ray lines Si XIII, S XV, and S XVI (Markert et al. 1983) and verified by radio observations (Kenney and Dent 1985). This geometry can be the result of either an asymmetrical supernova explosion or the asymmetrical outflow of the pre-supernova (if the outflow density of the pre-supernova was enhanced in the equatorial plane, the reverse wave arose here earlier and was expressed more distinctly).

These are the basic observational manifestations of outflow of the precursor in young supernova remnants. There is no exhaustive theory of remnant evolution taking into account radiation and mass loss of the precursor although some steps have been made.

Figure 14.5 gives a general representation of the evolutionary tracks of SNRs in a homogeneous interstellar medium for the standard density $n_o = 1$ cm$^{-3}$ and for a density characterizing extended cavities formed by the wind and multiple supernovae ($n_o = 10^{-2}$ cm$^{-3}$). The stages indicated are: free expansion, adiabatic expansion, and adiabatic expansion with the predominant evaporation of dense clouds and radiative cooling. One can see that the transition to the adiabatic stage for a remnant in a rarefied medium occurs for sizes reaching 20 pc and the formation of a radiative shell for sizes more than 200 pc or when the expanding remnant reaches the dense shell of the cavity. (Loops I, II, III, and IV are in all probability supernova remnants exploded in rarefied media (Lozinskaya 1991 and references therein). In particular, the Loop I which is the most studied is localized inside a cavity blown-out by stars of the Sco-Cen association (Davelaar et al. 1980). Although the size of these remnants reaches 200 pc, they are young objects in an evolutionary sense, still not having reached the stage of radiative cooling.

For a supernova envelope expanding into the material of the wind of the precursor with a spherically-symmetric density distribution of the form $\rho = \rho_o R^{-u}$ there is a self-similar solution of which the Sedov (1981) solution for an adiabatic remnant in a homogeneous medium is a particular case. The radius $R_s$ and the expansion velocity $v_s$ are described by the equations

$$R_s = (E_o/\rho_o)^{1/(s-u)} t^{1/(s-u)}$$

$$v_s = \frac{2}{s-u} (E_o/\rho_o)^{1/(s-u)} t^{-(3-u)/(s-u)} ,$$

where $E_o$ is the initial kinetic energy of the supernova explosion.

Any steep density gradient of the ambient gas, for example, with the presence of a circumstellar shell, induces multiple secondary shock waves. According to the calculations of Itoh and Fabian (1984) even one steep gradient between the wind of a red supergiant and the homogeneous interstellar gas initiates a secondary, reflected inward shock wave, which reheats the wind gas, and, probably one more inward-propagating shock wave in the ejecta and an outward-reflected one, in the wind. Although there is no adequate model taking into account the existence of multi-layered and multiple near-stellar shells, analytic (Shull et al. 1985; Chevalier and Liang 1989) and numerical solutions of the problem of a supernova explosion inside one shell or cavity exist (Dickel and Jones 1985; Falle 1988; Ciotti and D'Ercole 1989; Tenorio-Tagle et al. 1990). In the last study the calculations were made for an explosion in a cavity swept-out

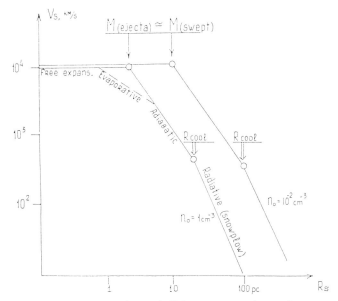

Figure 14.5. The evolution of SN remnants in a homogeneous medium for a standard density $n_0 = 1$ cm$^{-3}$ and for a density $n_0 = 10^{-2}$ cm$^{-3}$. The stages indicated: free expansion, adiabatic expansion, and adiabatic expansion with a dominant evaporation of dense clouds and radiative cooling.

by a wind typical for a WR star and encompass the period up to the stage of radiative cooling of the shell. The calculations show the generation of multiple direct and reverse shock waves which leads to a complex structure and complex distribution of the velocity field in the expanding remnant, and changes the course of evolution. The evolution of the supernova shell in a cavity of small size differs especially sharply from the standard remnant in a homogeneous medium if the shock waves induced by the supernova expansion go beyond the shell swept-up by the wind (Tenorio-Tagle et al. 1990).

An adequate comparison of the numerical models of evolution with actual observations of supernova remnants is a quite involved problem. To verify by actual observations the process of collision of the supernova envelope with the shell blown-out by the wind is still difficult. Secondary shock waves lead to variations of the X-ray and optical emission of the remnant in the course of evolution, changing the conditions of heating and radiative cooling of gas behind the front (Tenorio-Tagle et al. 1990).

It is still more complicated to compare the results of radio observations with the theoretical evolutionary course. In the presence of a pulsar the synchrotron radiation of the remnant is determined by the interaction of the cloud of relativistic particles and magnetic field injected by the pulsar with the expanding supernova shell, on the one hand, and the interaction of the latter with the ambient gas perturbed by the action of radiation and the wind of the precursor, on the other. The evolution of the radio remnant was first considered by Shklovsky (1960). In the framework of this simple model expansion of the cloud of relativistic particles with a magnetic field led to a dependence of the surface radio brightness $\Sigma$ on the linear diameter D and he proposed a quite productive method of estimating the distance to supernova remnants according to the empirical dependence $\Sigma(D)$. Taking into account the injection of relativistic particles and magnetic field by the pulsar, Pacini and Salvati (1973) and Reynolds and Chevalier (1984) carried out analytical and numerical modeling of the evolution of the pulsar plasma cloud interacting with the supernova envelope. The next step should be a model of the interaction of the pulsar with the supernova envelope and interstellar gas. Finally one should extend the chain still further, i.e., take into account the action of the pre-supernova because the multiple secondary shock waves in the ejecta and swept-up gas are undoubtedly important for amplifying the magnetic field and the acceleration of relativistic particles, i.e., they can strongly change the synchrotron radiation of remnants.

Contemporary radio observations with high angular resolution show at least three types of radio remnants: plerion, shell, and combined (plerion surrounded by a shell); the shell objects are often characterized by an axially symmetric barrel-form morphology. Due to expansion of the central cloud of relativistic plasma and a decrease of energy pumping by the pulsar the radio emission of a plerion is observed for a comparatively short time, probably less than $10^4$ yr. Therefore, one can suppose that plerions are changed initially into combined ones and then purely shell remnants while the formation of a radio shell around the plerion connected with amplification of the field frozen into the interstellar gas depends directly on the density of the latter, i.e., on the action of the precursor wind. For example, for a SNII explosion in the cavity around an OB association one can expect to observe very young combined remnants at a stage when the radio shell is formed due to amplification of the field in the convective layer at the boundary of the ejecta and the gas of the precursor wind (red supergiant) or at the boundary of the pulsar wind and the ejecta, and then in the course of a long time one should observe a plerion devoid of the shell. The detection of such plerions of large size in cavities around OB associations may serve as a verification of the proposed scenario.

The formation of axially symmetric shells may also be connected with an asymmetric distribution of circumstellar density due to the action of the precursor; in any case it was successful in explaining the morphology of the best-known object of this type G296+10 (Bisnovatii-Kogan et al. 1990).

Finally, observations of the peculiar object CTB 80 (Fesen et al. 1988b) showed still another aspect of the interaction of a pulsar with a supernova remnant. A fast-moving energetic pulsar may overtake the retarding shell of the supernova remnant, so old that it is already invisible in the optical, X-ray and radio ranges and reanimate it, bringing fresh relativistic particles into the magnetic field frozen into the compressed gas of the shell.

In particular, in the case of CTB 80 an extended IR shell-remnant was detected whose age is about $10^5$ yr. The pulsar with a cloud of relativistic plasma is a plerion situated at the western boundary of the shell and reanimated its synchrotron radio emission due to injection of relativistic electrons. Apparently in the complex G 5.3–0.9 one is observing the more advanced stages of this interaction of a plerion with a central pulsar, penetrating through the retarding supernova remnant (Shull et al. 1989). Finally, one can pose the question consistently: what will happen if a fast moving pulsar, still continuing to inject relativistic particles, overtakes the shell swept-up by the stellar winds of the massive star or super shell around the OB association?

The question mark suits well the end of our article. The unsolved questions in the problem of supernovae and stellar winds in the interstellar medium are now much larger than 5, 10, or 20 years ago. However, the flow of new information is growing accordingly in this rapidly developing region of astronomy giving rise to hope and making our science so charmingly interesting.

## References

Abbot, D. C. (1982) *Astroph. J.* **263**, 723.

Avedisova, V. S. (1971) *Astron. Zh.* **48**, 894.

Bandiera, R. (1987) *Astroph. J.* **319**, 885.

Barlow, M. J., and Hummer, D. G. (1982) in IUA Symp. No. 99, Wolf-Rayet Stars: Observations, Physics, Evolution, C. W. H. de Loore and A. J. Willis (eds.) Reidel, Dordrecht, 387.

Bisnovatii-Kogan, G. S., Lozinskaya, T. A., and Silich, S. A. (1990) *Astroph. Space Sci.* **166**, 277.

Blinnikov, S. I., Chugay, N. N., and Lozinskaya, T. A. (1988) *Soviet Scientific Rev. Sect. E: Astroph and Space Phys. Rev.* **6**, 3, 195.

Bochkarev (1987) *Nature* **332**, 518.

Bruchweiler, F. C., Gull, T. R., Henize, K. G., and Cannon, R.D. (1981) *Astroph. J.* **251**, 126.

Bychkov, K. V., and Pikel'ner, S. B. (1975) *Pis'ma Astron. Zh.* **1**, 29.

Bychkov, K. V., and Lebedev, V. S. (1979) *Astron. Astrophys.* **80**, 167.

Cappa de Nicolau, C. E., and Niemela, V. C. (1984) *Astron. J.* **89**, 1398.

Cappa de Nicolau, C. E., Niemela, V. C., and Arnal, E. M. (1986) *Astron. J.* **92**, 1414.

Cappa de Nicolau, C. E., Niemela, V. C., Dubner, G., and Arnal, E. M. (1987) *Rev. Mex. Astron. Astrophys.* **14**, 2, 611.

Chanan, C. A., Helfand, D. I., Reynolds, D. P. (1984) *Astroph. J.* **287**, L23.

Chevalier, R. (1977) in Supernova, D. N. Schramm (ed.), Reidel, Dordrecht, p.53.

Chevalier, R., Kirshner, R. P., and Raymond, J. C. (1980) *Astroph. J.* **235**, 186.

Chu You-Hua (1981) *Astrophys. J.* **249**, 195.

Chu You-Hua (1990) in Wolf-Rayet Stars and Interrelations with Other Massive Stars and Galaxies, IAU Symp. No 143, Reidel, Dordrecht.
Chu You-Hua, Treffers, R. R., and Kwitter, K. W., (1983) *Astroph J. Suppl.* **53**, 937.
Chu You-Hua, Jacoby, G. H., and Arendt, R. (1987) *Ap. J. Suppl.* **64**, N.3.
Ciotti, L., and D'Ercole, A. (1989) *Astr. Ap.* **215**, 347.
Cowie, L. L., Songalia, A., and York, D. G. (1979) *Astroph. J.* **230**, 469.
Davelaar, J., Bleeker, A. M., and Derrenberg, A. J. M. (1980) *Astron. Astroph.* **92**, 231.
Deenefeld, M. (1982) *Astron. Astrophys.* **112**, 215.
Dickel, J. R., and Jones, E. M. (1985) *Astroph. J.* **288**, 707.
Dopita, M. (1987) in Star Forming Regions, IAU Symp. No 115, M. Peimbert and J. Jugaku (eds.), Reidel, Dordrecht.
Dopita, M. A., Lozinskaya, T. A., McGregor, P. J., and Rawlings, S. J. (1990) *Astrophys. J.* **351**, 563.
Dopita, M., and Lozinskaya, T. A. (1990) *Astroph. J.* **359**, 419.
Dopita, M. A., Mathewson, D. S., and Ford, V. L. (1985) *Astroph. J.* **297**, 599.
Dufour, R. J., Parker, R. A. R., and Henize, K. G. (1988) *Astrophys. J.* **327**, 859.
Dyson, J. E. (1981) in Investigating the Universe, F. D. Kahn (ed.), Reidel, Dordrecht, p. 125.
Falle, S. (1981) *MNRAS* **195**, 1011.
Falle, S. (1988) in Proc. IAU Coll. No. 101 The Interaction of Supernova Remnants with the Interstellar Medium, Reidel, Dordrecht.
Feitzinger, J. V. (1987) in Star Forming Regions, IAU Symp. No. 115, M. Peimbert and J. Jugaku (eds.), Reidel, Dordrecht.
Fesen, R. A., Becker, R. H., and Blair, W. P. (1987) *Ap. J.***313**, 378.
Fesen, R. A., Wu, C. C., Leventhal, M., and Hamilton, A. J. S. (1988a) *Ap. J.* **327**, 164.
Fesen, R. A., Shull, J. M., and Saken, J. M. (1988b) *Nature* **334**, 229.
Goudis, C. (1982) *Astrophys. Space Sci. Libr.* **90**, 1, Reidel, Dordrecht.
Hamilton, A. J., Sarazin, C. L., Szymkowiak, A. E., and Vartanian, M. H. (1985) *Astroph. J.* **297**, L5.
Hester, J. , (1987) *Astrophys. J.* **314**, 187.
Itoh, H., and Fabian, A. C. (1984) *MNRAS* **208**, 645.
Jewitt, D. C., Danielson, G. E., and Kupferman, P. N. (1986) *Astroph. J.* **302**, 727.
Kenney, J. D., and Dent, W. A. (1985) *Ap. J.* **298**, 644.
Kirshner, R. P., and Chevalier, R. A. (1978) *Astron. Astrophys.* **67**, 267.
Kirshner, R. P., Morse, J. A., Winkler, P. F., and Blair, W. P. (1989) *Astroph. J.* **342**, 260.
Kwitter, K. B. (1984) *Astrophys. J.* **287**, 840.
Leitherer, C., Chavarria, K. C. (1987) *Astron. Astrophys.* **175**, 208.
Lozinskaya, T. A. (1982) *Astrophys. Space Sci.* **87**, 313.

Lozinskaya, T. A. (1983) *Pis'ma Astron. Zh.* **9**, 469.
Lozinskaya, T. A. (1990) in Wolf-Rayet Stars and Interrelations with Other Massive Stars in Galaxies, IAU Symp. No. 143, Reidel, Dordrecht.
Lozinskaya, T. A. (1991) Supernovae and Stellar Wind: Interaction with the Interstellar Gas, (1986) Moscow, Nauka; (1991) Second edition, AIP, New York.
Lozinskaya, T. A., and Repin, S. V. (1990) *Astron. Zh.* **67**, 1152.
Lozinskaya, T. A., and Sitnik, T. G. (1988) *Pis'ma Astron. Zh.* **14**, 240.
Maeder, A., and Meynet, G. (1989) *Astron. Astrophys.* **210**, 155.
Markert, T. H., Canizares, C. R., Clark, G. W., and Winkler, P. F. (1983) *Ap. J.* **268**, 134.
McCray, R. (1988) in the Supernova Remnants and the ISM, IAU Coll. No 101, R. S. Roger and T. L. Landecker (eds.), Reidel, Dordrecht, p. 447.
McKee, C. F., Van Buren, D., and Lazareff, B. (1984) *Astroph. J.* **278**, L115.
Midleditch, J., Pennypacker, C. R., and Burns, M. S. (1987) *Astroph. J.* **315**, 142.
Pacini, F., and Salvati, M. (1973) *Astroph. J.* **186**, 249.
Pikel'ner, S. B. (1954) *Isvestia Krymskii Astrophys. Obs.* **12**, 93.
Pikel'ner, S. B. (1956) *Astron. Zh.* **33**, 785.
Pikel'ner, S. B. (1961) *Astron. Zh.* **38**, 21.
Pikel'ner, S. B. (1968) *Astrophys. Lett.* **2**, 97.
Reynolds, D. P. (1985), *Astroph. J.* **291**, 152.
Reynolds, S. P., and Chevalier, R. A. (1984) *Astroph. J.* **278**, 630.
Reynolds, R. J., and Ogden, P. M. (1979) *Astron. J.* **87**, 306.
Romani, R. W., Reach, W. T., Koo, B. C., and Heiles, C. (1990) *Ap. J.* **349**, L51.
Sedov, L. I. (1981) Similarity and Dimensional Analysis in Mechanics, Nauka, Moscow.
Shklovsky, J. S.(1960) *Astron. Zh.* **37**, 222.
Shklovsky, J. S.(1976) Supernovae and Related Problems, Nauka, Moscow.
Shklovsky, J. S.(1978) *Astron. Zh.* **55**, 726.
Shklovsky, J. S.(1984) *Pis'ma Astron. Zh.* **10**, 723.
Shull, P., Dyson, J. E., Kahn, J. E., and West, K. A. (1985) *MNRAS* **212**, 799.
Shull, M., Fesen, R. A., Saken, J. N. (1989) *Astroph. J.* **346**, 860.
Smith, A. M., Cornett, R. H., and Hill, R. S. (1987) *Astrop. J.* **320**, 609.
Tenorio-Tagle, G., Bodenheimer, P., Franco, J., and Rozyczka, M. (1990) *MNRAS*, in press.
Torres, A. V., Conti, P. S., and Massey, P. (1986) *Ap. J.* **300**, 379.
Van Buren, D. (1985) *Astroph. J.* **294**, 567.
Van Buren, D., and McCray, R. (1988) *Astroph. J.* **329**, L93.
Van der Hucht, K. A., Williams, P. M., and The, P. S. (1987) *Q. J. Roy Astr. Soc.* **28**, 254.
Weaver, R., McCray, R., Castor, J., et al. (1977) *Astroph. J.* **218**, 377.
Wendker, M. R. (1984) *Astron. Astrophys. Suppl. Ser.* **58**, 291.

# 15

# Joseph Shklovsky and X-ray Astronomy

## Herbert Friedman
*U.S. Naval Research Laboratory*

## INTRODUCTION

My close friendship with Joseph Shklovsky began at the time of the 1958 General Assembly of the International Astronomical Union in Moscow. Although his reputation as a leading figure in astronomy was already well established through his publications, only a handful of astronomers from the western world had met him face to face. The meeting was an occasion of euphoria for Shklovsky who was filled with the need to talk science with all the colleagues from distant countries who came to the great meeting. He roamed about the Hall of Columns at the opening ceremonies, grasping lapels to read the names on our identification tags, and exclaimed with pleasure when he discovered who we were. He bubbled with good humor and joked about how young the American scientists were; Americans traveled abroad at a much earlier age than their Soviet peers who traveled rarely and on pitifully meagre allowances when they were permitted.

Throughout the length of the meeting Shklovsky arranged informal seminars at Moscow University to maximize the scientific exchange. In that brief stretch of time I learned that Shklovsky and I had remarkably similar backgrounds. We were born within a week of each other and came from deeply religious backgrounds. Although neither one of us any longer practiced the orthodox Jewish faith, out cultural ethos was deeply ingrained. He felt strong concern about the survival of Israel; while logically pessimistic he constantly sought reassurances that he might be wrong. In 1958, I was already concentrating my research efforts in X-ray astronomy and Shklovsky was intrigued with the future potential of the field. The synergysm in our research and ethnic interests led to a close lasting friendship for the rest of his life.

My last meeting with Shklovsky was in the summer of 1984 on the occasion of the COSPAR meeting in Graz. Tom Donahue, my wife and I took Joseph for a drive into the countryside on a warm, sunny Sunday and stopped for lunch at the edge of a lake. Joseph announced that it was his birthday and we toasted his good health. He appeared very relaxed and remarked that it was the best birthday he had ever had. Returning to

Graz, he expressed a dreary view of the Soviet future; it would take another fifty years for any significant change for the better! In Moscow that winter, he was hospitalized for a thrombosis in his leg. Surgery released the blood clot to his brain and he died. How sad that he did not live to the era of Glasnost.†

## SOLAR X-RAY ASTRONOMY

In the immediate aftermath of World War II astrophysics was poised to examine the cosmic scene with a rush of brilliant new insights, and Shklovsky was a leader in the remarkable generation that pointed the way. With keen intuition he sensed every important development early on and came up with startling new explanations.

His candidate (1944) and doctoral (1949) dissertations were concerned with the physics of the solar corona. Astronomy had been focused on relatively commonplace low energy processes, characterized by temperatures of a few thousand degrees, but Shklovsky was intrigued with higher energy phenomena that implied million degree plasmas. At the time, the far reaching outer atmosphere of the sun was a source of great puzzlement. From a variety of indirect evidence it was inferred that the solar coronal temperature exceeded 1,000,000 K even though the temperature of the solar disk was no more than 6000 K.

Pioneering work on the corona had been initiated as early as 1938 by Bengt Edlen through spectroscope studies in his laboratory at Uppsala, Sweden. He obtained spectra of the helium-like and hydrogen-like ions of oxygen and nitrogen stripped of all but their last one or two electrons. The spectrum lines fell in the soft x-ray range from 10-30 angstroms. Edlen also solved the mystery of the coronal green line, thought to be some strange new element not found in the periodic table. Instead, Edlen showed that the green line came from iron that had been stripped of 13 of its normal complement of 26 electrons.

Shklovsky seized on Elden's spectra to devise a theory of ionization equilibrium in the corona as distinct from thermodynamic equilibrium. He postulated that the coronal ionization and line emission was caused by electron impact followed by electron capture with emission of continuum radiation. To explain the high temperature of the corona, Shklovsky proposed in 1944 that hydromagnetic waves could be a source of energy dissipation in the corona. As early as the late 1940s Shklovsky had resolved the differences in solar chromospheric and coronal radiation and the relationships of thermal and nonthermal components of solar activity. He also correctly predicted the strong emission of x-rays and extreme ultraviolet light in the superheated coronal plasma of highly stripped ions. In this model the total flux in emission lines exceeded the free-free radiation.

---

† My personal recollections of Joseph Shklovsky are recounted in the introductory chapter to his collection stories, "Five Billion Vodka Bottles to the Moon" published by the W. W. Norton Co., New York, 1991.

Between 1949 and 1960 Shklovsky's ideas were largely confirmed by the direct observation of solar x-rays from rockets by my group at the Naval Research Lab. In 1960 we finally obtained line resolved spectra that fit the detailed theory. By the time that G. Elwert at the University of Tubingen was developing his refined theoretical models of solar x-ray emission Shklovsky was moving on to galactic astronomy.

## GALACTIC X-RAY ASTRONOMY

When I met Shklovsky in Moscow at the 1958 General Assembly of the IAU, he was already intrigued with the possibility of observing galactic x-rays. Clearly, if all stars radiated x-rays at the intensity level of the Sun it would be virtually hopeless with the x-ray detectors we had available at the time to detect sources deep within the galaxy. But Shklovsky explained the radio and optical emission of the Crab Nebula as synchrotron radiation and suggested that the continuous spectrum could extend into the x-ray region.

While studying the x-ray emission from solar flares in the 1956 to 1958 period my colleagues and I had detected evidence for what appeared to be extra-solar system x-rays. When I reported those observations at the IAU Assembly, Shklovsky was excited about the possibility that we had in fact detected galactic x-rays. He strongly urged me to pursue x-ray astronomy studies of the galaxy and in particular of the Crab Nebula.

In 1962 convincing evidence was obtained by Rossi, Giacconi, Gursky and Paolini, at the American Science and Engineering Corp., of galactic x-ray emission. Their x-ray detector had a very wide acceptance angle and they could only surmise that the major flux came from the general direction of the galactic center. In 1963 my colleagues and I at the Naval Research Lab flew a more sensitive detector with sufficiently tight collimation to isolate the brightest x-ray source, Sco X-1, and the Crab Nebula. I suggested at the time that the x-ray object could be a neutron star radiating thermally at a temperature of about 10-million K. Such an explanation had some merit for a supernova remnant in the Crab but Sco X-1 was a complete puzzle since the only optical object within the x-ray error box was an undistinguished 13th mag blue star.

I remember presenting the NRL results to a Princeton colloquium attended by J. Robert Oppenheimer who responded with excitement to the possibility that his theoretical neutron star had been detected. Shklovsky was equally enthusiastic. We were aware that 1964 would be the occasion for a lunar occultation of the Crab. If an x-ray observation could be made during the course of the occultation it would give positive identification of the x-ray source with the Nebula and in principle could detect a high level of x-ray emission from a central neutron star. We had already initiated preparations for a rocket flight coincident with the occultation when I received an urgent communication from Shklovsky stressing the great importance of carrying out the observation. He remarked that his Soviet colleagues were not technically prepared for such an experiment but he felt sure that our rocket astronomy program was capable of doing it. Fortunately we succeeded, because the next opportunity would not come for another ten years. The occultation revealed no neutron star but it mapped the extent of the x-ray nebula. Five years later, without benefit of an occultation, we detected the 33 msec x-ray pulsations of the neutron star.

By 1966 optical astronomers at Caltech, Tokyo, MIT and AS&E had identified Sco X-1 with a nova-like star and theorists were proposing accretion "wind" models to produce the x-ray emission. But where did the wind originate? There was no supergiant blue star in the source vicinity. Geoffrey Burbidge postulated that an invisible neutron star was delivering the wind to the blue companion star. Shklovsky soon turned the picture around and proposed that matter was transferred from a more or less normal star onto a neutron star. The idea was new to western scientists although x-rays from an accretion process had been proposed in principle in 1965 by Zeldovich and Novikov at the USSR Space Research Institute in Moscow. To pursue a search for binary sources of x-rays clearly required detectors aboard satellites. When the UHURU mission was launched in 1970 x-ray binaries were quickly discovered and the field of x-ray astronomy took a great leap forward.

For the rest of his life Shklovsky remained deeply involved with x-ray astronomy. X-ray pulsars were a major discovery that Shklovsky coupled to his broad interests in supernovae and their remnants. The extension of x-ray astronomy to extragalactic objects like AGNs and quasars and the general x-ray background radiation occupied Shklovsky's attention and he continued to be a major theoretical contributor to the field.

## SOME THOUGHTS FOR THE FUTURE

Celestial x-ray observations are now being made with both imaging reflecting telescopes and coarsely collimated large area proportional counters. The great success of the Einstein Observatory reflecting telescope has been followed by the German ROSAT telescope and a new start is being made in the NASA program on AXAF (Advanced X-Ray Astrophysics Facility) which is being designed to exceed the sensitivity of the Einstein Observatory by a factor of fifty and to double its spectral range. In the category of large area, non-focusing detectors, considerable success was achieved by HEAO-1 (High Energy Astronomical Observatory), the European EXOSAT (X-Ray Observatory Satellite) and the Japanese series of x-ray observatories culminating in the currently orbiting GINGA. Several other x-ray instruments of intermediate scale have been flown in recent years to provide a substantial yield of interesting data. NASA still plans to launch an X-Ray Timing Explorer (XTE) scaled to a size very similar to the Japanese Ginga.

My personal interest in the past decade has been directed to the possibilities of x-ray timing measurements with a very much larger detector than any of those methods mentioned above. XLA, for X-Ray Large Array, is a concept for a modular assembly that could be built up to a size as great as 100 square meters. Some astronomers think it is a pipe dream but it is not altogether unrealistic. At the end of the Apollo program, it was thought that left over hardware could be used to carry a huge payload into earth orbit or to establish a lunar base with potential for scientific research. My proposal was to first put a very large x-ray proportional counter in orbit, to be followed by erecting a still larger x-ray array on the surface of the moon. For the moon base I estimated 1985 as a reasonable target; that was certainly overly optimistic, but now there is semi-serious talk of doing just that.

XLA is a much more sophisticated concept than anything I had in mind twenty years ago. A number of scientists, led by Peter Michelson of Stanford University and Kent Wood of the Naval Research Laboratory have contributed to the present design concept. XLA is not intended for a lunar base; it is conceived as a free-flyer platform for low earth orbit or as an attached payload to be assembled at a space station. I feel certain that the XLA mission would have great appeal to Shklovsky. His colleagues at IKI urged me to consider flying an XLA on their enormous Energia rocket, but the present times do not appear to be right for such a collaboration.

I will attempt to describe briefly some of the scientific potential envisioned for an XLA of 100 square meters aperture.

## X-RAY SOURCE STUDIES WITH A VERY LARGE AREA DETECTOR

### 1. Millisecond X-Ray Pulsars

The progenitors of millisecond radio pulsars are believed to be low mass x-ray binaries (LMXB) in which the neutron stars are spun up by accretion torques produced by mass flow during the x-ray emitting phase. Support for this model comes from the observation of five binary radio pulsars with periods ranging from 1.6 msec to 6.1 msec. Attempts to explain quasi-periodicities in several x-ray sources also invoke fast spin rates for the accreting neutron stars.

The expected maximum rotation frequency of the neutron star depends on the limiting Kepler frequency at the star's Alfvén radius. The mechanism should be quite common but observed msec pulsars are very rare. The Japanese X-Ray observatory, Ginga, with an area of 4000 $cm^2$ places upper limits on the pulse fraction from many LXMBs at only a few tenths of one percent.

For msec pulsations to appear in an LMXB, the neutron star magnetic field must be sufficiently strong to distort the flow of accreting mass. The weakness of the magnetic field that is required for the fast spin is also the explanation for the weakness of pulsations. For the 20 brightest LMXBs, XLA could sense pulse fractions as low as 0.01 percent. If these pulsations were measured, the information would contribute greatly to understanding the evolution of LMXBs.

## 2. Quasi-Periodic Oscillators

The phenomenon of quasi-periodic oscillations has been observed in several galactic bulge sources by the European EXOSAT, the Japanese HAKUCHO, and NASA's HEAO-1 but the process is still not well understood. Variations of brightness of several percent are not perfectly periodic; the frequency drifts over the range between 6 and 50 Hz. Unlike the precise pulsars the quasi-periodic sources are sloppy clocks. One suggested model is a beat frequency mechanism in which the quasi-periodicity reflects the difference in frequency of clumped plasma at the inner edge of the accretion disk.

QPO to date have been detected in the brightest LMXBs at a pulse fraction of greater than 1 percent. Individual oscillations have been observed clearly only in data from the Rapid Burster where the QPO strength is enormous (20-30% compared to the more typical 1-5%). GINGA and EXOSAT data have not been sufficiently precise to distinguish between various proposed mechanisms: fixed oscillation period with phase jumps, an on/off duty cycle, frequency modulation, or possible combinations of these. The fractional QPO that would be observable with XLA is about one-fifteenth of what is now achieved - about 0.1 percent for the brighter sources. This sensitivity is in the theoretically interesting domain.

Accretion flows are the most difficult phenomena to model in LXMBs and QPO can provide essential information on the environment. For bright sources XLA could resolve individual oscillations shorter than 10 microsec and provide spectra several times per oscillation. Individual blobs of matter might be observed as they spiral through the Alfvén surface to impact the neutron star.

## 3. X-Ray Bursts

Short-lived bursts of x-rays are observed in LMXBs. It appears that accretion onto the surface of a neutron star builds up to a runway thermonuclear ignition, but many features are not clearly understood. Is the burning material hydrogen or helium? Why are some bursts multi-peaked and what accounts for the short burst repetition intervals in others? Is the Eddington limit violated in some bursters? Numerical codes of the thermonuclear process predict variations of black body temperature on millisecond time scales. XLA could check the models by following the spectral evolution on the required time scales both during the rise and decay times.

## 4. Gravitational Redshifts

A gravitational redshift in emission lines or an absorption edge at the surface of a neutron star would provide an accurate measure of its mass. An absorption feature observed during a few seconds of the bursts from three different sources has been interpreted as a helium-like iron line from the stellar surface, gravitationally redshifted by

1.6. This interpretation requires an iron abundance in the neutron star atmosphere about 100 times cosmic abundance. The minimum detectable line width with EXOSAT was about 300 eV; XLA would be able to see 10 eV widths in a few seconds and would provide reliable determinations of gravitational redshifts for many neutron stars.

## 5. Neutron Star Seismology

Helioseismology, the study of global modes of oscillation in the solar atmosphere and the corresponding phenomenon in white dwarfs, is an advanced discipline of stellar astrophysics. Neutron star seismology has not been observationally confirmed, but there is ample reason to search for its manifestation. Small movements of the surface of a neutron star should give rise to faint transient signals that could be picked up by XLA.

At least four modes of neutron star oscillation are predicted: pressure modes with periods less than 0.2 msec; gravity modes with periods greater than 50 msec; shear modes with periods less than 2 msec; and interfacial modes with periods between tenths and tens of msec. Some modes may be excited when x-ray or gamma-ray bursts occur. Detection of such modes of oscillation should be diagnostic of the structure and composition of neutron star envelopes just as helioseismology probes the internal structure of the Sun.

Searches for vibrational modes have been made in connection with x-ray bursts, but no positive evidence has yet been found. The implication is that the level of modulation must be below one percent. XLA would detect fast neutron star vibrations down to the 0.1 percent level of modulation for transients and the 0.02 percent level for persistent effects.

## 6. Black Holes

Evidence for the existence of stellar mass black holes has been found in several binary systems, of which Cygnus X-1 is the best example. From the orbital parameters of the binary system a mass of about 9 $M_\odot$ is indicated. Rapid incoherent binary temporal fluctuations and spectral signatures can be attributed to an accretion disk extending close to the surface of a compact object without encountering a neutron star magnetosphere.

Tests for variability in Cygnus X-1 with data from HEAO-A1 showed evidence of an excess of variability noise over that expected from a Poisson distribution near 3 msec. With standard models of a viscous accretion disk it was concluded that the observed 3 msec activity occurred within 5 black hole radii of the center of the accretion disk. It should be understood that the evidence for 3 msec bursts was not based on the observation of single bursts, but on a statistical aggregate derived from long strings of data. With XLA it would be possible to observe the time profiles and spectra of single bursts. Analysis of individual events would lead the way to an improved theory of the structure of the inner disk.

## 7. Milliarcsec Angular Resolution

XLA combined with an occulting edge such as the moon or an artificial occulter placed at a distance could provide milliarcsec spatial resolution. For example, consider the possibilities of imaging fine detail in the Crab Nebula. Because of its prototypical status as a plerion, its high brightness, its central pulsar and its positioning for lunar occultation, the Crab is an especially interesting target for the XLA.

The details of the structure of the nebula have been mapped with the Einstein Observatory to a resolution of about 3 arcsec. XLA in the lunar occultation mode can provide detail about 100 times finer. The optical wisps in the nebula move back and forth on arcsec scales. MHD models of nebular structure invoke a strong shock propagating into a pulsar wind zone to a radius of about 10 arcsec. XLA with lunar occultation could provide subarcsec detail in the shock zone.

The Crab Pulsar is by far the brightest non-binary x-ray pulsar. Precise timing observations with XLA could detect internal neutron star mechanical modes with 20 times faster timescales and an order of magnitude higher sensitivity than has been obtained in previous studies. The interpretation in terms of neutron star superfluidity and equation of state would be especially interesting in resolving questions of crust/neutron superfluid coupling.

The above examples are just a sample of many applications of an XLA to fundamental problems in x-ray astronomy. While there are no early prospects of the application of larger than one or two square meter detectors to x-ray timing there is nothing fundamentally or technically forbidding about scaling up toward 100 square meters.

## CONCLUSION

It is sad to know that Joseph Shklovsky will no longer spark the imagination of astronomers around the world with his penetrating insights into every aspect of contemporary astrophysics. I feel especially privileged to have enjoyed his close friendship and to have derived much scientific inspiration from him.

# 16

# Quasars and Active Galactic Nuclei

B. V. Komberg

*Astro Space Center, Lebedev Physical Institute, Moscow*

## 1. INTRODUCTION

A. The role of the active nuclei phenomenon in the evolution and formation of galaxies, and the processes of enrichment and ionization of the intergalactic medium, in the refinement of the model of the universe, and in the clarification of very distant objects was not recognized immediately. The sources of this recognition which began in the middle of the '40s included such well-known scientists as F. Zwicky, K. Seyfert, V. Baade, R. Minkowskii, E. M. Burbidge and J. Burbidge, M. Schmidt, A. Sandage, J. Oort, H. Arp, V. A. Ambartsumyan, Ya. B. Zeldovich, I. S. Shklovsky, V. L. Ginzburg, S. B. Pikel'ner, B. A. Vorontsov-Velyaminov, B. E. Markaryan, et al. In the beginning the conclusions were based on data about nuclei of Seyfert galaxies (SG) and radio galaxies (RG) /1-3/ and their observational features summarized in his epic paper at the Solvay Congress in 1958 of V. A. Ambartsumyan /4/ who, apparently, was one of the first to understand the significance of active nuclei for the problem of energy supplied to stellar systems. Therefore, it is not amazing that the quasars discovered by M. Schmidt in 1963 (radio strong quasi-stellar objects, QSO) and quasi-galaxies discovered by A. Sandage in 1965 (much weaker radio quasi-stellar galaxies, QSG) were recognized immediately as a verification of the point of view on the birth of galaxies from nuclei which at the stage of quasi-stellar objects (QSO—both the quasars and quasi-galaxies) are still "naked". However, at the beginning of the '70s observations verified that QSOs were surrounded by extended nebulae reminding one by their basic characteristics of gigantic galaxies,† to be sure with a series of anomalies. They succeeded in measuring the red shifts of some of these nebulae which turned out to be close to the red shifts of the QSOs themselves. From this fact the conclusion was made that QSOs are very active nuclei in distant galaxies of stellar systems relatively rich in gas /6,7/.

As soon as QSOs were put in the series with active galactic nuclei (AGN), the question naturally arose about the place of QSOs among galaxies with active nuclei (GAN). Here different suppositions were made:

---

†This was first shown by J. Kristian /5/ for the near ($Z < 0.5$) QSOs.

1. Each form of GAN evolves along its own path and such objects as Seyfert galaxies, radio galaxies, and QSOs are not connected.
2. The observed forms of GAN are in fact different evolutionary stages of objects of a single (or similar) type.
3. The intermediate case when different GAN types and their different stages of evolution are observed. For example, quasi-stellar sources strong in the radio evolve with time into radio galaxies and those radio-weak QSG evolve into Seyfert galaxies.

Because the third case is the most general, it requires a careful analysis of observations relative to the properties of different types. It is possible for this purpose to put the observed different GAN into a single scheme. It is true in this case that one must answer a whole series of questions on the cause of features both of AGN models and their evolution: for example, it is necessary to know the light function of GAN, the degree of anisotropy of radiation of active nuclei (for introducing corrections in orientation), the direction of their evolutionary course, etc. There is still no single opinion among astrophysicists on all these questions. For example, some authors considered QSOs as the initial bright stage in the evolution of AGN and others the final stage. Others hold the point of view that QSOs are a relatively short-lived and repeating phase of activity in more or less normal nuclei of massive galaxies /8/.

Thus, an understanding of the place of QSOs among other AGN may shed light on the nature of this phenomenon. By the way, the reverse assertion is also true, but it, in our view, is more difficult to implement at the present time.

Just in the first years after detecting QSOs, these most energetic, most massive, most compact, most restless, and most far from known AGN, models were proposed, capable, in principle, of explaining the basic observational features /9–11/. We enumerate these models in order of popularity at the present:

1. A massive black hole with an accretion disk.
2. A compact massive stellar cluster in whose center a black hole possibly forms.
3. A rotating supermassive star with strong turbulent motions and large-scale magnetic fields.

In succeeding years these models were developed and refined. Schemes were proposed for the evolution of massive black holes with the accretional increase of their masses /12/. The electrodynamics of black holes were considered /13/. Problems of the formation in the center of compact clusters of massive black holes were discussed due to the loss and dissipation of the momentum with the gas ejected during interaction and evolution of stars of the cluster. The probability of formation of double black holes was estimated in the center of a system by merging of two or more galaxies into a single complex /14,15/. After the detection in 1979 of the peculiar double star system SS-433 /16/ which according to a whole series of signatures reminded one astonishingly of a miniature active nucleus, interest intensified in the problem of doubleness of AGN along with a model of a massive pulsar /17,18/. Although up to now the question of an AGN model has not been solved finally, an analysis of observational data already allows one at present to clarify some of the necessary conditions without whose fulfillment it is impossible to have significant activity in nuclei. One of these conditions is, apparently, the presence in central regions of galaxies of gas in a quantity sufficient for satisfying an accretion rate 0.01–1 $M_\odot$/yr for a time $10^{6-7}$ yr. The question arises about possible paths of the appearance of gas in the central kiloparsec, especially if one is concerned with galaxies of early morphological types where the most active nuclei are observed, but in the galaxies themselves there is usually little gas.

In recent years convincing evidence has appeared for using amplification of the activity of nuclei and the rate of star formation in interacting and (or) merging galaxies. As observations show, this relates to QSOs situated in central regions of tight groups of galaxies with evidence of interaction. These facts are at odds with the thought that it is just the interaction and (or) merging of galaxies that serves as the initial manifestation in the central parts of systems of a large quantity of gas which, in turn, enables the manifestation of active nuclei /19,20,21/ and the accompanying phenomena of activity, in particular, induced star formation /22/.

Thus, at the present the question on the role of the interaction or merging of galaxies in the process of AGN formation is a central one.[†] A possible answer to it does not help us in the choice of a model of an active nucleus, but it, undoubtedly, will enable an understanding of processes corresponding to supply of the "central" machine. Moreover, the processes of interaction and merging of galaxies may have an indirect influence on the whole combination of problems connected with searches for "young" galaxies. The point is that the presently detected far ($Z > 1.5$) galaxies may turn out not to be young (i.e., not initial), but rejuvenated in the course of interaction and (or) merging /25,26/. To clarify the nature and evolution of QSO and AGN it is necessary to answer a series of other questions.

B. 1. It would be interesting to find the evolution of the chemical composition in QSOs at different Z (a remark on this property was noted in /27/).[‡] However, spectra in very far QSO and radio galaxies turned out to be metal-rich which unequivocally shows that at the time of their activity in any case, massive stars were already able to evolve and enrich the gaseous medium with heavy elements. This means, firstly, that their age

---

[†]See, e.g., /23,24/.

[‡]In /28/ the evolution of the chemical composition (with a maximum at $Z \sim 1.5$) was found in clouds of gas corresponding to formation of the absorption lines Mg II and CIV in QSO spectra.

is >$10^{6-7}$ yr, i.e., the age of QSOs is in no way less than the age of massive stars which were able to enrich the gaseous envelope of QSOs with heavy elements. Secondly, the relative youth of QSOs does not in general presuppose the youth of "host galaxies." In any case, direct spectroscopic studies of nebulae surrounding QSOs and host radio galaxies reliably evidences their stellar composition (e.g., /29/). Therefore, the question arises how to combine the relative youth of the QSO phenomena with the more or less normal ages (~$10^{10}$ yr) of host galaxies? For a resolution of this problem it was supposed by the author in his series of studies that there exist two QSO populations (in analogy with stars of different populations). We shall return to this problem.

One can introduce an additional series of arguments that QSO are an early phase in the evolution of active nuclei as was supposed, e.g., in /30/. Firstly, this supposition provides a natural means of explaining the evolution of more powerful quasars toward less powerful active nuclei. Secondly, the total evolution time of objects with quasar-like nuclei may be significant which provides a means of understanding the observed variability of properties of the galactic systems surrounding them (from more peculiar around QSOs to less peculiar around radio and Secret galaxies). Thirdly, one obtains a reasonable explanation of the fact that in the past the spatial density of QSOs was significantly larger (for $Z \sim 2.5$ approximately $10^3$ times). Fourthly, at early stages of formation the system had a large gas reserve which, probably, enabled an increased activity of nuclei.

2. The lifetimes of compact nuclear radio galaxies in which ~10% of the total radio power is concentrated can hardly exceed $10^7$ yr. This follows from their small sizes (~10 pc) and from the form of the radio spectra. However, with extended radio components the situation is different. Their great extent (to several Mpc), large luminosity (~90% of the total), and steep radio spectra for weak magnetic fields ($\leq 10^{-5}$ G) evidence an age exceeding $10^9$ yr (e.g., in /31/). This means that the radio properties originating at this stage of quasars do not disappear with them, but accompany their evolutionary descendants (radio galaxies) in the form of extended radio components. However, it has been clarified that radio galaxies may be divided into two types with different properties. (The first such delineation was proposed by Fanaroff and Riley in /32/.) It is interesting that different types of radio galaxies have different optical properties:

1. FR II are classical double sources ($P^{1.4GHz}_{tot} > 10^{25}$ W/Hz) with steep radio spectra with an increase of radio brightness at the periphery and a significant flux of emission from hot spots. They have strong radio nuclei ($P^{5GHz}_{N} > 10^{24}$ W/Hz) with strong variability, super-light motions of the double components, and one-sided narrow and long radio ejecta in which the magnetic field is extended along the radio axis. They are situated in the centers of groups of galaxies. In the optical they have strong activity both in the continuous spectrum and in lines. The resolved lines are

16. QUASARS AND ACTIVE GALACTIC NUCLEI                257

very broad and have complex profiles, already forbidden and shifted to the blue (the Fe II lines are weak). A proportionality is observed between the radiated powers in the line (O III) from the nucleus and from extended emission regions. Host galaxies have peculiar morphologies and a blue color characteristic for strongly interacting systems.

2. FR I are irregular forms ($P^{1.4GHz}_{tot} < 10^{25}$ W/Hz) with less steep radio spectra and a decrease of radio brightness at the periphery without significant hot spots. They have weak radio nuclei ($P^{5GHz}_{N} < 10^{24}$ W/Hz), and two-sided wide and not too long radio ejecta with the magnetic field orthogonal to the axis. They are situated in the central regions of clusters. The activity of their nuclei in the optical is 10–100 times less in comparison with FR II. The resolved lines are narrow with smooth profiles (the Fe II lines are strong). The host galaxies are red and less distorted by interaction.

Among astrophysicists there is no generally accepted opinion yet relative to the cause of the differences of properties in radio galaxies of types FR II and FR I. Some consider that the cause is the difference of parameters of magnetohydrodynamic jet flows (ejecta) from active radio galaxy nuclei and others that the cause is different properties of radio galaxy surroundings, i.e., in different properties of the intergalactic medium in galactic clusters and outside them (see, e.g., /33,34/). It is also impossible to completely exclude the influence of evolutionary changes connected with a weakening of the activity of nuclei in radio galaxies and an increase of the medium density around central galactic clusters with time. In fact, one can suppose that radio galaxies of the type FR I in clusters of galaxies are far evolved radio galaxies of the FR II type in which the activity of the nucleus is already weakened and the extended radio components are pressed by "cooling flows" to the parent galaxies.

Although the radio nuclei of quasars in the radio ejecta are, on the average, ~10 times more powerful than for radio galaxies, in their global radio structure they also are divided into types similar to FR II and FR I which are designated in /35/ as D2 and D1. It is interesting that for both radio galaxies and quasars the transition from one type to another occurs for approximately identical $P_{tot}$.

Apparently, all this may strengthen the conclusion of using evolutionary transitions from quasars of types D2 and D1 to radio galaxies of types FR II and FR I, respectively. Moreover, starting from the differences of host galaxies, one may conclude that there are different evolutionary tracks of objects with strong and weak radio emission and they can hardly transfer from one to another. An active nucleus at the early stage of its development either is a strong radio source (quasar) or is not (quasi-galaxy) and in further evolution goes either to a radio galaxy of another type or to a normal galaxy with an active nucleus, respectively. It is interesting that strong radio emission is formed only in nuclei of peculiar spheroidal systems which have an unusual excess of gas.

C. The detection of QSOs which due to their large luminosity expanded the boundaries of the observed metagalaxy to $Z \sim 5$ /36/ (Fig. 16.1), however, did not solve the question about the properties of young galaxies. On the one hand, known QSO have more or less normal chemical composition which means they are not initial objects. On the other hand, their accompanying spatial density grows sharply to $Z \sim 2.5$, but for large Z this growth slows or even a decrease is observed /37/. This makes the order of the day the question of searches for $Z > 3$ objects, being either a new population, or the progenitors of the observed QSO. On the other hand, the galaxies found for $Z \sim 2$–$3$ (and not QSO) are also already far evolved stellar systems in which, although the fraction of young stars is large, there is no visible deficit of heavy elements /38/.

All these facts force us to consider young galaxies as massive systems in which for about a billion years a significant fraction of gas transfers to stars with a standard mass function which are little observed. More probably, one can suppose that galaxies with $Z \sim 2$–$3$ are formed in the merging of smaller complexes in the composition of groups of dwarf galaxies in which star formation already occurred for $Z > 4$–$5$. With this scenario the observed galaxies with $Z \sim 2$–$4$ are not young, but revived due to star-formation bursts connected with processes of interaction and merging. It is possible that active nuclei, i.e., QSO and initial dwarfs and secondary gigantic galaxies will have considerably different properties: basically, we now observe QSO II[†] that are active nuclei in secondary galactic systems. A transition is required to the identification of weak ($m_B > 25$) objects at $Z > 4$–$5$ to solve the problem about the possible existence of weakly active nuclei in initial dwarfs (QSO I, i.e., quasars of the first generation).

In this connection, searches of far galaxies are at the present one of the basic problems of astrophysics. Here large attention should be given to a careful analysis of the properties of weak radio sources among which very far galaxies should be present. The question is whether on the basis of these or other known features of radio sources one can distinguish those which are connected with far galaxies among them /39,40/.

In this case it should be emphasized especially that the epic corresponding to the maximum spatial density of QSOs ($Z^* \sim 2$–$3$) may turn out to be distinguished also in another relation. The point is that for these $Z^*$, apparently, very intense merging occurs in rich groups and core clusters of initial small mass galaxies into massive ones, mergers. This process should be accompanied by an increased frequency of QSOs in systems and strong IR emission from heated dust which in addition may serve as a pump for maser emission on OH molecules /41/. In this connection the question arises about the possible contribution of radiation from the epoch with $Z^* \sim 2$–$3$ to the background radiation in the wide range of wavelengths shifted by $(1+Z^*)$; this range may extend from the radio to the optical /42/.

---

[†]QSO II are quasars of the second generation.

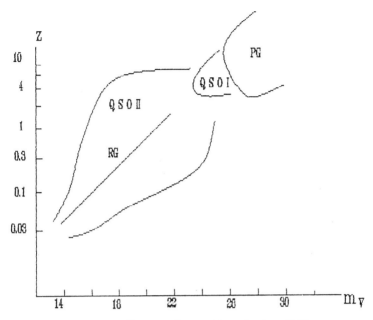

Figure 16.1. Hubble diagram for radio galaxies (RG), quasars of the I and II populations (QSO I and QSO II), and protogalaxies (PG).

## 2. HISTORY OF QUASAR DETECTION

A. After the detection at the beginning of the '60s of the first quasars amongst compact radio sources and then optical quasars as the brightest, the farthest, and the most variable of presently observed metagalactic objects a series of complex problems arose among astrophysicists. Firstly, would it be necessary to investigate the relation of QSOs to galaxies in their groups and clusters? It already became clear at the beginning of the '70s that QSOs are a very active stage in the evolution of the nuclei of massive galaxies. The QSOs intense in the radio turned out to be connected with spheroidal host galaxies[†] and those weak in the radio with spheroidal or disk galaxies (a situation similar to the one observed previously in the association of radio sources with galaxies). It became clear at the beginning of the '80s that QSOs belong in the composition of tight groups of

---

[†]Host galaxies is a term used for galaxies with strong active nuclei.

interacting galaxies, being the nuclei of their most bright central members and that a correlation exists between the luminosities of QSOs and their host galaxies. The scale found of QSOs of their spatial distribution (~10 Mpc) is similar to the scale of distribution of groups and falls between the corresponding scales for galaxies (~5 Mpc) and galaxy clusters (~15 Mpc). Significantly, it was necessary to construct a luminosity function of QSOs at small Z in order to compare them with similar functions for other galaxies (e.g., Seyfert, Markaryanovskii, and radio galaxies). It would also be necessary to follow the change of the form of the QSO luminosity function with a growth of red shifts in order to investigate the degree of influence of cosmological effects and for estimating the rate of evolution for different luminosities. All of this would help estimate the lifetimes of QSOs in comparison with other GAN as well as to understand their place among other objects with active nuclei. Thirdly, with established activity signatures, it would be desirable to have even a qualitative classification of AGN, allowing one in the future to also apply quantitative criteria. For this case it would be necessary to put special attention on the role of selection effects, especially in the case of the anisotropic character of radiation from AGN when the axis inclination of radio ejecta in relation to the line of sight should strongly influence the observed properties. Fourthly, it would be necessary to make some type of assumptions relative to the paths of evolution of AGN, how do they originate and how their properties change with time. A priori, it would be impossible to reject the possibility of their recurrence, i.e., a temporal repetition of the properties of AGN. And finally, starting from observations it would be desirable to clarify the necessary conditions required for the appearance of active nuclei. This would help make a choice between different models of AGN because there should exist a definite connection between these understandings.

Twenty-five years of QSO research showed that the assault of the problem of "active galactic nuclei" using quasars did not succeed and requires a long siege with the incorporation of data on the nearest active galaxies. However, QSOs turned out to be indispensable as markers of the universe, which, because of their exceptional brightness allow one to obtain information on the epoch ($Z^* \sim 2$–$3$) in which the intensive birth of massive systems with very active nuclei occur. In addition, QSOs are widely used for "deep probes" of the metagalaxy bearing information about the objects and media occurring at the threshold of the line of sight. Special interest in recent years was caused by a study of tight pairs or groups ($\Delta\Theta < 20''$) of QSOs, some of which are possibly the result of massive invisible gravitational lenses situated between the observer and the QSOs. Although the majority of them, most probably, are true pairs of galaxies with active nuclei (e.g., /43,44/).

B. As often occurs in science, ten years may occur between the detection of one or another phenomenon, and the comprehension and understanding of its place among other phenomena. This occurred with the phenomenon of active galactic nuclei which were already observed 50 years ago, but were not understood at that time. We are speaking of observations of objects with emission lines (OEL) begun already by E. A. Fath (USA) in 1908 (see the history of the problem in /45/). The greatest contribution to this problem was brought by Carl Seyfert (USA) in 1943 by obtaining spectra of the bright nuclei in galaxies: NGC 1068, NGO 1275, 3516, 4051, and 4151, and NGC 7419. In their spectra he detected strong and wide emission lines of $H_\alpha$, [OIII], $H_\beta$, etc. and galaxies with these nuclear spectra subsequently obtained the designation Seyfert galaxies. In 1956 G. Aro (Mexico) showed that there is a correlation between the existence in the spectra of objects of emission lines and the blueness of their light. On this basis a method was proposed of searching for OEL using three-color (UBV) photometry. From the beginning of the '60s in Byurakan Beniamin Egishevich Markaryan began to carry out on the one-meter Schmidt using an objective prism sky scans to find objects with noncoincident color and spectral characteristics. Although from the beginning this study was planned for a search of hypothetical "bodies" of a nonstellar nature, already from 1965 Markaryan and his coworkers began to publish lists of "Markaryan galaxies." If their first review dealt with galaxies to $17^m.5$, then the second begun at the end of the '70s extended the limit to $19^m.5$. It was clarified that Markaryan galaxies constitute 20% of the S galactic field. About 85% of galaxies with an ultraviolet excess have emission lines in their spectra, but ~10% of these relate to the Seyfert type. Here it should be noted that weak emission lines, manifested in higher dispersion scans than scans with an objective prism, already exist in ~90% of the nuclei of S galaxies, i.e., weak emission is a quite general phenomenon. In 1936–1940 F. Zwicky and T. Humason carried out a scan on an 18-inch telescope of high latitude areas to $15^m$ and detected small blue stars which they considered dwarf stars of the corona of our galaxy with a small metal enrichment. They noted that the number of these objects grew in transition to weaker stellar values and planned to carry out a scan on the 48-inch telescope with $m_{lim} = 20$. Unfortunately, such a scan was not carried out in those years. However, even among their bright small blue stars Humason and Zwicky detected "blue compact galaxies" (BCG), about 50% of which as clarified only in the beginning of the '70s have emission spectra and 3% are of the Seyfert type. As became clear in the '60s, some bright quasars were among them.

In 1958 V. A. Ambartsumyan /4/ gave a talk at the XIth Solvay Congress on the possible role of activity of nuclei in galactic evolution. In this talk he assembled the evidence existing at that time on the appearance of the phenomenon of activity in the nearest galaxies and radio galaxies. He advanced the hypothesis on the existence of kernels of stars and galactic nuclei of "prestellar matter" (or, the so-called "D-bodies") which lead to processes of ejection and the division of objects into dispersing components. And although during the last thirty years this point of view did not obtain special development, however, the idea itself of the activity of nuclei turned out to be extremely productive. This was verified again in the discovery in 1963 of quasars which, from the point of view of the hypothesis of Ambartsumyan, should be active nuclei in a pure form.

C. The term quasars, quasi-stellar sources (QSS), was introduced for a designation of compact radio sources (CRS) identified in the optical with star-like objects. The history of the detection of QSS is quite instructive both from the point of view of overcoming dogmas in science and from the point of view of the role of radio astronomy in manifesting new types of objects. At the end of the '40s radio astronomers detected several discrete sources of radio emission in the Cygnus, Virgo, Centaurus, Taurus, and Cassiopeia constellations. In these years a part of them were identified with gigantic E galaxies situated in the central regions of clusters and groups of galaxies. These radio sources obtained the name radio galaxies (Cygnus A, Virgo A, and Centaurus A). Another part of the discrete radio sources were identified with supernova remnants: the Crab Nebula in the Little Dipper, the remnant in Cassiopeia, etc. In 1953 I. S. Shklovsky put forth the brilliant idea on the synchrotron nature of the radiation of the Crab Nebula in the range from the radio to the optical and in 1954 he extended it to radiation of the ejecta from the M87 nucleus (Virgo A) /46/. A verification of this hypothesis was the detection of optical polarization both in the Crab (V. L. Ginzburg and I. M. Gordon) and in M87 (V. Baade). On the basis of synchrotron theory in a series of studies in the '50s and '60s I. S. Shklovsky estimated the energetics of relativistic particles in magnetic fields and radio galaxies /47/. Analyzing the problem of the identification of extended extragalactic radio sources, V. Baade and R. Minkowskii put forth the hypothesis in 1954 of the collisional (catastrophic) nature of radio galaxies. In this case they started from the fact that the radio source Cygnus A was identified in the optical with a "colliding" galaxy. It was clarified that this hypothesis could hardly relate to a series of collisions. More probably, it is similar in cross section to the dust band of the radio galaxy Centaurus A which, if it collided, occurred in the distant past. Many astrophysicists were against the collisional nature of radio galaxies including J. S. Shklovsky who, together with P. N. Kholopov, identified the radio galaxy Fornax A. He supposed that the strong radio emission from radio galaxies was connected with the occurrence in the past of very intense sources of relativistic particles from multiple SN ($\sim 10^4$/yr). In this active period of the galaxy, the "supernova" or "exploding" phase one could have $M_V =$ $-(26-27)$. Shklovsky proposed to search for these "eruptive" far galaxies using the strong radiation from them in optical lines. Thus, already in the sixties, Joseph Samuelovich in fact predicted that far radio galaxies could be seen as extended intense emission regions /48,1/.

However, a series of discrete less bright radio sources remained unidentified and the suspicion arose that they could be connected with stars of our galaxies and the term "radio star" appeared. However, to reliably identify a radio source with a star would require an angular resolution of several arc seconds. At that time such accuracy could be achieved only by lunar occultation, successfully applied by Australian radio astronomers. At the beginning of the '60s coordinates were obtained of several compact radio sources of the 3C catalog ($F_{178\ MHz} > 9$ Jy). At the end of the '60s at the 107th Meeting of the

American Astronomical Society, A. Sandage proved the identification of radio source 3C 48 with a $16^m$ star surrounded by a red nebula with angular size 6"×12". Strong lines were visible in the spectrum of this star which could not be identified with known stellar lines /49/. In 1963 the radio sources 3C 196, 3C 286, and 3C 273 were also identified with "stars" /50/. The student of J. Oort, Martin Schmidt, working in the USA used the lines and their spectra for the identification. In particular, he obtained the spectrum of the object 3C 273 identified with a $12^m.8$ star with a blue 20" ejecta reminiscent of the ejecta from the nucleus of the radio galaxy Virgo A (NGC 4486) also known by the photographs of G. Curtis from the beginning of the century. In the radio range 3C 273 consisted of a compact source (B) and an extended ejecta (A). The lines in the spectrum of the radio source 3C 273 initially were not identified. However, in 1963 Schmidt succeeded in solving this riddle /51/. He guessed that the lines in the spectrum of 3C 273 were the well-known spectroscopic signatures of the Balmer series of hydrogen, but shifted to the red part of the spectrum by $\Delta\lambda = \lambda_o z$. It followed from this that with all the lines one could calculate the red shift $Z = (\lambda - \lambda_o)/\lambda_o$ which turned out to be 0.158. From this Schmidt made the conclusion that 3C 273 is an extragalactic object, i.e., Z is a cosmological red shift. The guess of Schmidt served as the key to deciphering the spectra of other QSS. Thus, for example, for 3C 48 the red shift was 0.367 /52/. Temporarily these objects obtained the name "interlopers," interfering, as it was thought, because they were small and wandering among radio stars. However, it was soon clarified that not only strong radio sources had significant red shifts, but also a series of blue "small stars," not being in general radio sources.

In 1965 A. Sandage made the stunning discovery /53/ that all blue high velocity small stars weaker than $16^m$ are radio quiet quasars, quasi-staller galaxies (quasi-galaxies). According to his estimates their surface density at $19^m$ is about 4/□° which exceeds by a factor of 500 the density of radio quasars at the same $m_{lim}$. Studies of Sandage received strong criticism from a series of astronomers. In particular, from T. Kinman and F. Zwicky /54/ who considered that the majority of objects measured by Sandage are stars of our galaxy. They turned out to be correct for 90%, but all the same 10% of the objects discovered by Sandage turned out to be a new radio quiet population of quasars. They obtained the name quasi-stellar galaxies or quasi-galaxies and in density they exceed the population of QSS by 50 times. It was clarified that objects of this sort in small numbers already occurred in the scans of Humason and Zwicky /55/ (the object HZ 46 with Z = 0.045) and Aro-Leiton (the object Ton 202 with Z = 0.366). Thus, in principle, the discovery of quasi-stellar objects (QSO) could have occurred much earlier than 1963–1965. However, without the prompting of radio astronomy this rare type of object would not have been identified amongst the much more numerous population of weak stars of the galactic corona. According to the opinion of D. Weedman /56/ there is still another reason of a psychological nature. Astronomers were attracted to the idea that extragalactic objects could only be extended formations of the type of galaxies. They could not imagine that one could encounter "starlike" extragalactic objects, even the basic role of galactic nuclei was discussed only after 1958.

After the recording of the spectra of the first quasars the majority of astrophysicsts unreservedly took the point of view of their extragalactic nature, i.e., agreed with the cosmological nature of red shifts in their spectra. However, in principle, there are other possible explanations of the red shift: these are gravitational and Doppler. For example, J. Burbidge and Halton Arp took the point of view of a local nature of QSOs with quasars situated considerably closer than follows from the red shifts. In attempts to verify his point of view (see, e.g., H. Arp (1986) (IAU Symp. N 124), Halton Arp, one of the most remarkable observers of the twentieth century, continued to introduce more and more examples of connections between QSO and galaxies having completely different red shifts. A clear example of this situation is the case of the galaxy NGC 4319 (Z = 0.005) and the QSO Mrn 205 (Z = 0.072), connected, according to the opinion of H. Arp by a light filament /57/. It is true that other observers considered the observed "bridge" as a purely projection effect. The careful analysis carried out by Arp of examples, undoubtedly, enabled a deeper understanding of the nature of QSOs, and the argument of the opinions of proponents of different directions did not give astrophysicists laurel crowns.

Among those who met the discovery of a new cosmological population of objects, quasars, with enthusiasm were the leading Soviet astrophysicsts: V. A. Ambartsumyan, J. S. Shklovsky, Ya. B. Zeldovich, V. L. Ginzburg, S. B. Pikel'ner, et al. We already mentioned the role of Ambartsumyan. Immediately after the discovery of QSOs, J. S. Shklovsky proposed to seek variability in the optical range for them. This was done for the quasar 3C 273 in 1963 by A. S. Sharov and Yu. N. Eframov with the old plates of the Shternberg Astronomical Institute /58/. In 1965 Shklovsky /59/ put forward the idea about the existence of "mini" quasars in the nuclei of Seyfert galaxies. In the same year G. B. Sholomitskii, a student of Shklovsky's, discovered variability in the QSS CTA 102 in the radio range /60/. After the detection of five QSOs, Ya. B. Zeldovich tried to estimate the rate of their cosmological evolution and then, together with I. D. Novikov gave a lower limit to the mass of QSOs /61/,[†] starting form the assumption that the mass of QSOs should be sufficient for their maintenance from break-up due to the force of light pressure. V. L. Ginzburg together with L. M. Ozernoy and S. I. Syrovatskii /62/ discussed the problem of the mechanism of optical radiation of the object 3C 273 "B." Soviet observers also contributed to the study of AGN and quasars. Among them one can list: E. A. Dibai, M. A. Arakelyan, V. I. Pronik, I. I. Pronik, K. K. Chuvaev, E. E. Khachikian, V. F. Esipov, V. M. Lyutyi, V. Yu. Terbizh, V. L. Afanas'ev, Yu. N. Pariiskii, S. Ya. Braude, and N. S. Kardashev.

One can find more information on the history of the discovery of QSOs from the monographs and studies, for example, of J. Greenstein, Margaret and J. Burbidge, M. Elvis, and V. Wilknes.

---

[†]For a flux $\sim 3 \times 10^{47}$ erg/s one obtains $M \geq 3 \times 10^9$ $M_\odot$.

## 3. THE HYPOTHESIS OF SEVERAL QSO POPULATIONS

A. 1). We have already talked enough about the fact that QSOs are a short-lived ($<10^7$ yr) very bright phase in the evolution of giant galactic nuclei. Thus, for example, according to /63/ there is a direct dependence between the luminosity of the host galaxy and the average energy output of the nucleus. However, what can one say about the conditions for formation of the largest galaxies and the properties of their evolution? It is appropriate to start from two important observational aspects. First, hopes for the existence at large Z of gigantic galaxies in which bursts of star formation would have encompassed all their mass in a short interval of time ($\leq 10^{8-9}$ yr) were not verified /64/. Secondly, it was clarified (e.g., /65/) that with the normal form of the initial mass function of stars formed in protogalaxies one could not explain the observed dependence (V–R) = f(Z) for distant E galaxies. It is possible that one can relate these difficulties in the framework of the standard theory of formation of large galaxies from gaseous massive protogalaxies to the limitation of $\Delta T/T < 10^{-5}$ in the inhomogeneity of background relic radiation in the radio range on scales of angular minutes /66/.

All these facts led to the approach to the problem of galaxy formation from several other positions. These were namely: it was assumed that the stage of a massive gaseous protogalaxy, in general, did not exist. Gigantic galaxies arise in the merging of smaller formations ($M < 10^{10}$ $M_\odot$) in which the process of star formation has already occurred. Thus, the merging process and the intensification of star formation connected with it in a gas compressed by shock waves could extend in time. Therefore, one succeeds to a considerable degree in decreasing the brightness of a young giant galaxy. This means that a young galaxy in an extended not too bright form with a low surface brightness could be detected at small red shifts (Z ~ 1–2) although the merging process itself of initial components with masses from $10^6$ to $10^9$ $M_\odot$ into a single conglomerate could also start for Z > 5. In particular, this picture follows from the theory of cold dark matter (CDM) which was proposed for an agreement with the observed pattern of the large-scale distribution of galaxies. In the framework of this theory (e.g., /67/, /68/) the phenomenon of galaxy formation is viewed as a continuous process in which star formation begins for Z = 10 in small-mass clouds, but the maximum luminosity of mergers is reached for Z = 2–3. Thus, simultaneously with the growth of the mass of the total formation due to accretion the mass of the central body of the nucleus and its luminosity also reaches a maximum for Z = 2–3. This agrees well with the observed evolution of the QSO light function (see, e.g., /69/, /70/).

2). After the detection by Sandage of radio quiet quasi-stellar sources it became clear that QSS strong in the radio (i.e., quasars) comprise altogether several percent of a more numerous population of radio weak quasi-stellar galaxies. Thus, it was clarified that QSS and quasi-stellar galaxies had different rates of cosmological evolution (e.g., /71/). Differences in the rates of cosmological evolution were also noted for QSS with steep ($\alpha_r > 0.2$) and flat ($\alpha_r < 0.2$) radio spectra /72/ This is connected either with the different lifetimes of these radio sources and (or) with a difference in the properties of the surrounding medium /73/. Attempts were also made to divide QSOs into groups according to properties of optical continuum /74/ and line /75/ spectra.

In recent years considerable attention was given to dividing QSOs according to properties of their surrounding host galaxies which, as it was clarified, differ in optical spectra /76/, colors /77/, and radio and X-ray properties /78/. In a series of studies this division was connected with the identification of QSS with elliptical galaxies (EG) and quasi-stellar galaxies with plain systems of the type of Seyfert galaxies. Thus, in all probability, these differences in properties of host galaxies from QSOs extend to more normal galaxies: radio galaxies are identified with E galaxies, but plain systems, on the contrary, are not strong radio sources.

A somewhat different aspect of the problem of two types of QSOs was considered in /79, 80/. It was proposed in these studies that the existence of two types (or populations) of QSOs might explain the observed form of the dependence of the number of QSOs in Z (or the form of the change of the spatial density of QSOs in Z) which has a maximum somewhere in the range $Z^* = 2.5-3$. (It was noted in /81/ that the value $Z^*$ may increase with increasing QSO luminosity). The point is that the appearance of a maximum in the rate of formation of QSOs at $Z^*$ can be understood if one makes the quite natural assumption about the connection of the phenomenon of the activity of nuclei with processes of interaction or (and) merging of galactic structures (e.g., /82,83/). According to similar logic the existence of two QSO populations was admitted in /79/:

1. QSOs of the first population (QSO I) are active nuclei in single, not very luminous, small mass, and far ($Z > 4$) galaxies. They should be dominated by evolutionarily young objects if they exist in optical counts of QSOs with $m_B > 25-26$.
2. QSOs of the second population (QSO II) are very active nuclei in luminous, massive, and not too far ($Z < 3$) systems that are the products of merging of less luminous members in groups and clusters. These in some sense secondary objects will dominate in optical counts with $m_B < 23$ and in radio counts (Figs. 16.1–16.3).

Estimates of the characteristic merging time ($\tau_{mr}$) of members into groups and central regions of clusters and into a single giant galaxy (merger) carried out by numerical simulation (e.g., /84, 85/), show that $\tau_{mr} \sim \kappa\tau_{cr}$, where $\kappa = 1-3$ and $\tau_{cr}$ are crossing times in systems equal in order of magnitude to $\sim 10^9$ yr. These $\tau_{mr}$, depending on the model of the universe used, are in units of red shifts in the range $\Delta Z = 2-3$. Thus, it becomes understandable in principle why during the epoch of galaxy formation distant from us by $Z^{**} \cong 5-6$, the epoch of QSO II formation is observed for $Z^* \cong 2-3$. For $Z > Z^*$, i.e., at epochs when galaxy merging into gigantic systems had still not occurred, QSO II should not have been observed. If this is really the case, then in a model of the

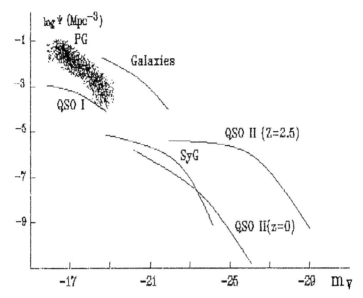

Figure 16.2. Luminosity function in the optical for normal galaxies, Seyfert galaxies (SyG), quasars of the I and II populations (QSO I and QSO II) for different Z, and protogalaxies (PG).

universe with $q_0 = 0.5$ the temporal delay $\Delta t$ for different values $Z_{\text{gal. form.}}$ and $Z_{\text{QSO II}}$ are (for $H_0 = 100$):

| $Z_{\text{gal. form.}}$ | $Z_{\text{QSO II}}$ | $\Delta t$ (yr) |
|---|---|---|
| 10 | 3 | $6.6 \times 10^8$ |
|  | 2 | $1.1 \times 10^9$ |
| 5 | 3 | $3.7 \times 10^8$ |
|  | 2 | $8.2 \times 10^8$ |

Thus, the quantity $\Delta t \cong 5 \times 10^8 - 10^9$ yr is obtained completely acceptably for the theory of galaxy merging because it turns out to be comparable with the crossing times for groups and poor clusters of galaxies:

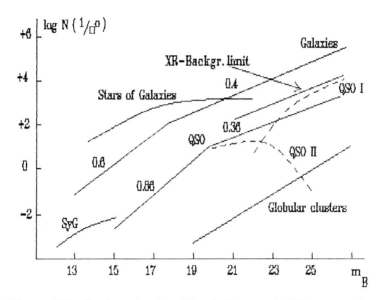

Figure 16.3. Surface density (N) of objects of different types as a function of observed stellar magnitude ($m_B$). The numbers on the lines indicate their slopes. The arrow indicates the limiting value of the surface density of QSO I, obtained with data on X-ray background radiation.

$$\tau_{cr} \cong R/G \cong 500 \text{ kpc}/1000 \text{ km/s} \sim 5\times10^8 \text{ yr.}$$

We assume that for $Z > Z^*$ only QSO I will be basically observed, although it should be noted here that their existence is more probable than necessary. This is connected with the fact that the formation of active nuclei in initially gas rich, but not very massive galaxies is completely natural and does not contradict the presently existing observational data (e.g. /86/). However, a final answer on the problem of the existence of the hypothetical QSO I should be given by future optical studies at the level $m_B > 25$, where the expected surface density may reach $10^3/\square°$ /79/ (Fig. 16.3). This problem is completely resolvable using the Space telescope. Even with observations with large telescopes on the Earth using the new generation of light receivers (such as PZS), results have already been obtained of galaxy counts to $m_r = 26$ /87/. Their number reaches $1.5\times10^5/\square°$ and corresponds to a slope of the curve log N(m) characteristic for galaxies with m > 22. It is interesting that QSO counts at different Z give a similar dependence,

## 16. QUASARS AND ACTIVE GALACTIC NUCLEI

but with a number of objects at $1/\square°$ which is about 20 times smaller /88/ (See Fig. 16.3).

3). The assumption of the existence of two QSO populations puts some limitations on their properties and on the properties of systems in which they arise from the point of view of agreement with observational data. In our opinion /79/ the most serious limitations are: 1) data relative to the background radiation and 2) the relative avoidance of relatively near QSO II occupied by rich clusters of galaxies.

a. According to the estimates of /89/ the magnitude of the background radiation in the energy range 2 keV is $\sim 1.16 \times 10^{-29}$ erg/cm$^2$ s Hz. For estimates of the contribution to the X-ray background of the radiation from QSO the dependence N(<m) observed for them was used or $\phi(M_B)$ which for a given spectral index between the optical and X-ray ranges ($\alpha_{OX}$) can be introduced into the dependence $N(> S_{XR})$. According to /90/ $N(> S_{XR}) \cong 2.7 \times 10^{-16} \, S_{XR}^{-1.53}$ which for the limiting flux from HEAO "B" $\sim 2.6 \times 10^{-14}$ erg/cm$^2$ s lies in the range 0.5–3.5 keV $(6.3 \pm 2.6 \times 10^4)$ of X-ray sources per steradian. The estimates carried out show that from QSO with $m_B < 20$ the contribution to the X-ray background is ~25% which corresponds to 20 sources per square degree for $\alpha_{OX} \cong 1.5$. The radio emission of QSO has, as a rule, $\alpha_{OX} \cong 1.3$ and is, on the average, three times more powerful than the X-ray sources. However, due to their small projected density, the basic contribution to the background is given by non-radio-emitting QSO.

It is clear from what has been said /79/ that data on the background radiation did not contradict the existence of objects of the QSO I type, although to be sure they give some limitations on the form of their electromagnetic spectrum and the rate of evolution of their projected density in the past. Overall, one may allow the following parameters of spectra and the slope of the optical dependence for QSO I and QSO II:

|  | QSO I | QSO II |
|---|---|---|
| $\alpha_{ro}$ | <0.5 | >0.5 |
| $\alpha_{OX}$ | >1.55 | <1.55 |
| b | <1.5 | >2 |

The search for QSO I in the radio range is apparently not very promising. This follows both from data of direct counts of weak radio sources (e.g., /91/) at the level $S_v < 0.01$ Jy and from limitations on their spatial density following from data on observations of the background in the radio range. It also follows directly from studies of QSO and their light functions both in the radio and in the optical that for $Z \cong 1$ the fraction of QSS strong in the radio decreases by almost 10 times in comparison with the situation at $Z \cong 0$. This is most probably connected with the faster evolution of the radio luminosity in comparison with the optical (this relates especially to QSS with flat spectra in the radio) or with the change of conditions necessary for the formation of strong radio sources at high Z.

b. Concerning the entry of QSO into galaxy clusters, as clarified in recent years (e.g., /92/), the probability of this occurrence grows with Z. However, there are well-known cases of the entry of QSO into near clusters. For example, QSO 3C 206 (Z = 0.2) is situated in the center of a cluster of richness R = 1 /93/. In our opinion, all these facts may be explained in the framework of our assumptions relative to several QSO populations. In fact, in the central parts of rich clusters the merging of small mass members into massive mergers will occur at earlier epochs ($Z^* \cong 2-4$) at which the maximum spatial density of QSO II will be reached. The formation of mergers in centers of groups and poor clusters having dynamic properties different from rich clusters will occur at later epochs right up to the present. In these epochs in centers of rich clusters mergers of the II population may be formed (with the merging of mergers of the I population) whose active nuclei may be observed as QSOs of the third population. It is possible that the QSO 3C 206 is related to them.

B. 1. Can one make any kind of conclusions on the existence of objects of the QSO I type today, without waiting for information on direct counts of starlike objects at a level $25-26^m$ with deep scans? In our opinion /79/ this possibility exists. It is true that it depends on the reliability of our assumptions relative to the evolutionary properties both of the QSOs themselves and their host galaxies, i.e., on the reliability of assumptions about the variability of observed properties of QSO with time. It is quite probable that the differences in properties of radio galaxies of types FR I and FR II may be connected with properties of galaxy merging processes in clusters and groups, respectively. In this case it is also necessary to take into account that there should be a strong influence of the morphology of members participating in the merging process on the properties of the central formation (merger) formed in the process of multiple mergers. When many gas-rich S galaxies participate in the merging (as is characteristic for poor irregular clusters in groups of galaxies), then a gE galaxy is formed with a stronger radio source of the type FR II. In the central parts of rich regular clusters more gas-poor spheroidal galaxies participate in merging and although very massive cD galaxies may be formed, in their radio luminosity they are less than gE galaxies, i.e., here radio galaxies of type FR I arise. And probably the important thing here is not only the quantity of gas brought into the merger, but also the fact that the galaxies in merging also carry a contribution in the form of magnetic fields amplified in winds of spiral galaxies. It is possible that just these fields serve as a trap for the generation of regular strong magnetic structures required for the properties of radio emission of strong radio sources. It is not excluded that in the merging of a large number of S galaxies the large specific angular momenta which they carry are important. It is possible that the fast rotation of the central parts of radio galaxies is connected with just this fact /94/. This, in turn, facilitates the amplification of magnetic fields using the hydrodynamic dynamo mechanism /95/.

2. We consider now the possible evolutionary transformation of QSO I. It is already

clear that from the determination of QSO I as objects not having a stage of rapid merging that they could hardly be strong sources of thermal IR radiation. However, BSO in the initial galaxy as clarified, apparently, did not have a large quantity of gas (e.g., /64/). The dust in them should also not be large. Thus it is hardly possible that objects of the type QSO I could be revealed as the result of deep IR scans. This also relates to a search of QSO I in the radio range since they are most probably weak radio sources. (This is evidenced both by the identification of objects corresponding to the thickening of the log N–log S curve at the weakened ($S_V <$ mJy) and to the limitation on the radio background. Therefore, we give basic attention to the question which of the observations in the optical at the present time of galaxies with one or another signature could be related to far highly evolved host galaxies connected in the past with QSO I.

It was shown in studies, for example /96/, that in regions of "voids" of neutral hydrogen there is not more than $3 \times 10^{10} \, h_{100}^{-2} \, M_\odot$. However, it was shown in /97/ that in spectra obtained on IUE in the UV range absorption lines of the ions C IV, Si IV, and L$\alpha$ are visible in the gap from far QSO. If one assumes that these lines are formed in clouds with $R = 100 \, h_{100}^{-1}$ kpc, then the gas mass is estimated as $10^9 \, h_{100}^{-2} \, M_\odot$. In this case the column density turns out to be $\sim 3 \times 10^{14}$ cm$^{-2}$ (for Si IV and C IV) and $\sim 10^{19}$ cm$^{-2}$ (for H I).

In a series of studies searches were applied inside "voids" of galaxies with low surface brightness. For example, in /98/ with the Uppsala catalog of galaxies no tendency was detected to a blurring by them of voids. The authors of /99/ came to a similar conclusion and showed that 58 galaxies with $m_B < 15.5$ and $m_B = -(17-13)$ are situated along walls of "voids." However, some galaxies inside voids were detected all the same (e.g., /100/). The presence in the void of Bootes galaxies of the Markaryan type were also indicated by the authors of the Markaryan survey /101/. Although in /102/ the presence of a void was verified in Bootes (D = 62 Mpc, center coordinates: $\alpha = 14^h \, 50^m$, $\delta = +46$, $v_r = 15{,}500$ km/s), it was noted at the same time that there was a presence in it of some number of emission galaxies. The properties of these eight galaxies was studied in detail in /103/. It was clarified that for absolute stellar quantities $m_v = -(21-19)$ these galaxies have relatively small sizes ($\sim 10$ kpc) and are strong sources in the line [OIII] ($W_{[OIII]} \sim 10$ A). Such strong radiation in this line is encountered only in $\sim(5-7)\%$ of the galactic field in which one could have expected to detect about 70 within the size limits of this void. However, 10 times fewer were detected and no emission galaxies were detected in general. This fact has not yet found a satisfactory explanation. Starting from theoretical assumptions, in a series of studies (e.g., /67, 104/) the conclusion was made about the possible formation at late epochs inside voids of dwarf galaxies of low surface brightness. However, the emission galaxies detected in Bootes are not dwarfs either according to their luminosity or their surface brightness.[†]

---

[†] In /105/ in 21 cm a dwarf galaxy ($m_v = -15$, $R_{21 \, cm} = 10$ kpc, $M_{H \, I} \cong 2 \times 10^8 \, M_\odot$, $L_{21 \, cm}/L_{"v"} = 1.5$) was detected in one of the voids.

It is presently impossible to give a final answer on the question whether the galaxies detected in voids are posterity of QSO I. Additional studies of the properties of these galaxies in Bootes and a search of similar objects in other known deserts are required for this conclusion. However, if it were so then the interesting and important conclusion would follow on the existence in different epochs of different types of perturbations. For $Z^{**} \sim 6-4$ these perturbations should lead to the formation of not too massive galaxies (e.g., /68,106/) whose nuclei could show themselves as quasi-galaxies of the first population. For $Z < 4$ larger scale perturbations should exist leading to the formation of large-scale structures in which the probability of interaction and merging of separate initial galaxies into larger systems was large. The maximum rate of the merging process, apparently, occurred at $Z^* \cong 2.5-3$, where the largest spatial density of active nuclei of these mergers is observed, i.e., the QSO II population (see /79/).

Thus, a study of the properties of galaxies encountered in large-scale voids, possibly, can help shed light on the question of the possible existence of quasi-galaxies of the I population. In conclusion we note that the far evolved descendents of quasi-galaxies I should have a series of specific properties caused by the fact that evolution "in the pure form" should be observed in them, i.e., not distorted by processes of interaction and merging and, as a result, without bursts of star formation induced by them. Therefore, in all probability, in these "pure" galaxies more gas could have been preserved in them and, possibly, the process of star formation would be extended in comparison with systems where star formation had a more rapid (induced by mergings) character. Undoubtedly, this should affect the chemical composition of these galaxies toward their impoverishment in metals.

In general, searches of distant descendants of QSO are an important problem both from the point of view of cosmology (e.g., /107/) and from the point of view of clarifying the evolutionary properties of QSO themselves (e.g., /86/, /108/) (see Table 16.1).

## References

1. Shklovsky, I. S. (1962) *Usp. Fix. Nauk.* **77**, 1.
2. Shklovsky, I. S. (1964) *Vestnik AN SSSR* **2**, 22.
3. Vorontsov-Velyaminov, B. A. (1978) Extragalactic Astronomy, Nauka, Moscow.
4. Ambartsumyan, V. A. (1958) *Izv. AN Arm. SSR* **11**, 9.
5. Kristian, J. (1973) *Astrophys. J.* **179**, L61.
6. Wyckoff, S., Wehinger, P. A., and Gehern, T. (1981) *Astrophys. J.* **247**, 750.
7. Wampler, E. J., Robinson, L. B., Burbridge, E. M., and Baldwin, J. A. (1978) *Astrophys. J.* **198**, L49.
8. Weedman, D. W. (1976) *Q.H.R. Astr. Soc.* **17**, 227.
9. Zeldovich, Ya. B., and Novikov, I. D. (1967) Relativistic Astrophysics, Nauka, Moscow.

Table 16.1. Possible Evolutionary Scheme of GAN Taking Into Account Merging of Galaxies and Formation of QSO II and QSO I [79].

| Type of interaction | Morphology | Star formation and IR luminosity | Spectral type 0-I wide allowed | Spectral type II average | Spectral type III narrow-forbidden | Radio properties Type | Radio properties $L_2$ erg/s | Radio properties $P_2$ W/Hz |
|---|---|---|---|---|---|---|---|---|
| Multiple merging in rich groups in central parts | Basically SG in rich groups | ← Growth of IR luminosity / Growth of star formation rate | OSS II (SS) $[10^{-9}]$Mpc$^{-3}$ | NG (radio) | NLRG | PR II (gEG) | $\geq 10^{44}$ | $>10^{25}$ |
|  | SG + EG in scattered clusters |  | QSS((( SE) $[5\times 10^{-9}]$ Mpc$^{-3}$ | BLRG $[3\times 10^{-8}]$ | $[3\times 10^{-7}]$ | FR I (cD) | $>10^{43}$ | $<10^{25}$ |
|  | Basically EG in rich compact clusters |  | QSS II (EE) $[10^{-8}]$ Mpc$^{-3}$ | NG (no radio) | cD, gE | radioactive nuclei | $<10^{42}$ | $<10^{21-24}$ |
| Low-number mergings in poor groups | S + S |  | QSG (SS), SG / Weak RS corresponding to thickening of log N − log S | SG II | SG III | ← Growth of $L_{radio}$ of nucleus | $10^{37-40}$ |  |
|  | S + E |  | QSG (SE) $[10^{-7}]$ Mpc$^{-3}$ | LINERS (?) $[3\times 10^{-5}]$ | EG$_{pec.}$ $[3\times 10^{-5}]$ |  | $10^{34-38}$ |  |
| Galaxy fields or in voids (without interaction) | SG |  | QSGI(S) $[10^{-6}]$ Mpc$^{-3}$ | Micro SG dw SG | Extended star formation |  | Weak radio sources |  |
|  | EG |  | QSGI(E) $[10^{-5}]$ Mpc$^{-3}$ | $[3\times 10^{-5}]$ | $[3\times 10^{-4}]$ |  |  |  |
| Length of active phase (yr) |  |  | $<3\times 10^7$ | $3\times 10^8$ → Evolution | $3\times 10^9$ |  |  |  |
| Z of the population |  |  | >4 QSO I / <4 QSO II | from 2 to 0 |  |  |  |  |

10. Burbridge, E. M. (1967) *Annual Rev. Astron. Astrophys.* **5**, 399.
11. Ginzburg, V. L., and Ozernoy, L. M. (1977) *Asdtrophys. Sp. Sc.* **48**, 401.
12. Hills, J. G. (1975) *Nature* **254**, 295.
13. Novikov, I. D., and Frolov, V. P. (1986) Physics of Black Holes, Nauka, Moscow.
14. Roos, N. (1988) *Astrophys. J.* **334**, 95.
15. Heckman, T. H., Smith, E. P., Baum, S. A., et al. (1986) *Astrophys. J.* **311**, 526.
16. Margon, B. (1982) *Science* **215**, 247.
17. Sorrell, W. H. (1981) *Nature* **291**, 394.
18. Sil'chenko, O. K., and Lipunov, V. M. (1985) *Astrophys. Sp. Sc.* **117**, 293.
19. Sanders, D. B., Soifer, B. T., Elias, J. M., et al. (1988) *Astrophys. J.* **325**, 74.
20. Byrd, G. G., Sundelius, K., and Valtonen, M. (1987) *Astron. Astrophys.* **171**, 16.
21. Hernquist, L. (1989) *Nature* **340**, 687.
22. Lavery, R. J., and Henry, J. P. *Astrophys. J.* **330**, 596.
23. Shlosman, I., Begelman, M. C., and Frank, J. (1980) *Nature*, in press.
24. Hernquist, L. (1989) Preprint IASSNS-AST. 32.
25. Sanders, D. B., Soifer, B. T., Ellias, J. H., et al. (1988) *Astrophys. J.* **328**, L41.
26. Barnes, J. E. (1989) *Nature* **338**, 123.
27. Thakura, R. K., and Dasgupta, A. (1982) *Astrophys. Sp. Sc.* **85**, 277.
28. Alvensleben, U.F.V., Krugel, H., Fricke, K. J., and Loose, H. H. (1989) *Astron. Astrophys.* **224**, L1.
29. Chambers, K. C., and McCarthy, P. J. (1990) *Astrophys. J.* **354**, L9.
30. Komberg, B. V., and Sunyaev, R. A. (1971) *Astron. Zh.* **48**, 235.
31. Kellermann, K. I., and Owen, F. N. (1987) Galactic and Extragalactic Radio Astronomy, G. L. Vbrashuur and K. I. Kellermann (eds.), Springer-Verlag, Berlin, p. 563.
32. Fanaroff, B. L., and Riley, J. M. (1974) *Monthly Nat. RAS* **167**, 31.
33. Komissarov, S. S., and Ovchinnikov, I. A. (1988) Levedev Inst. Prepr. 120.
34. Bicknell, G. V., de Ruiter, H. R., Fanti, R., et al. (1990) *Ap. J.* **354**, 98.
35. Kapahi, V. K. (1981) *Astrophys. Astron. (Indian)* **2**, 43.
36. Schneider, D. P., Schmidt, M., and Gunn, J. E. (1989) *Astrophys. J.* **98**, 1951.
37. Schmidt, M., and Gunn, J. E. (1987) *Astrophys. J.* **316**, L1.
38. Djorgovski, S., Spinrad, H., Reldelty, J., et al. (1987) *Astron. J.* **93**, 1307.
39. Chambers, K. C., Milley, G. K., and Van Brengel, W.J.M. (1988) *Astrophys. J.* **327**, L47.
40. Komberg, B. V. (1989) *Astron. Zh.* **66**, 710.
41. Baan, W. A. (1986) *Nature* **315**, 26.
42. Burdyuzha, V. V., and Komberg, B. V. (1990) *Astron. Astrophys.* **234**, 40.
43. Komberg, B. V. (1989) *Astrofizika* **30**, 406.
44. Meylan, G. (1990) *New Scientist* **10/11**, 32.
45. Wasilevski, A. Z. (1983) *Astrophys. J.* **272**, 68.

46. Shklovsky, I. S. (1955) *Astron. Zh.* **32**, 215.
47. Shklovsky, I. S. (1961) *Astron. Zh.* **37**, 945.
48. Shklovsky, I. S. (1962) *Astron. Zh.* **39**, 591.
49. Matthews, T. A., Balton, J. G., Greenstein, J. L., et al. (1961) *Sky and Tel.* **21**, 148.
50. Matthews, T. A., and Sandage, A. R., (1963) *Astrophys. J.* **138**, 30.
51. Schmidt, M. (1963) *Astrophys. J.* **197**, 1040.
52. Greenstein, J. L., and Matthews, T. A. (1963) *Nature* **197**, 1041.
53. Sandage, A. R. (1965) *Astrophys. J.* **141**, 1560.
54. Kinman, T. D. (1965) *Astrophys. J.* **142**, 1241.
55. Humason, M. L., and Zwicky, F. (1947) *Astrophys. J.* **105**, 85.
56. Weedman, D. W. (1976) *Q. J. RAS* **17**, 227.
57. Sulentic, J. W., and Arp, H. C. (1987) *Astrophys. J.* **319**, 687.
58. Sharov, A. S., and Efremov, Yu. N. (1963) *IBVS* N 23.
59. Shklovsky, I. S. (1965) *Astron. Cir* N 332.
60. Shlomitskii, G. B. (1965) *Astron. Zh.* **42**, 673.
61. Zeldovich, Ya. B., and Novikov, I. D. (1964) *DAN SSSR* **158**, 811.
62. Ginzburg, V. L., Ozernoy, L. M., and Syrovatskii, S. I. (1964) *DAN SSSR* **154**, 557.
63. Gehren, T., Fried, J., Wehinger, P. A., and Wyckoff, S. (1984) *Astrophys. J.* **278**, 11.
64. Tinsley, B. H., and Larson, R. B. (1979) *Monthly Not. RAS* **186**, 503.
65. Wyse, R.F.G., and Silk, J. (1987) *Astrophys. J.* **319**, L1.
66. Bulaenko, E. V., and Pariiskii, Yu. N. (1981) Proc. Formation of Large Scale Structure of the Universe, Tallinn.
67. Dekel, A., and Silk, J. (1986) *Astrophys. J.* **303**, 39.
68. Silk, J., and Stalay, A. S. (1987) *Astrophys. J.* **323**, L107.
69. De Roberts, H. (1985) *Astron. J.* **90**, 998.
70. Koo, D. C., and Kron, R. G. (1988) *Ap. J.* **325**, 92.
71. Petrosian, V. (1973) *Astrophys. J.* **183**, 359.
72. Schmidt, M. (1976) *Astrophys. J.* **209**, L55.
73. Stocke, J. T., Perrenod, S. C. (1981) *Astrophys. J.* **245**, 375.
74. Capps, K. W., Sitko, M. L., and Stein, W. A. (1982) *Astrophys. J.* **255**, 413.
75. Steiner, J. E. (1981) *Astrophys. J.* **250**, 469.
76. Boronson, A., Persson, S. E., and Oke, J. B. (1985) *Astrophys. J.* **293**, 120.
77. Modes, M. (1985) *Astron. Astrophys.* **152**, 271.
78. Marshall, H. L. (1987) *Astrophys. J.* **326**, 84.
79. Komberg, B. V. (1984) *Astrofizika* **20**, 73; (1982) Space Research Inst. Prepr., N 727; (1989) *Proc. SAO* **61**, 134.
80. Veron, P. (1986) *Astron. Astrophys.* **170**, 37.
81. Schmidt, M., Schneider, A., and Gunn, J. (1986) *Astrophys. J.* **306**, 401.

82. Heckman, T. M., Smith, E. P., Baum, S. A., et al. (1986) *Astrophys. J.* **311**, 526.
83. Byrd, G. G., Sundeliks, B., and Valtonen, M. (1987) *Astrophys. J.* **171**, 16.
84. Cooper, R. G., and Miller, R. H. (1982) *Astrophys. J.* **254**, 16.
85. Kool, N., and Norman, C. A. (1974) *Astron. Astrophys.* **76**, 75.
86. Anderson, S. F., and Morgan, B. (1987) *Nature* **327**, 125.
87. Hall, P., and MacKay, C. D. (1984) *Monthly Not. RAS* **210**, 979.
88. Afanas'ev, V. L., Dodonov, S. N., Lorenz, H., and Terebizh, V. (1987) IAU Symp. N 121, p. 49.
89. Marshall, F. E., Boldt, E. A., Volt, S. S., et al. (1981) *Astrophys. J.* **95**, 7.
90. Maccacaro, T. (1982) *Astrophys. J.* **253**, 504.
91. Lilly, S. J., and Kongair, M. S. (1982) *Monthly Not. RAS* **199**, 1053; (1984) *Monthly Not. RAS* **211**, 833.
92. Smith, E. P., and Heckman, T. M. (1990) *Astrophys. J.* **348**, 38.
93. Ellington, E., and Yee, H.K.C. (1989) *Astron. J.* **97**, 1539.
94. Jenkins, C. K. (1984) *Monthly Not. RAS* **207**, 361.
95. Zeldovich, Ya. B., Ruzmaikin, A. A., and Sokolof, D. D. (1983) Magnetic Fields in Astrophysics, Gordon and Breach, New York.
96. Krum, N., and Brosch, N. (1984) *Astron. J.* **89**, 1461.
97. Borsch, N., and Condhallka, P. M. (1984) *Astron. Astrophys.* **140**, 43.
98. Bothum, G. D., Beers, T. C., Mould, Z. R., and Nuckra, J. P. (1986) *Astrophys. J.* **308**, 510.
99. Thuan, T. X., Gott, J. R., and Schneider, S. E. (1987) *Astrophys. J.* **315**, 95.
100. Sanduleak, N., and Pesch, P. (1982) *Astrophys. J.* **258**, L11.
101. Balzano, V. A., and Weedman, D. W. (1982) *Astrophys. J.* **255**, 1.
102. Krischner, R. P., Oemler, A., Schechter, P. L., and Stectman, A. (1987) *Astrophys. J.* **314**, 493.
103. Moody, J. W., Krishner, R. P., MacAlpine, G. M., and Gregory, S. A. (1987) *Astrophys. J.* **314**, L33.
104. Melott, A., Einasto, J., Saar, E., et al. (1983) *Phys. Rev. Lett.* **51**, 935.
105. Henning, P. A., and Kerr, F. J. (1989) *Astrophys. J.* **347**, L1.
106. Pritchet, C. J., and Hartwick, F.D.A. (1987) *Astrophys. J.* **320**, 464.
107. Borra, E. F. (1983) *Astrophys. J.* **273**, 55.
108. Edwards, G., Borra, E. F., and Hardy, E. (1985) *Astrophys. J.* **289**, 446.
109. Yoshii, Y., and Takahara, F. (1988) *Astrophys. J.* **326**, 1.
110. Kellermann, K. I., and Wall, J. V. (1986) NRAO Prepr. N 190.
111. Windhorst, R. A. (1984), Thesis, Sterrewacht, Leiden.

# 17
# A Lucky Chance for Cosmology

## V. N. Lukash and I. D. Novikov
*Astro Space Center, Lebedev Physical Institute, Moscow*

## 1. INTRODUCTION

Cosmology is one of few branches of astrophysics that J. S. Shklovsky and S. B. Pikel'ner rarely turned to. Nevertheless, they had to their credit some original works in the field. This paper is to commemorate the memory of these scientists. The two prominent astrophysicists put particular attention to the most puzzling problems and suggested their own original solutions. Up to his very last days J. S. Shklovsky showed great interest in the current problems of the physics of the Very Early Universe.

Nowadays we are coming to a point when a confrontation of the two well-known facts in observational cosmology that brought about the numerous theoretical speculations in the last decade will be settled. The facts are: (i) the Universe around us has a well-developed structure (clusters and superclusters of galaxies, voids, QSOs, large peculiar galaxy velocities etc.); (ii) cosmic microwave background radiation reveals a black body spectrum and seems to be highly isotropic which means that there is no any significant structure at very large scales and at great red shifts.

The structure that one observes implies nonlinear perturbations ($\delta = \delta\rho/\rho \sim 1$) at least at scales of $\sim 10$ $h^{-1}$ Mpc, while at the scale of the present horizon $\sim 3 \times 10^3$ $h^{-1}$ Mpc, the perturbations in the total matter density should be no more than $\delta \lesssim 1-2 \times 10^{-5}$ because a higher perturbation level would be in contradiction with the upper limits for the quadrupole and higher multipole anisotropy of $\Delta T/T$ obtained in the RELIC I experiment. Combined with the absence of small scale perturbations (at angles $\theta \sim 10'$, $\Delta T/T < 3 \times 10^{-5}$ which are causally related to the present structure, this gives us direct evidence of an evolutionary origin of the observed structure of the Universe. Gravitational instability perhaps is operative to trigger the evolution, and the cause or seeds are the small perturbations of density and gravitational potential (against a homogeneous Friedmann expansion) which remain in the postrecombination epoch as relics of the earlier evolution of the Universe.

A headache of modern cosmology is to find out such a model of galaxy origin which on the one hand would be able to explain the observed structure of the Universe (match the experiment) and on the other hand would arise naturally as a result of the evolution of the Very Early Universe (match the theory). Thus, the model becomes a bridge that links theory and experiment. By now, one could find a dozen models of this type in the literature. This variety of models was worked out during the last decade in course of the smooth progress of experiments. However, recent developments in experiments have changed the situation. Important discoveries (the absence of spectral distortions in the cosmic microwave background (CMB) at a level of ~1%, the Great Attractor, new constraints on the gravitational wave density obtained through pulsar timing, etc. caused the rejection of many models. Furthermore, in coming years two important groups of experiments will be performed which will practically leave no room between the estimations of $\delta\rho/\rho$ from $\Delta T/T$ on the one hand, and from observations of large scale structure on the other hand for most of the models (including the standard CDM one). All this allows us to say that today we are witnessing the crisis of the market of cosmological models of the origin of galaxies. With a certain degree of optimism, one might hope that this crisis will lead to the construction of a true model and determination of its fundamental properties such as the spectrum of primordial density perturbations at large scales ($1 \gtrsim 10\ h^{-1}$ Mpc) and the composition of dark matter. Finding out the two important properties of the postrecombination Universe will provide a great impetus to the development of theories of the Very Large Universe, in particular inflationary theories, whose crucial test, as is well known, is the postinflationary cosmological perturbations.

This paper is arranged in the following way. In the next section (Sec. II) we will give our view on which of the forthcoming observations will have the most impact on the progress of theory. In Sec. III we discuss the possibilities that modern theory gives for the construction of a model of formation of the large scale structure of the Universe.

This review does not pretend to treat all the related topics that are now under discussion in the literature. We rather give here our view on the important forthcoming and already initiated changes in cosmology.

## II. EXPERIMENT

### 1. Confrontation

We see a tremendous importance for the modern cosmology of two groups of experiment nowadays: $\Delta T/T$ observations on large scales ($\theta > 6°$) and the Great Attractor (GA) confirmation. There are two reasons for this attention: theoretical and practical.

The first reason stems from the fact that these two experiments confront each other. If $\Delta T/T$ upper limits which we have up to now force us to lower primordial perturbation amplitudes on scales larger than a few degrees of arc ($1 > 30'\ h^{-1}$ Mpc), then GA needs for its existence high enough cosmological perturbation amplitudes on scales 1 ~10–100 $h^{-1}$ Mpc. For most theories of galaxy formation the gap between these two requirements is negligible. Say, within Gaussian perturbation theories any reasonable assumption for GA to be a more or less standard phenomenon in the visible Universe inevitably leads to $\Delta T/T$ prediction levels capable of current detection (Hnatyk et al. 1991). It leads to great optimism to obtain a large scale primordial perturbation spectrum directly from these two experiments with a high degree of accuracy.

Another idea is based on the possibility of finishing these observations in the nearest future. $\Delta T/T$ observations on scales $\theta > 7°$ are carried out by the COBE (which is launched) and RELIC II (to be launched in 1993) missions. A necessary sensitivity ~3–5 × $10^{-6}$ for cosmological $\Delta T/T$ detection is very reachable by both of them, but even the level of ~$10^{-5}$ which is expected for COBE in half a year is very promising in view of GA existence. To have two independent satellite experiments in $\Delta T/T$ with very different techniques at the same angular scales and at similar periods of time is really a striking and exciting thing for cosmology.

On the other hand, GA confirmation is also very close to us. Up to now ~30–40% of the total volume of GA has been measured. If we take into account the zone of avoidance in optics (due to Milky Way dust) which occupies about 30%, then we see that half of the GA volume is still to be measured. However, nobody seems to doubt the GA existence. The fact is that all available data now do not contradict the reality of the GA, not to speak of the observation of different groups being in accord with each other. Because of the vital importance of this phenomenon for cosmology, we dwell upon this subject in a bit more detail.

## 2. Great Attractor—A Challenge to Theory

The GA is a large scale phenomenon in the galactic peculiar velocity field (Dressler et al. 1987, Linden-Bell et al. 1988, see also recent publications of Mathewson et al. 1990, Dressler and Faber 1990, Burstein et al. 1990). One can say that it is a systematic (coherent) distortion of the Hubble diagram in the direction $l = 307°$, $b = g°$ on the celestial sphere (see Fig. 17.1). Two points are important for the GA notion:

1. Coherent flow of galaxies and clusters, and
2. A large scale correlation length ($l \geq 40$ $h^{-1}$ Mpc) for this coherent velocity field.

The later means that the typical scale of the mean velocity field is larger than the averaging scale to be used for getting such a field. The location of the GA, a region towards which galaxies infall, is at a distance $r_{LG} \sim 40$ $h^{-1}$ Mpc from the Local Group (LG).

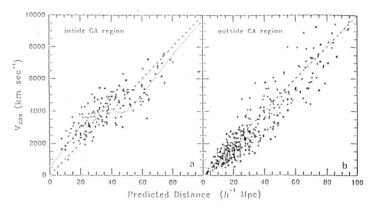

**Figure 17.1.** Hubble diagram in the direction of the GA (the GA center is at a distance ~40 $h^{-1}$ Mpc) and outside the GA region (from Dressler and Faber 1990).

Two puzzles immediately arise in connection with the GA: very high peculiar velocities (~600 km s$^{-1}$ on the LG radius from the GA center and ~$10^3$ km s$^{-1}$ on the Centaurus cluster radius, $r_{Cen}$ ~ 10 $h^{-1}$ Mpc) and, what is much more important, coherent flow on scales ~ 40 $h^{-1}$ Mpc. If the r.m.s. velocity ~600 km s$^{-1}$ is a result of positive and negative density fluctuations which, in principle, can be found in many theories providing for sufficient power of primordial perturbations on large scales, then coherent flow (GA) requires a high peak of a certain shape which becomes a crucial test for the theory.

The next important idea about the GA is that, according to observations, the central part of the GA (within the Centaurus radius) is certainly in the non-linear dynamical region now ($\delta \geq 3$). All this requires an explanation from the point of view of current cosmology.

Further on we follow the paper of Hnatyk et al. (1991) who considered the GA origin as a result of evolution of a high peak of density of the adiabatic growing mode of Gaussian perturbations in a two-component medium: dark matter (N-body approximation) and hydrogen-helium primordial plasma (hydrodynamic approach). Numerical simulations were carried out beginning from z = 10 (where perturbations are small) and up to z = 0 to be normalized here to the Faber and Burstein (1988) observational fit:

$$r_{LG} = 4300\ (\pm 300)\ \text{km s}^{-1}$$

$$u_{LG} = 535\ (\pm 70)\ \text{km s}^{-1}$$

$$n_{LG} = 1.4 \ (\pm 0.5)$$

$$r_c \cong 0.3 \ r_{LG}$$

The latter two quantities are just the velocity gradient index $n = \ln u/\ln r$ in the vicinity of the LG and the core radius of the GA (Halfwidth of density profile). Small cases of $n_{LG} \leq 1$ backed by Aaronson et al. (1989) observations of clusters promote more flat profiles than those based on elipticals and spirals, and both possibilities are included.

The following principal steps were taken.

(i) First, some primordial perturbation spectrum at the postrecombination era (say $z > 10$) was assumed.

In fact, seven different models were considered: three neutrino and the standard CDM models as limiting cases, and three hybrid spectra: CDM + X (Bardeen et al. 1986, handmade modification with extrapower on scales $10 < 1 < 100 \ h^{-1}$ Mpc), HC (hot and cold particle phenomenological spectrum) and DI (double inflation phenomenological spectrum proposed by Turner et al 1987). All the model parameters and corresponding spectra, standardly normalized with the help of the correlation function of galaxies, are presented in Table 17.1 and Fig. 17.2.

(ii) Second, the spectrum dependent mean density profile of high peaks was derived according to the Doroshkevich (1970) and Bardeen et al. (1986) statistical theory.
The orientation-averaged density profile about a peak of height $\nu >> 1$ is closely approximated by a very simple formula having a transparent physical meaning:

$$\delta(r) = \frac{\nu}{\sigma_o} \xi(r) \qquad (1)$$

where r measures distance from peak center, $\sigma_o = \xi(0)^{1/2}$ is a r.m.s. density perturbation, and $\xi(r)$ is a correlation function of density.

(iii) Third, the peak shape (1) was taken as the initial condition at $z = 10$ and then numerically simulated to $z = 0$ where normalization to the peculiar velocity on the distance of the LG was made.

(iv) Fourth, the realization probability of a number of GAs within the cosmological horizon was estimated.

Two points should be clarified before we present the results of these considerations. The initial shape (1) depends on two parameters: $\nu$ (peak-height) and $R_f$ (Gaussian filter which correlates with the GA core radius $r_c$). After normalization (third step), we are left with the one-parameter family of initial conditions (GAs) which have a very different realization probability (fourth step). Also, the Gaussian filter $R_f$ for smoothing the random field perturbations on which the real peak shape is explicitly dependent needs a physical interpretation.

Table 17.1. The main parameters of cosmological models.†

|  | HDM(1) | HDM(3) | HDM(3) | CDM | HC | CDM+X | DI |
|---|---|---|---|---|---|---|---|
| h | 0.4 0.5 0.6 | 0.4 0.5 0.6 | 0.4 0.5 0.6 | 0.4 0.5 0.6 | 0.4 0.5 0.6 | 0.4 0.5 0.6 | 0.4 0.5 0.6 |
| $\Omega_b$ | 0.1 0.1 0.1 | 0.1 0.1 0.1 | <<0.1 | 0.1 0.1 0.1 | 0.1 0.1 0.1 | 0.1 0.1 0.1 | 0.1 0.1 0.1 |
| $\Omega_{CDM}$ | — — — | — — — | — — — | 0.9 0.9 0.9 | 0.5 0.5 0.5 | 0.9 0.9 0.9 | — — — |
| $\Omega_{HDM}$ | 0.9 0.9 0.9 | 0.9 0.9 0.9 | 1.0 1.0 1.0 | — — — | 0.4 0.4 0.4 | — — — | 0.9 0.9 0.9 |
| b | .73 .64 .57 | .48 .51 .53 | .49 .52 .55 | 1.7 1.7 1.7 | 1.7 1.7 1.7 | 1.7 1.7 1.7 | 1.7 1.7 1.7 |
| $\sigma_{og}$ | 2.0 2.0 2.0 | 2.0 2.0 2.0 | 2.0 2.0 2.0 | 2.2 2.4 2.6 | 1.3 1.3 1.3 | 2.1 2.3 2.5 | 2.7 2.8 2.8 |
| $Q*10^5$ | 2.0 1.3 0.9 | 6.0 3.8 2.7 | 4.6 3.1 2.3 | 1.0 0.8 0.6 | 1.4 1.0 0.6 | 1.2 1.0 0.8 | 2.7 1.6 1.1 |

†"b" is the biasing parameter, $\sigma_{og}$ is the r.m.s. density fluctuation on the galactic scale, and $Q \equiv (\Delta T/T)_2$ is the quadrupole r.m.s. amplitude of the microwave background anisotropy.

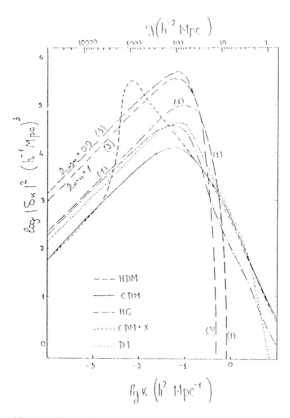

Figure 17.2. Primordial postrecombination power spectra for the cosmological models presented in Table 17.1 (from Hnatyk et al. 1991).

There is no general consensus in the literature about a $R_f$ interpretation. One can list at least three possibilities.

1. $R_f = R_{cutoff}$ is a perturbation spectrum cutoff scale which requires a new spectrum normalization procedure each time for a new $R_f$.
2. $R_f = R_s$ is a typical scale of objects with which one deals. According to this interpretation, one does not know which class of objects (cluster or supercluster) is responsible for the realization of the GA. trying small and large cases of $R_f$, one models narrow (massive cluster in the GA center) and broad (supercluster without distinguished center) peaks.
3. $R_f$ is a technical parameter for smoothing a "real" peak supposed to exist independently of $R_f$. In this case, one can compare observations with only scales $r > R_f$, and for smaller radii smaller filters are needed, etc.

Numerical simulation is certainly independent of these speculations, but the interpretation of the results should be done within some of these ansatzs. Following Hnatyk et al. (1990) we will assume the second and third possibilities for $R_f$.

The results can be divided into three groups: initial peak parameter, realization probability, and contraflows.

(a) Some parameters of the real initial peaks, which generate observational profiles at $z = 0$, are practically independent of the model or primordial spectrum chosen. Among them $\delta(0) = (1-2)(1+z)^{-1}$, $r_c = (10-15)h^{-1}$ Mpc, and others.

(b) However, the realization probability of such peaks is drastically different for different models. The higher n, the smaller the probability is of finding peaks ensuring the GA. If for the CDM-model these are 6–11 σ peaks which are extremely unprobable phenomena, then for hybrid models the situation is more promising (~3–6 σ), while in HDM-models GAs are met much more frequently (~1–2σ) (see Table 17.2). However,

Table 17.2. Parameters of the mean profile peaks which most closely approximate the FB observational profiles.†

|  | HDM(1) | HDM(3) | HDM(3) | CDM | HC | CDM+X | DI |
|---|---|---|---|---|---|---|---|
| $R_f$ (Mpc) | 13.5 | 11.0 | 12.0 | 12.0 | 11.0 | 8.0 | 12.5 |
| ν | 5.2 | 2.0 | 2.4 | 11.7 | 6.7 | 5.9 | 6.6 |
| $\delta_o$ | 0.15 | 0.16 | 0.15 | 0.14 | 0.15 | 0.14 | 0.14 |
| N | 41 | $9 \times 10^5$ | $5 \times 10^5$ | $10^{-22}$ | $2 \times 10^{-2}$ | 2 | $3 \times 10^{-2}$ |
| $n_{LG}$ | 1.7 | 1.6 | 1.6 | 1.4 | 1.5 | 0.7 | 1.6 |

†h = 0.5; $\delta_o \equiv \delta$ (r = 0), N is the number of GAs within the contemporary horizon (from Hnatyk et al. 1990)

even for a fixed spectrum model, the realization probability grows very sensitively for smaller Hubble parameter h and/or more flat velocity profiles (larger $R_f$). It is demonstrated in Fig. 17.3 with the example of the HDM (1) model.

A very important conclusion concerns the symmetry of the GA. The point is that very high peaks tend to be spherically symmetric. Say, in the CDM-model the GA should be a spherically symmetric object while the neutrino and hybrid model GAs are provided by 1–3σ peaks which are not at all spherically symmetric (we recall that Zeldovich pancakes form from 1–2σ peaks).

(c) All peaks are non-linear near the GA center at $z = 0 (\delta > 3)$. However, peaks with initial $\delta(0) > 1.5 \ (1+z)^{-1}$ develop a strong non-linearity with contraflows of dark particles and galaxies in the center and with a global shock wave in the baryonic intergalactic gas. The appearance of contraflows is not confused by the symmetry properties of the GA and directly related to the peak height, thus providing a very sensitive test of the theory. It is, in principle, a simple observational problem to find contraflows of galaxies (or their absence) in the GA center. Note, that recent observations of Dressler and Faber (1990) and Burstein et al. (1990) pointed out zero mean velocities of galaxies in the GA center (i.e., the absence of contraflows) which favors the moderate initial peak and agrees with the spherical assymetry of GA which seems to have been observed as well. If so, the GA is a very standard and representative phenomenon in the Universe.

## 3. ΔT/T and Large-scale Structure

We have paid a special attention to GA observations for the reasons explained above, but there are other experiments showing growing evidence of large structure activities in the Universe. We list briefly some of them.

First, we have the cluster-cluster correlation function which biased the galactic correlation function by a factor of 25. Recently, it was confirmed by Huchra (1990) on ~350 clusters.

Second, it is the large visible constructions in the Universe whose scales grow with the exploration of a larger sample volume. There are huge voids (~100 h$^{-1}$ Mpc), great filaments and pancakes like the Great Wall (~150 h$^{-1}$ Mpc), etc.

It is very important for theory to investigate cluster evolution because these objects give a characteristic scale in the theory and since they formed rather recently, they are close to us to be observed. There are not only direct cluster measurements but also very important indirect ones. Among them: ΔT/T observations on small scales (~1"–1') and quasar statistics.

If clusters had not evolved since z ~ 10, they would have hot gas causing small-scale fluctuations due to the Zeldovich-Sunyaev effect, $\Delta T/T \sim 10^{-4}$, which are not observed (Partridge 1990). Therefore early clusters did not generate hot gas and, thus, were not yet virialized. Recent observations of far distant quasars (z > 0.5) also evidence this.

## 17. A LUCKY CHANCE FOR COSMOLOGY 285

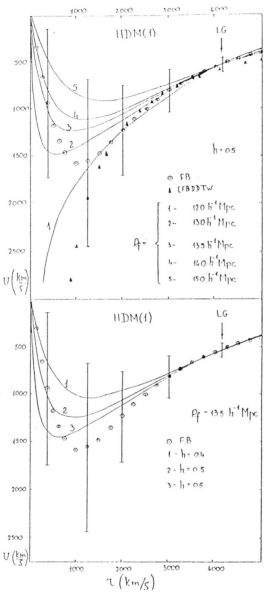

**Figure 17.3.** The GA peculiar velocities simulated in the HDM(1) model with one sort of massive neutrino for different $R_f$ and h parameters (from Hnatyk et al. 1991).

Far distant QSOs seem to be found in rich galaxies centered in clusters, they are brighter than nearby QSOs and die before $z \sim 0.5$ (Ellingson 1990; Yee 1990). The reason might be that these far distant clusters are only in the process of collapse and, because of that, they are cold and can supply enough material for black hole accretion (QSO). After virialization, the origin of hot gas and high galactic velocities destroy this channel and the quasar decays, converting into a radio-loud huge galaxy (short-lived) in the cluster. Nearby quasars occur mostly in groups, and thus are not so active (radio-quiet) and long-lived. This hypothesis, if true, could give very rich information about quasar and cluster physics, and thus, about the theory of galaxy formation. An independent test for this hypothesis could be extracted from quasar statistics: distant quasars ($z > 0.5$) might have a larger correlation function then nearby ones ($z < 0.5$).

All these large structures cannot be extended to the past because of another fundamental reason: they are upper limited by the absence of $\Delta T/T$ fluctuations.

We think that multifrequency observations of large scale $\Delta T/T$ fluctuations are the most important which cannot be smoothed over by secondary and re-ionization processes. In the near future one expects results from the COBE and RELIC II satellite experiments. Some Earth observations are also promising (on South pole, Canarie, etc.).

## 4. A Bridge to the Theory of the Very Early Universe

Current observations undoubtly evidence extra power in the primordial perturbation spectrum on scales $10 \, h^{-1} < 1 < 100 h^{-1}$ [Mpc] in comparison with the standard CDM-model. If the GA as a large scale coherent flow of galaxies is confirmed, then the CDM-model is untrue. Hybrid or HDM spectra are more appropriate and further observations will distinguish between them.

Thus, current theory must account for such primordial perturbations to which the observations point. Within random field perturbation theories, the goal is as follows: to find models which provide for high enough power on large scales as well as a sufficient perturbation amplitude on galactic scales in the postrecombination era. It should be something in between two limiting spectra: CDM (which is good for galaxy formation, but unable to explain the GA phenomenon) and HDM (which can account for the GA, but helpless in galactic origin).

The next section deals with how theory tries to solve this problem currently.

## III. THEORY

The most strict requirement to the theory of the Very Early Universe as seen now in cosmology is the necessity to form perturbations with extrapower on large scales. Thus, we will pay most attention to this topic here.

According to modern cosmological views, primordial perturbations are the result of two factors: postinflationary perturbations (generated before the beginning of the Hot Universe expansion) and the transfer function (saying how initial perturbations have changed to the postrecombination era). Both factors are responsible for the final appearance of perturbations. Here we study them separately.

First, we describe briefly the gravitational mechanism which forms the transfer function and then we list some possibilities to obtain non-flat (non-Harrison-Zeldovich) spectrums within the inflationary theories. For the sake of time we dwell mostly on adiabatic random field perturbations with some other possibilities mentioned at the end of this section.

## 1. Dark Matter

Evolution of perturbations is a matter of gravitational instability. The most delicate thing here is a point (in time) of complete balance between relativistic and non-relativistic particle densities which is called the equality epoch $t_{eq}$. Before $t_{eq}$ the Universe was dominated by relativistic particles (hot Universe) and expanded linearly in conformal time (a ~ $\eta$ ~ $t^{1/2}$); after $t_{eq}$ non relativistic particles dominated the expansion (cold Universe) which was quadratic in time ( a ~ $\eta^2$ ~ $t^{2/3}$), so during the equality epoch the Universe expansion accelerated. This was imprinted in the perturbation growth rate, leaving a typical scale in the transfer function as the scale of the horizon at the equality time $l_{eq}$. (For the standard CDM model it is about a supercluster scale).

Therefore, the scale of the equality horizon is of prime interest which is, in turn, sensitively affected by the dark matter composition in the real Universe. Among these components there are now dominating (maintaining critical density) and now non-dominating (e.g., relativistic weakly interacting particles). The heavy non-relativistic matter can be easily counted if one assumes that it consists of massive neutrinos (which are initially in thermal equilibrium) and relic cold particles:

$$\Omega_m + \Omega_r = 1 \qquad (2)$$

and thus we are left with one free parameter, say $\Omega_r \in$ (0.1) (the restmass m is unambiguiously determined by $\Omega_r$). The light component provides another free parameter

$$\nu = \frac{N_\nu}{N_\nu + N_\gamma} \in (0,1) \qquad (3)$$

which gives us the total amount of relativistic weakly interacting particles ($\nu$) in terms of all relativistic particles including photons ($\gamma$). The task is to probe transfer functions versus these two parameters.

Below, we emphasize two main results concerning this problem (Lukash 1989; Kahniashvili and Lukash 1990).

i. *v-scaling*: in a certain normalization of the wave vector k, where $k_{eq} = 1$ corresponds to the equality horizon scale, transfer functions C(k) both for adiabatic and isocurvature perturbations are practically independent of the v-parameter (see Fig. 17.4). Certainly real scales are v-dependent according to the background model connection:

$$\lambda = 36 \, k^{-1} h^{-2} (1-v)^{-1/2} \, \text{Mpc} \qquad (4)$$

The v-scaling effect is very useful in practice since it allows one to make any calculations for $v = 0$ (or $v = 0.4$ as for the standard CDM model) and then rescale all results with the help of Eq. (4) and other background relations. Generally, with v growing, all scales and perturbation amplitudes grow as well, so v cannot be larger than 0.8 to be in agreement with $\Delta T/T$ observations.

ii. *Two-component dark matter effect:* Fig. 17.5 demonstrates a set of adiabatic $\delta\rho/\rho$-spectra (in the case of Harrison-Zeldovich postinflationary perturbations) for different $\Omega_r$ parameters. The two limiting cases $\Omega_r = 0;1$ account for HDM and CDM models,

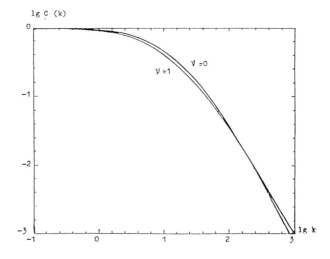

Figure 17.4. Adiabatic transfer function C(k) in the CDM model for two limiting cases of the parameter v. All the other curves (0 < v < 1) are placed in between these two (from Lukash 1989).

## 17. A LUCKY CHANCE FOR COSMOLOGY

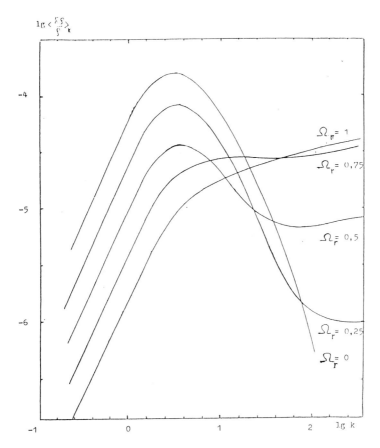

Figure 17.5. Spectra of adiabatic density perturbations in the two-component dark matter model $\Omega_m + \Omega_r = 1$ (one sort of massive neutrino + cold relic particles) for different parameters $\Omega_r = 0, 0.25, 0.5, 0.75, 1$ (from Kahniashvili and Lukash 1991).

respectively. It can be seen that some hybrid spectra with $0 < \Omega_r < 1$ are very promising as far as cosmological requirements are concerned: they have a supercluster-scale peak (larger than in CDM and less than in HDM models) and a galactic-scale noticeable amplitude (larger than in HDM and less than in CDM models). Further analysis is occurring.

Here we briefly outline the possibilities of dark matter to modify the primordial spectrum with minimum assumptions about the dark matter composition. Certainly, there are some other proposals in the literature (strong loops, domain walls, explosions, etc.), but all of them need additional hypotheses.

## 2. Inflation

Nowadays, inflation seems to be the most popular way for producing cosmological perturbations. There is a very simple definition which shows the principal difference of inflation from conventional Friedmann expansion: Freidmann expansion proceeds with deceleration ($\ddot{a} < 0$) while inflation means acceleration ($\ddot{a} > 0$). It causes Universe scales to expand faster than the horizon scale:

$$(lH)^{\bullet} > 1 \tag{5}$$

where $l = a/k$, $H = \dot{a}/a$, the dot is a derivative over the universal time and, "a" is a scale factor.

If inflation proceeds more than 60 e-folds, which is necessary for a spacially homogeneous and flat Universe to be set within the contemporary horizon,

$$H \Delta t_{inf} \geq 60, \tag{6}$$

then the large structure scales, which are found now within the horizon, appeared initially from inside the horizon to inflate first outside of it at the inflationary stage and then to reenter the horizon again at the Freidmann Universe. As for the primordial perturbations, such dynamics work as a parametric (nonadiabatic) amplification mechanism for them to be generated from some seed fluctuations located within the horizon at the inflationary stage. The role of these seeds is naturally played by quantum zero-point fluctuations of physical fields acting during inflation.

Recent investigations of many authors showed that the resulting spectrum of primordial perturbations is directly connected to the analytical properties of the effective potential of a scalar field responsible for the perturbations (see Hodges and Blumenthal 1989, and references therein). Here we illustrate this statement on the example of a single inflation field $\varphi$ with the simplest Lagrangian:

$$L = \frac{1}{2} \varphi_{,i} \, \varphi^{,i} - V(\varphi) \tag{7}$$

During inflation the space becomes very flat and only quantum fluctuations of uniform inflation can provide the seed fluctuations. The power spectrum of density perturbations when a scale reenters the horizon at the Freidmann stage is then simply related to the potential $V = V(\varphi)$ when the scale left the horizon at the inflationary stage (Guth and Pi 1982; Hawking 1982; Starobinsky 1982; Bardeen et al. 1983, etc.):

$$P^{1/2}(k) \equiv k^3 \left\langle \left(\frac{\delta\rho}{\rho}\right)^2_k \right\rangle \cong V^{3/2}/V' m_{pl}^3 \tag{8}$$

where $V' = dV/d\varphi$ and $\varphi$ is connected to the wavevector k as $k = H(\varphi)a(\varphi)$. Hodges and Blumenthal (1989) have shown this expression can be reversed: for any given spectrum P(k) one can derive the potential $V(\varphi)$ needed. However, some potentials for a non-flat spectrum appear as rather peculiar ones but, nevertheless, the principal possibility to obtain a broad variety of spectra is there (see Fig. 17.6). Obviously, Harrison-Zeldovich flat spectra are the most natural result of the inflation since they are obtained from a large class of smooth functions V slowly depending on $\varphi$. However, analytical peculiarity in V or V' which gives a new typical scale $k_V$ immediately reflected in the non-trivial behavior of P(k) around the scale $\sim k_V$.

We have just described one of the ways to generate non-Zeldovich spectra in chaotic inflation. It seems a rather simple and attractive one, but it is only half of the truth: the main problem is to justify the validity of this or that potential $V(\varphi)$ which is a matter for a future theory.

In the absence of a true theory of the Very Early Universe, there are many other proposals how to produce perturbations at the Freidmann expansion beginning. Among rather popular ones currently there are:

- Extended inflation (revised Guth scenario for the first order phase transition, La and Steinhardt 1989).
- Cosmic strings (especially in the light of recent and future pulsar timing observations).
- More complicated topological features like cosmic textures (Turok 1989).

This list can be continued, but it is not our aim.

In ending this Section we can conclude that there exist some possibilities to modify the standard Harrison-Zeldovich spectrum in the chaotic inflationary theory by choosing more sophisticated potentials than $m^2\varphi^2$ or $\lambda\varphi^4$. Thus, it is not the first time in cosmology that theory can propose a few ways to obtain the desired result (spectrum), but in each case the observational consequences are different and the truth can be revealed only by the confrontation of theory and observations.

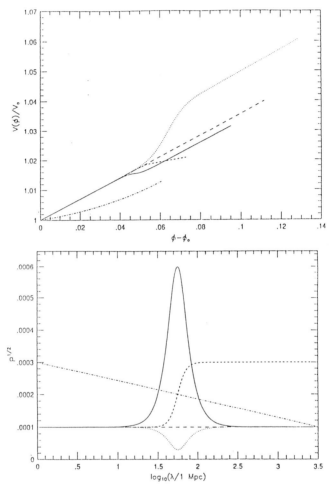

Figure 17.6. Postinflationary perturbation spectra (the Harrison-Zeldovich spectrum is a horizontal line) for different potentials $V(\varphi)$; $V_0$ at the shortest scale $\lambda = 1$ Mpc is taken to be $V_0 = 10^{-12} m_p^4$ (from Hodges and Blumenthal 1989).

## V. CONCLUSION

Recent observations of large scale features in the Universe, in particular, evidences for the GA existence indicate a conflict with the standard CDM model. The GA as a large scale phenomenon can be most likely formed only by gravitational instability which requires some extra-power of primordial perturbations on scales $\sim 10\text{-}100h^{-1}$ Mpc relative to the standard CDM spectrum. Such additional perturbations bring about greater $\Delta T/T$ amplitudes, predicted on angular scales $\theta > 1°$ which reach the level of the neutrino HDM models $\Delta T/T \sim 10^{-5}$ quite amenable to current detection. Therefore, the GA problem together with $\Delta T/T$ observations appear to play the key role in the direct detection of a postrecombination spectrum of primordial perturbations.

On the other side, the right postrecombination spectrum could easily form after the equality epoch in the two-component dark matter model (neutrino type + cold type particles) on the basis of the standard Harrison-Zeldovich postinflationary spectrum. This might prove to be a crucial test regarding inflation and dark matter problems.

## ACKNOWLEDGMENTS

This work was basically done in Moscow (August 1990) and finished partly in Moscow and partly in Spain (University of the Basque Country, Bilbao, September 1990) where one of the authors (V. N. L.) enjoyed hospitality. He is very grateful to the Basque Government for financial support and to Prof. A. Chamorro and the staff of the Department of Theoretical Physics for providing a scientific and stimulating atmosphere for work.

## REFERENCES

Aaronson M., Bothun G. D., Cornell M. E. et al. (1989) *Ap. J.* **338**, 654.
Bardeen, J. M., Bond, J. R., Kaiser, N., and Szalay, A. S. (1986) *Ap. J.* **304**, 15.
Bardeen, J. M., Steinhardt, P. J., and Turner, M. S. (1983) *Phys. Rev.* **D28**, 679.
Burstein, D., Faber, S. M., and Dressler, A. (1990) *Ap. J.* **354**, 18.
Doroshkevich, A. G. (1970) *Astrofizika* **6**, 320.
Dressler, A., and Faber, S. M. (1990) *Ap. J.* **354**, 13.
Dressler, A., Faber, S. M., Burstein, D., et al. (1987) *Ap. J.* **313**, L37.
Ellingson, E. (1990) in Superclusters and Clusters of Galaxies and Environmental Effects, G. Giuricin et al. (eds.).
Faber, S. M., and Burstein, D. (1988) *Proc. Pontific. Acad. Study Week* N27.
Guth, A. H., and Pi, S. (1982) *Phys. Rev. Lett.* **49**, 1110.
Hawking, S. W. (1982) *Phys. Lett.* **115B**, 295.
Hnatyk, B. I., Lukash, V. N., and Novosyadly, B. S. (1991) *Pis'ma Astron. Zh.*, N8.
Hodges, H. M., and Blumenthal, G. R. (1989) Preprint SCIPP 89/56.

Huchra, J. P. H. (1990) in *Superclusters and Clusters of Galaxies and Environmental Effects*, G. Giurcin et al. (eds.).
Kahniashvili, T. A., and Lukash V. N. (1991), in press.
La, D., and Steinhardt, P. J. (1989) *Phys. Rev. Lett* **62**, 376.
Lukash, V. N. (1989) in *Large Scale Structure and Motions in the Universe*, M. Mezzetti et al. (eds.), Kluwer Acad. Publ., Dordrecht.
Lynden-Bell, D., Faber, S. M., Burstein, D., et al. (1988) *Ap. J.* **326**, 19.
Mathewson, D. S. (1990) *Austral. J. Phys.*, in press.
Partridge, R. B. (1990) in *Superclusters and Clusters of Galaxies and Environmental Effects*, G. Giuricin et al. (eds.).
Starobinsky, A. A. (1982) *Phys. Lett.* **117B**, 175.
Turner, M. S., Villumsen, J. V., Vittorio, N., et al. (1987) *Ap. J.* **323**, 423.
Turok, N. (1989) *Phys. Rev. Lett.* **63**, 2625.
Yee, H., (1990) in *Superclusters and Clusters Galaxies and Environmental Effects*, G. Giuricin et al. (eds.).

# 18

# The Last Love

## I. L. Rosenthal
*Space Research Institute, Moscow*

### PROBABILITY OF THE FORMATION OF A METAGALAXY

1. Joseph Samuelovich Shklovsky was an extremely multifaceted person. He was attracted to painting, literature, music, and, of course, science. However, the time of Leonardo da Vinci was long in the past and therefore it was necessary for Joseph Samuelovich, basically, to concentrate on his constant attraction, astronomy, and in the last ten years astrophysics, where he was attracted to the study of the most exotic objects: quasars, pulsars, and the metagalactic gas. He also made important contributions to the study of the $3°$ cosmic isotropic radiation which he began to call relic radiation. However, in the latter years of his life in the time of our multiple conversations he was interested in the rapid progress in the theory of elementary particles and, especially, the unifying tendencies of quantum field theory. I "should" tell him the latest accomplishments: the theory of weak interactions and attempts at constructing a more unified theory including electromagnetic weak and strong interactions, and the role and necessity of Higgs particles.

In his first years his interest was to the new, important, and rapidly developing region of theoretical physics.

However, the tone and degree of his interest radically changed after he noticed the deep connection between the theory of elementary particles and cosmology.

Joseph Samuelovich understood well the tragic disparity between the splendid accomplishments of the Friedman theory of the evolution of the universe, a theory in whose framework all observed characteristics of the universe were interpreted, and the incompleteness and unclosedness of this model expressed above all in the existence of a singular state.

The apathy reigning among cosmologists at the beginning of the '80s could be expressed approximately as: "Well, so what? The singularity is 'the end of world,' the end of space, and the end of time," which did not satisfy Shklovsky and he often tried in our conversations to approach an understanding of this problem from physical (and not somewhat mystical) positions. The new inflationary cosmology, proposed by A. Guth, initially did not incite great interest in him. His piety toward Friedman was too great and on a first inspection it seemed that the new cosmology was extremely abstract and did not relate to reality which was the deciding criterion for Shklovsky.

It seems to me that a breakthrough in the opinion of Joseph Samuelovich of the new cosmology occurred after the excellent lecture of A. D. Linde in 1983 or 1984 at the Space Research Institute astrophysics seminar (organized by J. S. Shklovsky). The lecture demonstrated the close connection between the new cosmology, progress in the physics of elementary particles, and the successful solution of many problems of Friedman cosmology.

It is possible that our conversations played some role in the change in the relation to the uniqueness of canonical cosmology in which the problem of the uniqueness of the observed universe was demonstrated.

I introduced many physical arguments adopted from the physics of elementary particles and evidencing the exceedingly fine tuning of physical parameters to the fundamental fact: the existence of the basic elements of our Metagalaxy: atomic nuclei, atoms, stars, and galaxies. One (and in my opinion the most probable) of the possible interpretations of this amazing phenomenon was the hypothesis on the existence of an internal unstable physical vacuum from which from time to time metagalaxies arose with significantly different physical properties. Our Metagalaxy was an enormous fluctuation causing its complex structure.

This picture agreed well in general features with inflationary cosmology. Apparently, a similar argument had some influence on the ideas of Shklovsky on the universe. The last (as far as I know) article of Shklovsky, "What There Was When There Was Nothing" /1/, was connected with just these problems. In this article Shklovsky with insight and logic showed the existence of many metagalaxies, proposing to call all those existing in the old transcription by the "world." This fantastically interesting (and important) range of questions deeply interested Shklovsky. My conclusion is based on many telephone conversations on this subject which he carried on with me from the Hospital of the Academy of Sciences. In particular, he was interested in the quite nontrivial question, which proponents of inflationary cosmology to avoid or, at least, to encounter quite rarely. The essence of the problem is this: the expansion rate of the universe in the framework of this theory, at least, exceeds by millions of orders of magnitude the rate of light expansion. Shklovsky was interested by the actual problem: how could one compare such an assertion with the postulate of relativity theory about the limitation of the velocity of any motion to the velocity of light. There is at present no unique answer to this question. Later, in Section 2, I consider this problem in more detail.

We discussed this range of questions several times on the telephone. As I recall, Shklovsky still phoned from the hospital. Finally, it became clear, that the object of discussion was not a subject for telephone conversation and we agreed that I would come on the following day to him in the hospital for a more thorough discussion. However, fate would have it differently. The "following day" did not arrive. Joseph Samuelovich was postponed by the insult caused by his untimely death.

## 18. THE LAST LOVE

What amazes me? To the last minutes of his life Joseph Samuelovich believed in himself; he experienced unlimited interest in the central problems of science. This infinite interest on the threshold of a rendezvous with a different world gave our Scientist a remarkable personality.

2. We consider in more detail the problem of superlight velocities in the framework of the model of an inflationary universe.

Although the idea of solving this problem, apparently, has occurred to all authors known to me, the details of its realization differ considerably. Because nuances are important here, I must sometimes cite things word for word. The general idea of solving the paradox is this: models in exaggerated time cannot determine the inertial frame and therefore the idea of velocity has no sense, in particular, the velocity of light.

Thus, it is written in the book of I. D. Novikov /2/:

For this [measurement of the velocity of light] one must have the possibility although only in thought to connect the frame of reference with the observer in the form of an imagined hard carcass . . . then measuring the velocity of the body in relation to this carcass . . . we obtain the velocity of the body in relation to the observer . . . the imagined carcass should be solid and solidly connected with the observer. Otherwise the deformation of the carcass causes relative changes of its parts and the velocity of the body, measured in relation to the carcass in the place where the body moves, will not be the velocity in relation to the observer.

It is further maintained that in inflation conditions when a very strong gravitational field exists that such hard carcasses for particles separated by large distances are absent and, consequently, the idea of velocity loses its meaning.

According to A. D. Linde /3/ in the framework of the inflation model a transfer of information is absent between two points participating in expansion and only their relative motion is important: "The impossibility of observing superlight velocity relative to motion is connected with the impossibility of introducing a static reference system . . . " The conclusion follows from this on the absence of disagreement between results of observations (the velocity of light is the maximum observed one) and inflation theory. In these somewhat different and, in general, correct discussions there is an absence of the necessary strictness and determination which, by the way, is characteristic for expositions of the principal questions of relativity theory. Actually, an absolutely "solid" carcass is only a large idealization. In the framework of relativity theory absolutely solid bodies should not exist and, therefore, the introduction of a solid carcass for marking the frame of reference is not completely correct. The idea, of "a static reference system" is also unclear. It is known that the speed of light in vacuum is identical in all inertial reference systems including the ones in a state of motion. Therefore, in this case it would be very useful to clarify the sense of "the static state."

In my opinion, the interpretation of the paradox of the inflationary model is not connected with an abstract frame of reference, but concretely with the basic idea of relativity theory, an inertial reference system. Although several practically equivalent determinations of an inertial frame of reference exist, in our view the most successful is presented in the book /4/: an inertial frame consists of a central (point) reference body relative to which space is homogeneous and isotropic. Strong gravitational fields (and also nonpoint reference bodies) destroy the basic condition of this determination, isotropy and homogeneity, and consequently, the system of postulates of relativity theory including the speed of light limitation (see /5/). It is true that some incomplete representations about the condition of this limitation give multiple examples of the existence of superlight velocities in a system of microscopic bodies /6/. Sometimes it is maintained that these superlight velocities do not characterize information transfer. This assertion is not completely precise. One can demonstrate the situation arising here for the example of the effect of "a reflection" when the signal from the source rotating with an angular velocity w is transmitted a distance R to two points A and B of some macroscopic body. The velocity of signal transfer $v = wR$ may be arbitrarily large. In this process some limited information may be transferred. For example, if a ray from the source turns on a lamp at point A, and notification about this arrives at point B with a velocity v which can exceed the velocity c.

Thus, one may come to the following conclusion. In spite of some difference in approaches, in strong gravitational fields the relative velocity of motion of two material points may be arbitrarily large.

3. Shklovsky returned to the idea of the plurality of metagalaxies many times /1/. The first somewhat physical indications of this fundamental fact were obtained on the basis of the anthropic principle.[†] Its most popular and pictorial formulation was given by B. Carter /8/: "I think and therefore the world is such as it is" or if we go to the language of physics: "Physical laws should guarantee the existence of reasonable beings."

However, in spite of some successes and the use of the anthropic principle in the interpretation of cosmological regularities, similar formulations for a series of reasons are deeply unsatisfactory. Firstly, in our opinion, it is impossible on the basis of a physical principle to pose a quite indefinite (from a physical point of view) understanding of a reasoning being. Secondly, the anthropic principle is very weakly tied to microphysics, and in particular with experimental results obtained on accelerators. And finally, thirdly, on the basis of the anthropic principle much can be interpreted, but nothing can be predicted.

For these reasons a modification of the anthropic principle was proposed, that obtained the name of the expediency principle /9/: "Physical laws acting in our Metagalaxy are not only sufficient, but necessary for the origins and existence of basic stable states: atomic nuclei, atoms, stars, and galaxies."

---

[†]A quite complete exposition of this principle is given in the book /7/.

## 18. THE LAST LOVE

Not dwelling on the multiple applications of the expediency principle, we concentrate on an analysis, well studied on accelerators, of the distribution of elementary particles in mass /10/. The masses of approximately 300 elementary particles were determined with good accuracy and thus one can obtain an analytical representation of particles according to masses.

Qualitatively this distribution is characterized by two features: 1. It has a sharp maximum for $m \sim m_p$ ($m_p \sim 1$ GeV is the proton mass). Thus, in the interval ($\frac{1}{2}m_p - 2m_p$) the masses of more than 90% of known elementary particles are concentrated. 2. The distribution is unsymmetric relative to the maximum $m_{max} \sim m_p$.

As a consequence the masses of elementary particles are distributed within many orders of magnitude and the distribution can be conveniently approximated by presentation on a logarithmic scale. The analytic form of the normalized distribution of elementary particles has the form:

$$\frac{\Delta n}{\Delta \log(m/m_p)} \sim 30 \, (m/m_p)^{1.5}, \text{ if } m < 2m_p, \quad (1)$$

and

$$\frac{\Delta n}{\Delta \log(m/m_p)} \sim 300 \, (m/m_p)^{-2}, \text{ if } m > 2m_p. \quad (2)$$

The approximate equations (1), (2) were obtained without taking into account three particles: the electron ($e^-$) with a mass $\sim 10^{-3} \, m_p$, the X-boson (with a mass $m_X \sim 10^{15} \, m_p$), and the planckeon (with a mass $m_{pl} \sim 10^{19} \, m_p$). However, in agreement with modern experimental and theoretical ideas it is just these three particles that play the determining role in the structure of the Metagalaxy and unified field theory. This is well known; however, it is less known that the deviation of the masses of these three particles from distributions (1) and (2) is a necessary condition for fulfilling the expediency principle.

Actually, an increase of the electron mass by more than a factor of 2.5 would lead to the reaction

$$e^- + p \to n + \nu, \quad (3)$$

i.e., to the collapse of an atom of the element hydrogen necessary for the formation of galaxies.[†] It is impossible to allow a significant decrease of the X-bozon mass because with logarithmic accuracy the lifetime of the proton

---

[†]We recall that an era of neutral hydrogen with $Z \sim 1,000$ is necessary for gravitational condensation of matter in the galaxy.

$$t_p \propto m_x^4 \tag{4}$$

(see, e.g., /11/). If one thinks of decreasing the X-bozon mass by 3–4 orders of magnitude, then the time $t_p$ would be comparable with the lifetime of the Metagalaxy and all the matter, in a final analysis, would have been converted to photons and neutrinos. The lifetime of stars of the main sequence depends significantly on the mass $m_{pl}$ /12/:

$$t_s \propto m_{pl}^2 \tag{5}$$

Consequently, a significant deviation of the masses $m_e$, $m_x$, and $m_{pl}$ from the average value $\sim m_p$ of distributions (1) and (2) is necessary for fulfilling the expediency principle.

Using the approximations (1) and (2), one can estimate the probability that one particle (e') would have a mass less than $2.5\ m_e$ and two particles would have masses in the range $10^{15}-10^{19}\ m_p$. The first probability $\sim 10^{-6}$ and the second probability $\sim 10^{-45}$. Assuming both probabilities are independent, it is easy to estimate that the probability of the deviation of masses $m_e$, $m_x$, and $m_{pl}$ from average values $\sim 10^{-50}$.

How can one interpret this quite small number? In the framework of inflationary cosmology /3/ many universes and metagalaxies may originate with different quantum numbers although there is presently an absence of an expression for the law of their formation. Suppose that distributions (1) and (2) are universal in the sense that they reflect the law of formation of elementary particles for the origin of all metagalaxies or many universes. Then, the number $10^{-50}$ has the quite transparent sense of the probability of the formation of a metagalaxy with a complex structure similar to our Metagalaxy. Consequently, combining our conclusion with the basic idea of inflationary cosmology, one can come to the conclusion that there exists, at least, $10^{50}$ metagalaxies whose structure, as a rule, should be significantly simpler than the structure of our Metagalaxy.

## CONCLUSION

It follows from the arguments given in the last section that in the universe (world) there should exist at least $10^{50}$ metagalaxies (mini-universes) from which only one has a complex structure similar to ours. At first glance the conclusion on the existence of such a large number of metagalaxies is trivial because the universe should be infinite in space-time. In actuality, this postulate is not obvious in general. In the history of cosmology models existed of universes finite in space, but infinite in time (Einstein), finite in time and space (Friedman), etc. And at the present time the topology of the space-time of the universe is in no way obvious. Thus the number of $\sim 10^{50}$ mini-universes is the first experimental indication of the true scale of the universe (world). Furthermore, if one allows according to modern ideas that numerical values of fundamental constants formed initially are exaggerated, then the conclusion follows that the theory of the origin of the universe in metagalaxies should include variable fundamental constants. At the present time there are multiple attempts to create unified field theories and to construct a cosmology on their basis. However, all these take the fundamental constants as given (excluding the dimensionality of space; see, e.g., /13/). It follows from the arguments given that first of all it is necessary to interpret the amazing mass spectrum of particles and, consequently, a "true" unified theory should take into account the variability of mass spectra.† The same problem arises in a realistic cosmology. Therefore, it is clear that the problem of constructing a unified field theory and a theory of the origin and the evolution of the universe are equivalent.

In conclusion we make the following remarks. As noted, the distribution of elementary particles in mass is asymmetric and, apparently, cannot be represented as a Gaussian function. This evidences that the process of the formation of the mass spectrum does not have a purely statistical character. This assertion is completely consistent with the new direction in cosmology, attempts to describe the initial stage of the evolution of the universe as a stochastic process (see, e.g., /3/, /8-10; 18-22/), i.e., as a process combining statistical physics and dynamics.

The author is grateful to L. B. Okun' for a discussion of some of the problems discussed in this chapter.

## References

1. Shklovsky, I. S. (1984) *Zemlya i Vselennaya*, N4.
2. Novikov, I. D. (1988). How the Universe Originated, Nauka, Moscow, p. 119.
3. Linde, A. D. (1990) Physics of Elementary Particles and Inflationary Cosmology, Nauka, Moscow, p. 24.
4. Landau, L. D., and Lifshitz, E. M. (1988) Mechanics, Nauka, Moscow, p. 14.
5. Rosental, I. L. (1990) Mechanics as Geometry, Nauka, Moscow.
6. Ginzburg, V. L. (1987) Theoretical Physics and Astrophysics, Nauka, Moscow, Ch. 9.
7. Barrow, J. D., and Tipler, J. D. (1986) The Anthropic Cosmological Principle, Oxford Univ. Press, Oxford.
8. Carter, B. (1974) in Confrontation of Cosmological Theories with Observational Data, Reidel, Boston.
9. Rosenthal, I. L. (1980) *Usp. Fiz. Nauk.* **131**, 239.
10. Rosenthal, I. L. (1988) Big Bang, Big Bounce, Springer-Verlag, Berlin.

---

†It is impossible to say that the huge (in comparison with $m_p$) value of the masses $m_x$ and $m_{pl}$ have been completely unnoticed. This problem known as the hierarchy of masses was formulated many times; however, its solution was not found /14–17/.

11. Okun', L. B. (1981) Leptons and Quarks, Nauka, Moscow.
12. Dibai, E. A., and Kaplan, S. A. (1976) Dimensionality and Similar Astrophysical Values, Nauka, Moscow.
13. Linde, A. D., and Zelnikov, M. I. (1988) *Phys. Lett.* **215B**, 59.
14. Vysotskii, M. I. (1985) *Usp. Fiz. Nauk* **146**, 591.
15. Aref'eva, I. Ya., and Volovich, I. V. (1985) *Usp. Fiz. Nauk* **146**, 655.
16. Vainshtein, A. I., Zakharov, V. I., and Shifman, M. A. (1985) *Usp. Fiz. Nauk* **146**, 683.
17. Gell-Mann, M. (1985) Proc. of Conf. on Quantum Field Theory, R. Jakow, N. Khuri, S. Weinberg, and E. Witten (eds.), MIT Press, Cambridge, MA, p. 3.
18. Starobinskii, A. A. (1984) in Fundamental Interactions, Izd. MGPI, Moscow, p. 55.
19. Rey, S. J. (1987) *Nucl. Phys.* **284B**, 706.
20. Orotolan, A., et al. (1988) *Phys. Rev.* **38D**, 465.
21. Nambu, Y., and Sasaki, M. (1988) *Phys. Lett.* **205B**, 441.
22. Nambu, Y. (1989) *Prog. Theor. Phys.* **81**, 1037.

# 19

# Searching for Planetary Systems

Bernard F. Burke

*Massachusetts Institute of Technology, Cambridge*

## FERMI'S QUESTION

It could happen, at any time, that a lucky radio astronomer might stumble across the first messages from another planet, and would become famous in an instant. At the same time, the discovery would probably influence the course of civilization on earth in ways that cannot be calculated. At the very least it would answer Enrico Fermi's question 'Where are they?' The discovery might come that way, but it is an unlikely scenario. Most radio astronomy observations are taken with receivers that pre-process the information in such a way as to mask or reject the characteristic signature of coherent transmission. Even when the experiment is designed to recognize extraterrestrial signals, or accidentally might be well-suited to the job (pulsar searches, for example) the 'discovery space' that is investigated is usually small. Every radio astronomer should be alert to the possibility, but the chance of making the definitive discovery in the course of ordinary observations is much the same as the chance of buying a winning lottery ticket in an enormous game of chance where the lottery ticket's price is unknown, the numbers on the ticket are indistinct, and perhaps the honesty of the ticket seller can be questioned.

These complications were well understood by Joseph Shklovsky. Those of us who knew him well appreciate his contributions to the search program, perhaps as much by what he did not do as by what he did do. Joseph Shklovsky was an interesting combination of thought and action, of analysis and intuition, rationality and emotion. The central characteristics of his work, however, were his industry, inventiveness, and imagination. In these capacities, he was the leader of a small band of adventurers who had the courage to come forward to declare that the search for intelligent life in the universe was a useful scientific activity. Joseph contributed a critical component of his intellectual effort: the realization that a coordinated, large-scale search was needed because there was a low probability that other civilizations would be found by accident /1/. Nearly twenty years ago, his leadership was experienced in a tangible way when a varied group of scientists, many of them enthusiasts for a search program and some of them skeptics, gathered in Byurakhan, Armenia, to examine the search problem in its broadest context /2/. At that time, he was an optimist concerning the chances for finding civilizations, while in later years his opinion varied broadly, which was not surprising given the inherent uncertainty in our knowledge. He was steadfast, nevertheless, in his support for those who wanted to prosecute the search.

In the spirit of adventure that characterized Joseph Shklovsky's research life, let me examine another approach to the search for intelligent life in the universe. The direct searches for radio signals will ( and should) continue, but in any program of such vast implications, the promising alternate approaches must be examined. After all, there may be no radio transmissions to be detected, or the occurrence of civilizations that wish to do so is a rare event. If the chances for radio detection are small it would help the search program dramatically if there could be a way to target the stars that were most likely to have planets that harbor life. I propose, therefore, to outline a program that would start by looking for the solar systems of other stars, followed by determination of the chemical and physical properties of those planets, and, finally, to identify those planets that might harbor life in some form. The immediate answer to Fermi's 'Where are they?' is 'not here', but if the answer is 'out there', it would help to know where to start looking.

## TO FIND A PLANET

Planets have far less mass than their parent stars, reflect a relatively feeble fraction of the starlight, and so it is no surprise that a convincing example of another planetary system has yet to be found. The best case at present is a near planet: the companion of the star HD 114762 discovered by Latham and his collaborators /3/. The main star shows a quasi-sinusoidal shift in radial velocity that implies a mass of eleven Jupiter masses or more for the companion, deduced from the 84-day period and the 570-meter-per-second amplitude. Precision measurements such as this are one way of looking for planets; a related (but so far unsuccessful) technique is to look for the angular motion of the parent star about the barycenter.

Both techniques are indirect, and require great skill and ingenuity. Nevertheless, they appear to be the most feasible techniques for the short term, since the first experimental equipment has been built and put to use. Latham's discovery can serve as an example of the possibilities and limitations. The rms dispersion of the individual velocity deviation measurements, 400 m/s. allowed the 570 m/s amplitude to be determined to about ten percent accuracy by combing 280 observations. An accuracy improvement by about a factor of ten has been claimed by several groups, and the claims will probably be substantiated soon. There is an inherent limit imposed by the systematic motions in the stellar atmosphere. This limit has not been established yet, but 10 m/s can probably be taken as an optimistic limit.

The radial velocity measurements give the orbital eccentricity, the longitude of periastron, the period $\tau$, and the orbital velocity of the star $V_*$ about the barycenter. The stellar characteristics give an estimate for $M_*$, and from the orbital variables, a lower limit to the planetary mass, $M_p$, can be deduced. The orbital inclination $i$ remains unknown, and it is $M_p \sin i$ that is deduced, so the lower limit to $M_p$ is of the form

$$M_p \geq \left(\frac{V_*}{10 \text{ m/s}}\right)\left(\frac{M_*}{M\odot}\right)^{2/3}\left(\frac{\tau}{10y}\right)^{1/3} \text{ units of Jupiter mass.}$$

If 10 m/s accuracy can be reached, detection of a planet like Jupiter can just barely be achieved. The velocity varies as $(M_*/a)^{1/2}$, and the period varies as $a^{2/3}/M_*^{1/2}$, so the detectability deteriorates as the semimajor axis increases. Thus the method works best if the planet-to-star separation is small. The distance to the star does not affect the measurement and so all the bright stars are equally favorable subjects for study.

The search for planets by astrometric methods is presently limited by the effects of the earth's atmosphere, which limits an individual measurement to about $10^{-3}$ arc-second accuracy, with reasonable expectations for $10^{-4}$ arc-seconds in the near future, using the work of Shao and his group as a guide /4/. The astrometric method has the advantage that the orbital inclination can be derived from the measurements, and if the amplitude of the angular motion is $\theta_*$, the derived planetary mass for a star at distance D is

$$M_p = \left(\frac{\theta_*}{10^{-3}\text{a-s}}\right)\left(\frac{M_*}{M\odot}\right)^{2/3}\left(\frac{\tau}{10y}\right)^{-2/3}\left(\frac{D}{10 \text{ pc}}\right) \text{ Jupiter Masses.}$$

The method is distant-dependent, but it is reasonable to expect that a Jupiter-like planet in orbit about a nearby star (D = 10 pc) should be detectable. The detectable planetary mass diminishes as the semi-major axis increases, so the method is best suited to finding planets widely separated from the parent star.

The two methods have a certain complementarity, and their relative importance will be determined by the limits on velocity determination and not laboratory technique, and will be settled by the competition among groups over the next several years. Improvement is astronomy appears to lie in carrying the equipment to space, to overcome the atmospheric limitations. This will take time and money.

Both approaches will give interesting results, but suffer from two severe defects from the point of view of looking for life. The prospects for finding major planets are good, and astronomy, particularly, gives all the orbital elements, together with a definitive planetary mass. There might well be more than one large planet, though, and it will be necessary to Fourier analyze the deviations in order to disentangle the individual planetary parameters, and this could take many years. The most severe defect, though, is the inability to obtain physical and chemical information. Astronomy and Doppler measurements will find the systems, but one must then turn to other methods.

## TO SEE A PLANET

A simple pair of exercises shows that direct imaging of planets is a formidable problem. At the Byurakhan symposium in 1971 a group of us had a lively exchange on the subject /2, pp. 31–36/. The parent star is so bright that it dazzles the sight, so to speak. At elongation, the ratio of planetary flux to star flux at optical wavelengths, $S_p/S_s$, is approximately

$$S_p/S_s = (\eta/16)\,(d/a)^2$$

for a planet of mean albedo $y$, diameter $d$, and orbital separation $a$ from the star. Planets similar to those in our solar system would be of the order of 90 to 100 db fainter than the star, but even for nearby stars the angular separation is of the order of a second of arc. Diffraction and scattered light from the star dominate the background even for the best optical systems.

At infrared wavelengths, the situation is still extremely difficult. If one observes at the Planck maximum of the planet, its thermal radiation is still relatively feeble compared to that of the star:

$$S_p/S_s = 0.16(T_p/T_s)(d/D)^2$$

where $T_p$ and $T_s$ are the temperatures of the star and planet, and $d/D$ is the ratio of planetary to stellar diameters. For nearby stars, the planet is still of the order of 50 db fainter than the star, and since the diffraction disc of most reasonable telescopes is larger than the star-to-planet separation, the problem is still formidable.

Several layers of difficulty can be distinguished, using our own solar system as a model. The imaging of a giant planet like Jupiter, four or five astronomical units from the parent star, is the easiest task, as it is for the astronomical approach. Lesser giant planets like Neptune or Uranus are more difficult to image, and an earth-like planet, at one astronomical unit or so, presents an even greater challenge. There is no certainty, of course, that other planetary systems will resemble ours, but the general sequence from small planets made of refractory material, close to the sun, to giant planets formed of ices and light gases at greater distances seems to be a reasonable pattern. Since that is the only valid model we have, it gives us a reasonable starting point.

After the Moroz/Burke/Pariiskii exchange at the Byurakhan conference, the question of direct planetary detection lay fallow for years. In 1978–79 KenKnight (sic), followed by Bracewell and MacPhine /5/ proposed that a two-element optical Michelson stellar interferometer, carried by a spacecraft, could cancel out the starlight and detect a major planet like Jupiter. During a sabbatical year in 1984–85, in the course of considering prospects for optical aperture-synthesis interferometers in space, it became clear that there

## 19. SEARCHING FOR PLANETARY SYSTEMS 307

were no fundamental physical barriers to prevent detection of earth-like planets /6/, and this study was condensed into a brief publication /7/ that proposed the concept as a subject for discussion. That publication was accompanied by a different view of the problem, by Angel, Cheng, and Woolf /8/, and was followed by a vigorous exchange of views /9/. The issues have now been clarified: My examination was concentrated on the question of optical detection, while Angel, Cheng and Woolf examined the infrared detectability. From the viewpoint of signal-to-noise ratio, the infrared method appears to be favored; the crucial questions were: (1) Could a large (16-meter diameter) infrared telescope be expected to be launched into space? (2) Could optical methods be improved to allow medium-sized (2 meter) optical telescopes to be figured with a surface accuracy of a hundredth of a wavelength or better? At the time of the exchange, there was no clear answer to either question.

The principal conclusion still remains firm: it is feasible to contemplate direct detection of planets, even earth-like planets. A space-based instrument appears to be essential, because the earth's atmosphere perturbs the image quality so severely. The original debate over whether a monolithic, 16-meter infrared telescope is better or worse for planetary detection than an interferometric array of ten optical telescopes of 1.5 or 2 meter size has been overtaken by the realization that infrared observation can be carried out by interferometers also. For either wavelength range, the principle technical question is how to design an accurate array of 2- to 8- meter telescopes, mounted on a structure some ten to twenty meters in diameter, with the phases controlled to a small fraction of a wavelength. Whether the array is composed of, for example, four 8-meter telescopes, nine 4-meter telescopes, or seventeen 2-meter telescopes is a technical question that would be settled by cost considerations, since detection of earth-like planets could be accomplished with any of these choices. The larger the collecting area, of course, the easier the detection job becomes.

The long-range goal is to determine which stars possess planets that offer conditions that are favorable for supporting life. Since there is no firm theory for the origin and development of life, one can adopt a conservative strategy, designing a system that could give reliable detection of an earth-like planet belonging to any star within ten parsecs of the earth. There are about 50 stars not very different from our own sun within this range, so such a strategy would give a reasonable sampling of planets. Jupiter-like planets would be detectable at somewhat greater distances.

The technology is largely demonstrated. Aperture-synthesis interferometry is a well-developed technology in the radio domain /10/. The radio methods carry over completely into the infrared and optical regimes, except for signal-to-noise considerations (at radio wavelengths $h\nu \ll kT$ and the noise obeys Raleigh statistics, while as progressively shorter wavelengths are used, quantum noise dominates /11/). There are a number of ground-based optical interferometer projects already in an advanced state of development, and results have been given from at least two of these, the CERGA project in France /12/, and the NRL/MIT/JPL/SAO Mark III interferometer on Mt. Wilson in California

/13/. Further development of these techniques will proceed rapidly, and transferring the technology to space missions will follow. At least three international symposia have been held on the subject, outlining the preliminary concepts for space-based interferometers /14/.

## ONE STEP BEYOND

The discussion so far has been addressing the problem of planetary detection. The indirect methods may be the first to find a major planet, since the work is proceeding now at ground-based installations. The doppler method can find a Jupiter-like planet orbiting a solar-class star, and in principle can be successful for a star at any distance. in practice, there is a distance limit, because the method needs a large signal-to-noise ratio, but feasibility has been demonstrated for the case of HD 114762, at a distance of 30 pc, even though the companion is not quite a planet. The astrometric limit for ground-based equipment may include the possibility of finding smaller planets such as Uranus, out to a distance of 10 pc or somewhat worse. No one can yet rule out the possibility of life on such massive, cold, gaseous planets, but it would be hard to work up enthusiasm for a massive SETI program for such objects. Nevertheless, even the indirect evidence for other planetary systems would be a major scientific event, and the search would intensify for planets close in size and chemistry to that of earth.

The direct method, imaging the planets at infrared or optical wavelengths, appears to require space-based systems, which probably would be aperture-syntheses arrays of moderate-sized telescopes. The prospects for detecting the earth-like planets with such systems, however, are excellent, and certainly provide a strong justification for starting now, to prepare for systems that are unlikely to be ready for launch into space until after the year 2000.

If one accepts that the detection of earth-like planets is a goal that is so important that the search should start, an immediate question arises from the considerations that were outlined in the previous section, and the more extensive discussions in the references. The calculations show that the flux of photons from an earth-like planet within a distance of ten parsecs is small. At optical wavelengths, the flux is of the order of one photon per square meter per second. The infrared flux is greater, but still small: of the order of a hundred photons per square meter per second. The figure of merit for detecting such a weak signal is A/G, the ratio of total collecting area A to telescope gain G (borrowing the term that is commonly used in radio astronomy, but is unfamiliar to the optical community). The gain is normalized to unity on the instrumental axis, so G is the sidelobe suppression. It turns out that the principle cause of sidelobes will be the surface roughness of the individual telescopes that make up the array, and that an rms surface accuracy of a hundredth of a wavelength is required for the optical case. The infrared requirements are not quite so stringent, but still are formidable. Nevertheless, the art of producing mirrors of such quality is advancing rapidly.

## 19. SEARCHING FOR PLANETARY SYSTEMS

The total collecting area, A, is probably the principal limitation. A total effective area of ten square meters would probably allow detection of a planet like the earth (e.g. ten apodized telescopes of 2-meter diameter, but only having an effective area of one square meter because of their apodization). If such planets were to be found, however, an immediate question follows: What is the spectrum like? Finding the planet is only the first step. If the planet has an atmosphere (and one expects that to be the usual case), the atmospheric composition will be reveled by its spectrum. This leads to the conclusion that the larger the total collecting area, the better. The spectroscopic array has to have ten to a hundred times greater area than the detection array. An interesting policy question, then, is whether an array of several hundred square meter area is the logical first step. An array of seventeen 4-meter telescopes, for example, would qualify. Nine 8-meter telescopes might even be better. Could the massive resources for such an ambitious project be found? Can the project be carried out even if the resources were found?

A promising approach to the problem is already under discussion. A workshop was held in Albuquerque in 1989, building on several earlier studies, to consider the characteristics that a large aperture-synthesis array might have if it were to be constructed on the moon. The proposed title for the mission was LOUISA (Large Optical, Ultraviolet, and Infrared Synthesis Array) /15/. If the momentum could be established for building a permanent base on the moon, the building of an array such as LOUISA would be much easier to carry out. A permanent lunar base is, in itself, a project of such enormous magnitude that many uses should exist in order to justify the expenditure. Extensive lunar geographical studies is an example, and many types of astrophysical instruments and telescopes would also be natural candidates. A popular discussion of the astrophysical possibilities has already been published /16/, and a large-scale study of all scientific projects that might be conducted from a lunar base was held in Annapolis in 1990 /17/.

The prospect of building arrays of telescopes on the moon may be far more optimistic than one might expect. The instruments themselves might cost very little more than they would cost if built on the earth, since the lunar base allows them to be constructed and tended by the skilled personnel who reside at the lunar base. Estimates of the cost of transporting the equipment to the moon are uncertain, but if a proper system of space freighters can be developed to support the lunar base, they could be used to transport the telescopes also. The cost might well be comparable to the cost of the telescopes themselves. Nevertheless, the three basic segments of the program, the lunar base, the transport system, and the large aperture-synthesis instrument, would constitute the largest program ever undertaken in space. The lunar base would probably proceed independently from the astrophysical programs, depending upon considerations that scientists can only influence in a marginal way. Once the policy decision had been taken, however, the aperture synthesis array would, in all probability, follow as a natural consequence.

Finally, one must note that the project would be adapted naturally to a wide-world collaboration. The necessary resources are very expensive, and sharing the costs among many countries would bring the project within realistic reach. The array would be immensely powerful for many astrophysical programs, and all of the astrophysicists of the world would be welcomed as users of the instrument.

## THE SEARCH FOR LIFE

Once an aperture synthesis array, capable of performing spectroscopy at both optical and infrared wavelengths, has been constructed, the possible results from the instrument could lead to most interesting conclusions. Preston Cloud was the first to realize that the abrupt change in the geophysical record about two billion years ago signified a fundamental advance in the chemistry of living forms /18/. The earth's atmosphere today is rich of oxygen, far out of chemical equilibrium. The disequilibrium is maintained by life, and by oxygen-producing, chlorophyll-based life in particular. Cloud's proposal was that the sudden change in oxidation state of iron in ancient rocks was caused by the abrupt appearance of oxygen-producing life, and that the most probable agents were blue-green algae.

The process was stated more dramatically by Margulis and Lovelock /19/. They posed their Gaia hypothesis, that the atmosphere of our own plant is part of a feedback loop maintained by the earth's living system to support its continued existence. By extension, one might expect that living systems on other planets would be recognized by extreme disequilibrium in atmospheric conditions. We can see that this is the case for our own earth, and even though we may not represent the only route by which living systems can be sustained, the presence of major oxygen disequilibrium in any planetary atmosphere would be a clear signal. Gross disequilibrium of sulfur oxides, or even of reducing gases (with the caveat that volcanoes could be a non-organic cause!) could also be an indicator of some form of life. Long ago, Berkner and Marshall proposed that the shielding presence of ozone was a pre-condition for advanced terrestrial life /20/.

Tobias Owen proposed that detection of oxygen and ozone in the atmosphere of other planets would be a useful signature to look for when searching for evidence of life /21/. He proposed that the oxygen A-band, at 7600 Angstroms, might be the most feasible marker to look for. The ultraviolet absorption bands of ozone were, in his opinion, too likely to be confused with the sulfur dioxide bands, but this has been shown not to be the case if high-resolution spectroscopy can be employed /22/. Roger Angel and his associates have pointed out that there are prominent absorption bands of ozone in the infrared that appeared to be especially promising /8/. Furthermore, in the infrared domain there are prominent absorption features of carbon dioxide, methane, sulfur dioxide, and other molecules that give the promise of examining other planetary atmospheres in interesting chemical detail.

The ideal program, therefore, seems clear. In this decade (the 1900's), the first searches for giant planets belonging to nearby stars will be undertaken seriously by indirect methods, and preliminary experiments are even now under way. Space-based interferometers are under active study, and modest systems could be in orbit by the end of the decade. The most exciting prospect, however, is a large aperture-synthesis system, perhaps modelled on the Very Large Array (VLA) of the National Radio Astronomy Observatory /23/, even though its physical form might be dissimilar. The effective area would be at least a hundred square meters. In principle, the array could be built in earth orbit, perhaps in geosynchronous orbit or at a Lagrangian point. These locations, however, require nearly the same launch energy as for transport to the moon, and the systems will be difficult to service. A lunar location, on the other hand, has many practical advantages if, for national or international policy reasons, a permanent base is established on the moon.

## L'ENVOI

When the first suggestion was made to look for evidence of life by looking for oxygen, it appeared to be a purely academic concept. As technology developed, particularly during the decade of the 1980's, the suggestion has passed from the realm of the hypothetical to the realm of the possible. If planetary systems of other stars can be found (and they probably will be found), and if their atmospheres can be studied spectroscopically, a new age of planetary studies will begin. If even one planet shows a rich oxygen atmosphere, or evidence of some other gross chemical disequilibrium appears, the formal probability of finding intelligent life would immediately change from an unknown quantity to a very real possibility. If several such planets are found in the small sample that is available to us, even if those life forms do not or cannot communicate, we would know that life is a common phenomenon, and the search for signs of intelligent life could intensify. The importance of the project has been officially recognized /24/.

This study has necessarily been the result of working with many colleagues. I would like to single out several who have been particularly interested in the problem, and with whom I have had many fruitful discussions. These include Roger Angel, David Black, Michael Mumma, Michael Shao, Robert Stachnik, and a wider circles of associates who participated in the 1990 workshop 'Towards Observations of Planetary Systems (TOPS) /25/. Jack Burns, Nicholas Woolf and Steven Johnson should also be recognized, especially for their key role in running the LOUISA workshop.

Finally, there is the acknowledgement that comes too late. Joseph Shklovsky was an early inspiration in my career, with his fresh and enthusiastic approach to fundamental problems. His intellectual contacts were worldwide, even when his travel possibilities were constrained. It is appropriate, therefore, that a program that almost certainly must be international can be outlined in this memorial essay. Even if intelligent life is not found, the scientific rewards will fully justify the effort. After all, surprise and wonder are the best fruits of scientific endeavor.

# REFERENCES

1. J. S. Shklovsky and C. Segan (1968) *Intelligent Life in the Universe*, Dell, New York.
2. C. Segan (1973) *Communication with Extraterrestrial Intelligence*, MIT, Cambridge.
3. D. W. Latham, T. Mazeh, R. P. Stefanik, M. Mayor, and G. Burki (1989) *Nature*, **339**, 38.
4. M. Shao, M. M. Colavita, and D. H. Staelin (1986) *Proc.* **SPIE-628**, 250.
5. R. N. Bracewell and R. H. MacPhie (1979) *Icarus* **38**, 136.
6. B. F. Burke (1985) preprint 'Detection of Planetary Systems and the Search for Evidence of Life', MIT, Cambridge.
7. B. F. Burke (1986) *Nature* **322**, 340.
8. R. Angel, A. Cheng, and N. Woolf (1986) *Nature* **322**, 341.
9. B. F. Burke (1986) *Nature* **324**, 518.
10. A. R. Thompson, J. M. Moran and G. W. Swenson, Jr. (1986) *Interferometry and Aperture Synthesis in Radio Astronomy*, Wiley, New York.
11. B. F. Burke (1987) *Reflective Optics* **SPIE 751**, 50.
12. A. Layberie (1980) *Optical and Infrared Telescopes of the 1990's*, A. Hewitt (ed.), KPNO, Tucson.
13. D. J. Hutter et al. (1989) *Astrophys. J.* **340**, 1103.
14. (1987) *Bull. Am. Astr. Soc.* **16** *Workshop on High Angular Resolution Optical Interferometry from Space*, R. D. Reasenberg and P. B. Boyce (eds.), p. 747. (1987) Battelle *Cambridge Workshop on Imaging Interferometry*, European Space Agency *ESA* **SP-226**. (1985) *Colloquium on Kilometric Optical Arrays in Space* (Cargese Symposium).
15. (1991) *A Lunar Optical-UV-IR Synthesis Array*, J. Burns, S. Johnson, and N. Duric (eds.) NASA Conference Publication, in press.
16. J. Burns, N. Duric, G. J. Taylor, and S. Johnson (1990) *Sci. American* **262**, 42.
17. (1990) *Astrophysics from the Moon*, M. J. Mumma and H. J. Smith (eds.), Am. Inst. Phys., New York.
18. P. Cloud (1974) *Am. Scient.* **62**, 54.
19. L. Margulis and J. E. Lovelock (1974) *Icarus* **21**, 471.
20. L. V. Berkner and L. C. Marshall, (1965) *J. Atmos. Sci.* **22**, 225.
21. T. Owen (1980) *Strategies for Search for Life in the Universe*, M. D. Papagiannis (ed.), Reidel, Dordrecht.
22. B. F. Burke (1988) *'Biastronomy—The Next Steps*, G. Marx (ed), Kluwer, Dordrecht.
23. A. R. Thompson, B. G. Clark, C. M. Wade, and P. J. Napier (1980) *Astrophys. J. Supp* **44**, 151.
24. (1977) Lib. of Congress, Congressional Research Office, Science Policy Research

Division; Committee on Science and Technology, US House of Representatives, *Possibility of Intelligent Life in the Universe* **98-185 0**, US Govt. Printing Office, Washington.

25. A. Boss, B. F. Burke, M. J. Mumma, and H. J. Smith (1991) *Towards Observations of Planetary Systems (TOPS)*, NASA Conference Proceeding, in press.

# 20

## Space Radiointerferometry and Gravitational Waves

### V. B. Braginsky*, N., S. Kardashev, I. D. Novikov, and A. G. Palnarev

*Astro Space Center, Lebedev Physical Institute, Moscow and Moscow State University (*).*

We have shown in /1/ that the propagation of electromagnetic waves in a random gravitational wave field considerably differs from electromagnetic wave propagation in randomly inhomogeneous media. Their behavior is specific since gravitational waves are first transverse and second tensorial; thus they are "twice transverse" (the solution includes the factor $\sin^2 \theta$, rather than merely $\sin \theta$, as would be the case for certain vector transverse fields[†]) and the propagation velocity of gravitational waves is exactly equal to the velocity of light. Due to the cumulative effect of the above factors the squared phase dispersion does not grow with distance to the source of electromagnetic waves as is the case for randomly inhomogeneous media.

The paper /1/ discusses how gravitational waves affect the propagation of electromagnetic radiation from remote astrophysical sources and from active generators in the solar system.

One of the main tasks of observational cosmology is to get information about the possible existence in space of primordial gravitational waves (PGWs) generated at the very first moments of the Universe. Since PGWs are extremely weakly interacting with matter, they carry direct data about the birth of the Universe.

Much has been written about the problem of PGW detection (see, e.g., /1-30/). The most informative processes with respect to PGW detection are their interaction with electromagnetic radiation. This paper is devoted to a thorough analysis of these processes and to assessing the possibility to detect PGWs with VLBI space radio interferometry.

Thus the problem to which the article is devoted lies at the intersection of radioastronomy, the theory of the Early Universe and gravitational-wave astronomy. This is the reason why we include this article in this volume. J. A. Shklovsky's and S. B.

---

[†] Here $\theta$ is the angle between wave vectors of gravitational and electromagnetic waves.

Pikel'ner's contribution to radioastronomy is hard to overestimate and their contribution and interests in the theory of the Early Universe stimulated a lot of important research in this field. In addition, we recall Shklovsky's interest in the problem of gravitational radiation (see, e.g., the final section of his well-known book /65/ whose title is "Black Holes and Gravitational Waves").

For an arbitrary gravitational wave, with no sources and with the gauge condition for electromagnetic waves chosen as

$$A^i_{;i} = 0, \qquad (1)$$

where $A^i$ is the 4-vector of the electromagnetic field potential, and ";" is the covariant derivative (Latin letter indices take the values 0,1,2,3) Maxwell equations reduce to the following wave equations /31,32/

$$g^{ik} A^j_{;i;k} = 0 \qquad i,k = 0,1,2,3. \qquad (2)$$

The metric tensor for a weak field is written as

$$g_{ik} = \eta_{ik} + h_{ik}, \qquad g^{ik} = \eta^{ik} - h^{ik}, \qquad (3)$$

where $\eta_{ik} = \text{diag}(1,-1,-1,-1)$. Further, to the first order in $h_{ik}$ the indices are raised and lowered using a non-perturbed metric tensor $\eta_{ik}$. Then it follows from Eq. (2) that

$$\Box A^j + \hat{L}^j_m A^m = 0,$$

where $\Box = -\dfrac{\partial^2}{c^2 \partial t^2} + \Delta$, $\Delta$ is a conventional Laplacian, and the operator $\hat{L}^j_m$ is determined as

$$\hat{L}^j_m = -\delta^j_m h^{ik} \frac{\partial^2}{\partial x^i \partial x^k} + (h^{jk}_{,m} + h^{j,k}_m - h^{k,j}_m) \frac{\partial}{\partial x^k}. \qquad (5)$$

Eq. (5) was derived taking into account the gravitational wave gauge choice:

$$h^{ik}_{,k} = h^k_k = 0 \qquad (6)$$

and that $h_{ik}$ satisfies the wave equation:

$$\Box h_{ik} = 0. \tag{7}$$

Eq. (4) is easily generalized for the case when electromagnetic and gravitational wave interaction occurs in a homogeneous medium with the refraction index different from 1. To do that, it is sufficient that the light velocity in vacuum, c, in the $\Box$ be substituted with the phase velocity of the electromagnetic wave in a homogeneous medium, $c_\phi$. Further on, the operator $\Box_\phi$ is always either an operator with $c_\phi = c$ if vacuum is meant, or an operator with $c_\phi \neq c$ if it is the case with a homogeneous medium. Below we assume $c = 1$.

We consider the propagation of a flat monochromatic electromagnetic wave in the field of a flat monochromatic gravitational wave:

$$h_i^k = h\, \epsilon_i^k \exp(i\varphi_g). \tag{8}$$

Here h and $\varphi_g$ are the amplitude and phase of the gravitational wave; $\epsilon_i^k$ is the unit tensor orthogonal to the zero wave vector of the gravitational wave:

$$\kappa_i = \frac{\partial \varphi_g}{\partial x^i}, \quad \kappa_i \kappa^i = 0, \quad \epsilon_n^i \kappa^n = 0. \tag{9}$$

The nonperturbed electromagnetic wave is written as:

$$\overset{(0)}{A^j} = A_0 e^j \exp(i\varphi_e), \tag{10}$$

where $A_0$ and $\varphi_e$ are the amplitude and phase of the wave, $e^j$ is the unit space-like vector orthogonal to the electromagnetic wave vector $k^i$

$$k_i = \partial \varphi_e / \partial x^i, \quad e^i k_i = 0. \tag{11}$$

The non-perturbed vector-potential $\overset{(0)}{A^j}$ satisfies the wave equation $\Box_\phi \overset{(0)}{A^j} = 0$.

In this case the operator $\hat{L}_m^j$ reduces to a matrix. Note that:

$$\hat{L}_m^j A^m = -A_0 h b^j \exp[i(\varphi_g + \varphi_e)] \tag{12}$$

and Eq. (4) becomes

$$\Box_\phi A^j = A_0 h b^j \exp[i(\varphi_e + \varphi_g)] \tag{13}$$

where

$$b^j = -(\epsilon_{mn}k^m k^n)e^j + (\epsilon_n^j k^n)(e_m \kappa^m) + (\epsilon_n^j e^n)(k_m \kappa^m) - \kappa^j(\epsilon_{mn}e^n k^m). \tag{14}$$

The solution of Eq. (13) may be given as:

$$A^j = A_0 e^{i\varphi_e}(e^j + F b^j). \tag{15}$$

Here the scalar function F satisfies the equation

$$-2ik^n \frac{\partial F}{\partial x^n} + \Box_\phi F = h e^{i\varphi_g}. \tag{16}$$

If $\kappa \ll k$ and $v_\varphi = 1$ the solution of Eq. (16) taking account of boundary conditions at $z = 0$ becomes

$$F = \frac{h e^{i\varphi_g}}{2\kappa k(1-\mu)} [1 - e^{-i\kappa z(1-\mu)}], \tag{17}$$

where $\mu = \cos\theta$.

Generalizing the solution of Eq. (17) for the case when $\tilde{h}$ is a slowly varying function of z gives

$$F \cong \frac{h e^{i\varphi_g}}{2\kappa k(1-\mu)} [h(z) - h(0)e^{-i\kappa z(1-\mu)}]. \tag{18}$$

The condition of slow h variation is the following inequality

$$|h'/h| \ll \kappa(1-\mu) \tag{19}$$

We denote the characteristic distance

$$z_* = \min\{z, |h'/h|_{z=0}\} \tag{20}$$

Then the resonance condition can be written as:

$$\kappa(1-\mu)z_* \ll 1 \tag{21}$$

When the resonance condition (21) is satisfied, the function F grows linearly with the distance the electromagnetic wave passes. This turns out to be extremely important when the problem of electromagnetic wave passage in a random field of perturbations (see /1/) is solved

$$F \cong -\frac{ie^{i\varphi_g}}{2k} h(0)z_* \quad \text{at } 1 - (\kappa z_*)^{-1} < \mu < 1 + (\kappa z_*)^{-1}. \tag{22}$$

For further considerations it is convenient to introduce the following function:

$$\Psi(x,z) = \begin{cases} \dfrac{e^{ixz_*} - h(z)/h(0)}{x} & \text{at} \quad xz_* \gg 1 \\ iz_* & \text{at} \quad xz_* \ll 1 \end{cases} \tag{23}$$

Then for both limiting cases the function F can be written as:

$$F = \frac{e^{i\varphi_g}}{2k} h(0)\psi[\kappa(\mu-1),z]. \tag{24}$$

Further, let us consider the effect of relic gravitational waves on the propagation of electromagnetic radiation from distant sources with large red shifts $Z_s$. With due account taken of the adiabatic damping of gravitational waves caused by Universe expansion, the amplitude of the gravitational wave is a slowly varying function of the distance z the electromagnetic wave passes.

When the effect of the primordial gravitational waves on an electromagnetic wave is analyzed, it is convenient to turn to conformal time $\eta$, $d\eta = dt/a$, where a is the scale factor and to employ a wave equation in the curved spacetime which is the generalization of Eq. (2):

$$g^{ik}A^j_{;i;k} + R^j_n A^n = 0, \tag{25}$$

where $R^j_n$ is the Ricci tensor. The second term in (25) can be neglected if $k/H \ll 1$ and $\kappa/H \ll 1$, where H is the Hubble constant; (here and below the point means $\eta$-derivative). In fact for a zeroth approximation gravitational wave amplitude this term is of the order of $H^2 A_0$, whereas the term responsible for the adiabatic damping of electromagnetic waves for the zeroth-approximation gravitational wave amplitude is of the order of $H\kappa A_0 \gg H^2 A_0$. The first-order correction, in terms of the gravitational wave amplitude, to the Ricci tensor is zero

$$\delta R^j_n = 0. \tag{26}$$

It follows from (26) that

$$h^i_k = h_0 \frac{a_0}{a} \epsilon^i_k. \qquad (27)$$

Here, $h_0$ and $a_0$ is the wave amplitude and the scale factor for today. As $\kappa/H \lesssim 1$ the behavior of gravitational waves differs from the adiabatic law (see /16,33/). Substituting $h^i_k$ from (27) into Eq. (25) and

$$A^j = v^j/a^2, \qquad (28)$$

and using instead of Eq. (3) the metric

$$g_{ik} = a^2(\eta_{ik} + \frac{a_0}{a} h_0 \epsilon_{ik}) \qquad (29)$$

$$g^{ik} = 1/a^2(\eta^{ik} - \frac{a_0}{a} h_0 \epsilon^{ik}). \qquad (30)$$

(indices in the case of the vector $v^j$ and the tensor $\epsilon^j_k$ are raised and lowered with the help of the tensor $\eta^j_k$), we get instead of Eq. (4)

$$\Box v^j + \frac{a_0}{a} \hat{L}^j_m v^m = 0, \qquad (31)$$

where $\Box = -\frac{\partial^2}{\partial \eta^2} + \Delta$ and $\hat{L}^j_m$ is given by Eq. (5) with the quantity $h_0 \epsilon^{ik}$ instead of $h^{ik}$.

The solution of Eq. (31) is given by Eq. (24) if $h(0)$ is substituted for $h_0(1 + Z_s)$ and $h(z)/h(0)$ for $(1 + Z_s)^{-1}$

$$i\delta\hat{\varphi}_e \equiv i\delta\varphi_e + \delta A/A = (1 + Z_s)h_0 \frac{e^{i\varphi_g}}{2k} \Psi[\kappa(\mu-1),z]. \qquad (32)$$

In a random field of gravitational waves $\langle\delta\hat{\varphi}_e\rangle = 0$. Here and below $\langle \rangle$ denotes averaging over a random ensemble of gravitational waves. The correlation function for the nonpolarized PGW noise is written as

$$\Gamma(t,\vec{r},t',\vec{r}') \equiv \langle\delta\hat{\varphi}_e(t,\vec{r})\delta\hat{\varphi}_e^*(t',\vec{r}')\rangle = \frac{3}{4\pi}(1+Z_s)^2 H^2 k^2 \int_{a_*}^{\infty} \frac{d\kappa}{\kappa^3}\phi(\kappa) \cdot$$

$$\cdot \exp[i\kappa(\eta-\eta'-z-z')] \int_{-1}^{1} d\mu(1-\mu^2)^2 \Psi[\kappa(\mu-1),z] \Psi^*[\kappa(\mu-1),z'] \int_{0}^{2\pi} d\psi \exp[-i\kappa\sqrt{1-\mu^2}\Delta\rho\cos\psi]. \qquad (33)$$

Here the asterisk stands for complex conjugation and

$$\Delta\rho = \sqrt{(x-x')^2 + (y-y')^2}, \quad \kappa_* = 2\pi/\tau$$

where $\tau$ is the duration of the observation; as the literature on pulsar timing caused by cosmological gravitational waves often mentions (see /11/), this limitation on low frequencies is associated with the fact that the observation time is not sufficient for a wave with $\kappa < \kappa_*$ to undergo at least one oscillation and thus their effect cannot be distinguished from the systematic variation of any system parameters. As a result, the low frequency contribution to mean square quantities is zero. The quantity $\phi(\kappa)$ that enters Eq. (33) is the spectrum of gravitational wave background at the present moment, determined as:

$$\phi(\kappa)\frac{d\kappa}{\kappa} = \frac{1}{32\pi G}\left(\frac{3H^2}{8\pi G}\right)^{-1} 4\pi\kappa^4 <|h_0|^2> d\kappa. \tag{34}$$

The density of the gravitational wave background is expressed in terms of $\phi(x)$ as

$$\Omega_{GW} \equiv \varepsilon_{GW}/\varepsilon_{cr} = \int \phi(\kappa)\frac{d\kappa}{\kappa} \tag{35}$$

where $\varepsilon_{cr} = 3H^2/8\pi G$ is the critical density at which the dimensionless parameter of the mean density $\Omega = 1$. It follows from Eq. (34) that

$$<|h_0|^2> = \frac{3}{\pi}\phi(\kappa)/\kappa^5. \tag{36}$$

Relation (34) is taken into account in Eq. (33). From Eq. (33) for the correlation function we get the squared dispersion of the phase fluctuation putting $\eta' = \eta$, $x = x'$, $y = y'$, and $z = z'$:

$$\sigma^2 = \Gamma(t, \vec{r}, t, \vec{r}) = \frac{3}{2}(1 + Z_s)^2 H^2 k^2 \int_{\kappa_*}^{\infty} \frac{d\kappa}{\kappa^3}\phi(\kappa)I_\mu \tag{37}$$

where

$$I_\mu = \int_{-1}^{1} d\mu \, |\psi[\kappa(\mu-1),z]|^2 q(\mu). \tag{38}$$

Here $q(\mu) = (1-\mu^2)^2$.

Before doing integral (38), let us estimate the contribution from the angular time near the resonance (see Eq. (21)). As follows from Eqs. (22) and 23)

$$dI^{(R)}_\mu \cong \int_{1-(\kappa z_*)^{-1}}^{1+(\kappa z_*)^{-1}} q(\mu)d\mu \cong 2z_*^2 \left[ q|_{\mu=1}(\kappa z_*)^{-1} + 1/6 \frac{d^2q}{d\mu^2}|_{\mu=1}(\kappa z_*)^{-3} \right] =$$

$$2q|_{\mu=1}(z_*/\kappa) + 1/3 \frac{d^2q}{d\mu^2}|_{\mu=1}(\kappa^3 z_*)^{-1}). \tag{39}$$

It is evident from Eq. (39) why the factor of distance disappears in the dispersion relation for the case of random gravitational waves. As $v_\phi = 1$ for the latter, the coefficient of $z_*$ in Eq. (36) is proportional to $q|_{\mu=1}$, and $q|_{\mu=1} \cong (1-\mu^2)^2|_{\mu=1} = 0$ because of the gravitational wave transversity. Therefore, in this case, the contribution of resonance waves to the squared dispersion is inversely rather than directly proportional to the distance the EM wave passes. This is contrary to the situation with randomly inhomogeneous media for which $v_\phi \cong 0$ and $q|_{\mu \cong 0} \neq 0$.

Thus, we neglect below the contribution from resonant plane waves. In that case the integral (38) is easily taken and the dispersion (37) is

$$\sigma^2 = 4[(1+Z_s)^2 + 1]H^2k^2 \int_{\kappa_*}^\infty \frac{d\kappa}{\kappa^5} \phi(\kappa). \tag{40}$$

In practice, the value to be measured is the structure function determined as /22,23/

$$D(t,\vec{r},t',\vec{r}') \equiv <|\delta\hat{\phi}_e(t,\vec{r}) - \delta\hat{\phi}_e(t',\vec{r}')|^2> = \sigma^2(t,\vec{r}) + \sigma^2(t',\vec{r}') - 2\,\text{Re}\,\Gamma(t,\vec{r},t',\vec{r}'). \tag{41}$$

We now consider a space radio interferometer where the base is perpendicular to the direction toward the source while phase measurements are strictly synchronized, that is, t = t', ($\eta = \eta'$).

We calculate the transverse structure function with the help of E.q (41),

$$D(\vec{\Delta\rho}) \equiv D(t,\vec{r},t,\vec{r} + \vec{\Delta\rho}), \tag{42}$$

where $(\vec{\Delta\rho} \cdot \vec{r}) = 0$. We get from Eqs. (36), (41) and (42):

$$D(\vec{\Delta\rho}) = \frac{3}{2\pi}(1+Z_s)^2 H^2 \kappa^2 \int_{\kappa_*}^\infty \frac{d\kappa\phi(\kappa)}{\kappa^5} \int_{-1}^1 d\mu(1+\mu)^2 \left[ 1 - \frac{2\cos(\kappa z(1-\mu))}{1+Z_s} + \frac{1}{(1+Z_s)^2} \right] I_\psi, \tag{43}$$

where

$$I_\psi = \int_0^{2\pi} \left[1 - \cos(\kappa\Delta\rho \sqrt{1-\mu^2} \cos\psi)\right] d\psi. \qquad (44)$$

The integral $\psi$ reduces to the zero Bessel function $I_0(\alpha)$ /34/

$$I_\psi \cong 2\pi I_0(\alpha) \qquad (45)$$

where $\alpha = \kappa\Delta\rho\sqrt{1-\mu^2} = 2\pi(L/\lambda_g)\sqrt{1-\mu^2}$ ; here L is the base length of the interferometer. In asymptotical relations for long waves when $\alpha \ll 1$

$$I_\psi \cong \frac{\pi}{2} \kappa^2 L^2 (1-\mu^2), \qquad (46)$$

while in the short-wave limit when $\alpha \gg 1$ we get

$$I_\psi \cong 2\pi. \qquad (47)$$

In the asymptotical relations (46) and (47) the integral over $\mu$ is easily taken, which leads to

$$D(\Delta\vec{\rho}) \equiv D(L) \cong \frac{6}{5}k^2L^2H^2[(1+Z_s)^2+1] \int_{\kappa_*}^\infty d\kappa \frac{\phi(\kappa)}{\kappa^5} G(\kappa), \qquad (48)$$

where

$$G(\kappa) \cong \begin{cases} 1, & \kappa L \ll 1 \\ \dfrac{20}{3\kappa^2L^2}, & \kappa L \gg 1 \end{cases} \qquad (49)$$

Comparison of the structure function (48) with future radiointerferometry data will make it possible to derive limitations on the energy density of primordial gravitational waves at an arbitrary frequency $\kappa$ over a special range $\Delta\kappa \cong \kappa$ and to do so independently of the assumptions about the spectrum of cosmological gravitational waves. Indeed, it is evident from Eqs. (44) and (48) that the contribution to the structure function from waves with $\Delta\kappa \cong \kappa$ is equal to

$$D(\kappa,L) \cong \frac{k^2H^2}{\kappa^2} [(1+Z_s)^2+1] \, \Omega_{GW}(\kappa) \begin{cases} L^2, & 2\pi/\tau < \kappa \ll L^{-1} \\ 8\kappa^{-2}, & \kappa \gg L^{-1} \end{cases}. \qquad (50)$$

If the experiment provides a certain sensitivity in determining the structure function for phase shift (we denote it as $\delta$), then even if we don't have positive results in primordial gravitational wave measurements, we can, with the help of Eq. (50), obtain the following limitations on $\Omega_{GW}(\kappa)$

$$\Omega_{GW}(\kappa) < \frac{\delta^2 \kappa^2}{k^2 H^2 [(1+Z_s)^2+1]} \begin{cases} L^{-2}, & 2\pi/\tau < \kappa \ll L^{-1} \\ \kappa^2/8, & \kappa \gg L^{-1} \end{cases}. \tag{51}$$

A maximally severe limitation on $\Omega_{GW}(\kappa)$ is reached for wavelengths with a period of the order of the observation time and is written as:

$$\Omega_{*GW} \cong 2\delta^2 \left(\frac{T_H}{\tau}\right)^2 \left(\frac{\lambda_e}{L}\right)^2 [(1-Z_s)^2+1]^{-1} \cong$$

$$2 \times 10^{-6} \delta^2 \left(\frac{\tau}{1 \text{ year}}\right)^{-2} \left(\frac{\lambda_e}{1 \text{ cm}}\right)^2 \left(\frac{L}{1 \text{ A.U.}}\right)^{-2} \left[\frac{(1+Z_s)^2+1}{2}\right]^{-1}, \tag{52}$$

where $T_H = \frac{2}{3}H^{-1} \cong 20$ billion years is the Hubble time.

For an arbitrarily long gravitational wave, electromagnetic wave phase radiointerferometry yields the following limitations on $\Omega_{GW}(\kappa)$

$$\Omega_{GW}(\kappa) < \left(\frac{\delta}{3 \times 10^{-2}}\right)^2 \left(\frac{\lambda_e}{1 \text{ cm}}\right)^2 \left[\frac{(1+Z_s)^2+1}{2}\right]^{-1} \left(\frac{\lambda_g}{1 \text{ A.U.}}\right)^{-2}$$

$$\cdot \begin{cases} \left(\frac{L}{1 \text{ A.U.}}\right)^{-2}, & \lambda_g \gtrsim 3L \\ \left(\frac{\lambda_g}{1 \text{ A.U.}}\right)^{-2}, & \lambda_g \lesssim 3L \end{cases}. \tag{53}$$

The above described limitations on $\Omega_{GW}$ may be valid for some cosmological models predicting a specific type of gravitational wave spectrum. Thus, models based on inflation and phase transitions in the early Universe, within the wavelength range we are interested in, predict a flat spectrum of PGWs. Here /35–40/

$$\phi(\kappa) \cong \Omega_\gamma (t_{Pl}/t_{inf})^2 \tag{54}$$

where $\Omega_\gamma$ is the dimensionless density of relic radiation, $t_{Pl}$ is the Planck time, and $t_{inf}$ is the time from the beginning of Universe expansion till the beginning of its inflation. This time is $t_{inf} \cong t_{Pl}$ in quantum-gravitational inflation models and $t_{inf} \cong (10^3-10^4)t_{Pl}$ in models of inflation due to phase transitions at the Grand-Unification energy. To measure such a RGW spectrum with a space radio interferometer the accuracy of phase determination should be of the order of

$$\delta \leq 5 \frac{t_{Pl}}{t_{inf}} \left[\frac{(1+Z_s)^2+1}{2}\right]^{1/2} \left(\frac{\tau}{1 \text{ year}}\right) \left(\frac{\lambda_e}{1 \text{ cm}}\right)^{-1} \left(\frac{L}{1 \text{ A.U.}}\right). \quad (55)$$

When $Z_s \cong 3$ and $t_{Pl}/t_{inf} \cong 2 \times 10^{-4}$, the required $\delta$ is

$$\delta \cong 3 \times 10^{-3} \left(\frac{\tau}{1 \text{ year}}\right) \left(\frac{\lambda_e}{1 \text{ cm}}\right)^{-1} \left(\frac{L}{1 \text{ A.U.}}\right) \quad (56)$$

which seems quite realistic.

Models with ring cosmological strings predict the following for the range of interest: $\Omega_{GW} \cong 10^{-6}$, /41–47/, therefore in this case the accuracy with which the phase should be determined is written as:

$$\delta \cong 3 \times 10^{-2} \left(\frac{\tau}{1 \text{ year}}\right) \left(\frac{\lambda_e}{1 \text{ cm}}\right)^{-1} \left(\frac{L}{1 \text{ A.U.}}\right) \quad (57)$$

In fact real world requirements on $\delta$ should be an order of magnitude more stringent for the measurements to ensure the observed effect.

It is well known that the data on pulsar timing may currently yield the most stringent limitations on relic gravitational waves in the wavelength range of the order 1 to 10 light years (as to the 10 to $10^3$ Mpc range the strongest limitations can be obtained from the isotropy of relic (background) radiation, see, e.g., /36,48–54/.

The aim of a comparison made below is to determine the range of wavelengths within which space radiointerferometry may compete with, or even be better in its capabilities than pulsar timing. Fairly rough estimates will suffice for such an analysis.

In pulsar timing the accuracy with which wave amplitudes $h_0$ are measured depends on the following quantities: $\Delta t$ is the inevitable systems error in determining the time of arrival of radio pulses and $\tau_{PT}$ is the observation time. The quantity $\Delta t$ is now determined by the accuracy of the Solar system /28/ and for RSR 1937 + 21, it is of the order of tenths of a microsecond. The quantity $\tau_{PT}$ is of the order of tenths of a microsecond. The quantity $\tau_{PT}$ is of the order of 10 years. Therefore, the value measured is of the order of:

$$h_0 \sim \Delta t/\tau. \quad (58)$$

For the length of the gravitational wave $\lambda_g$ this imposes the following limitation on $\Omega_{GW}$

$$\Omega_{GW \, PT} \lesssim \left(\frac{R_H}{\lambda_g}\right)^2 \left(\frac{\Delta t}{\tau}\right)^2 \quad \text{for} \quad \lambda_g \ll \tau_{PT}. \quad (59)$$

The accuracy of $h_0$ measurements in terms of phase shift in the radiointerferometer is:

326     V. B. BRAGINSKY ET AL.

$$h_0 \cong \frac{\delta}{1+Z_s} \frac{\lambda_e}{L} \quad \text{for} \quad L < \lambda_g \ll \tau_{SI}, \qquad (60)$$

and the respective limitations on $\Omega_{GW}$ are

$$\Omega_{GW\;SI} \lesssim \left(\frac{R_H}{\lambda_g}\right)^2 \left(\frac{\delta}{1+Z_s}\right)^2 \left(\frac{\lambda_e}{L}\right)^2 \qquad (61)$$

Comparison of Eq. (58) with Eq. (60) or Eq. (59) with Eq. (61) shows that $\Omega_{GW\;SI}/\Omega_{GW\;PT} < 1$ when

$$\delta < (1+Z_s) \frac{\Delta t L}{\tau \lambda_e} \cong 2 \times 10^{-2} \left(\frac{1+Z_s}{4}\right) \left(\frac{\Delta t}{0.1\;\mu s}\right) \left(\frac{\tau_{PT}}{10\;\text{years}}\right)^{-1} \left(\frac{\lambda_e}{1\;\text{cm}}\right)^{-1} \left(\frac{L}{1\;\text{A.U.}}\right).$$

(62)

Thus, at fairly realistic $\Delta$ space radiointerferometry may compete with pulsar timing. It is essential to emphasize here that if 1 light year to 10 light years is the wavelength range within which the most stringent limitations can be derived for pulsar timing, the respective wavelength range for a space radiointerferometer is $\lambda_g \lesssim 1$ light year. In other words, not only could space interferometry be competitive with pulsar timing, it can also be complementary to the latter for the other wavelength range.

Another factor in favor of radiointerferometry is the one associated with adiabatic damping of relic gravitational waves. For distant sources with large red shifts the phase shift is determined by more intense relic gravitational waves near the source; then space radiointerferometry will provide nontrivial limitations on $\Omega_{GW}$ (or it may even lead to relic gravitational wave detection).

New possibilities should also be mentioned of optical interferometry which, reducing the length of the wave used, results in greater sensitivity of phase measurements by $\lambda_{radio}/\lambda$, i.e., by a factor of $10^5$.

If satellites of the Radioastron type are used to record low-frequency gravitational radiation with microwave interferometers to measure small variations in distance /46/, it is possible to achieve a sensitivity similar to the optimistic estimates for the relic gravitational wave background. We now estimate the requirements which must be imposed on such satellites and interferometers. Let the expected value of metric amplitude fluctuations be $<h_0^2>^{1/2} = 1 \times 10^{-18}$ in the frequency range $\Delta f_{GW} \approx f_{GW} \approx 10^{-3}$ Hz.[†] Then the requirements on the compensation level for nongravitational accelerations of satellites should be relatively milder. These accelerations should not be higher than

---

† According to /16/, $<h_0^2>^{1/2}$ is $\cong 10^{-(20-21)}/f_{GW}$ if $\Delta f_{GW} \sim f_{GW}$, while $<h_0^2>^{1/2} \cong (3-10) \times 10^{-20}$, according to /38/ at $f_{GW} \cong 10^{-4}$ Hz.

## 20. SPACE RADIOINTERFEROMETRY AND GRAVITATIONAL WAVES

$$2\pi^2 f_{GW}^2 L <h_0^2>^{1/2} \approx 3\times 10^{-10} \frac{cm}{c^2} \left(\frac{f_{GW}}{10^{-3}\,Hz}\right)^2 \left(\frac{L}{1\,A.U.}\right) (<h_0^2>^{1/2}/10^{-18}). \quad (63)$$

If three satellites are used to form three radiointerferometers with a common mcw-self-oscillator then the requirement to the relative stability of the oscillator frequency $\Delta\omega_0/\omega_0$ would not be too stringent either:

$$\frac{\Delta\omega_0}{\omega_0} \lesssim \frac{1}{2}\beta^{-1}<h_0^2>^{1/2} \cong 10^{-14}\left(\frac{\beta}{10^{-4}}\right)^{-1}\left(\frac{<h_0^2>^{1/2}}{10^{-18}}\right), \quad (64)$$

where $\beta$ is the fractional difference of distances between pairs of satellites.

The recorded phase-shift value is relatively large

$$\Delta\varphi_e \cong \frac{\pi}{\lambda_e} L <h_0^2>^{1/2} \cong 1\times 10^{-4}\,\text{rad}\left(\frac{\lambda_e}{1\,cm}\right)^{-1}\left(\frac{L}{1\,A.U.}\right)\left(\frac{<h_0^2>^{1/2}}{10^{-18}}\right) \quad (65)$$

To compensate for phase fluctuations caused by interplanetary plasma it is necessary that each satellite should be synchronized by a two microwave frequency link. Requirements on the dynamic range of the phase measuring device are not too severe: about 5 orders of magnitude since for $\lambda_e \cong 1$ cm and $L = 1.5\times 10^{13}$ cm the additional phase shift caused by the plasma is of the order of 10 rad /56/.

The signal-to-noise ratio is the most serious obstacle if we are to achieve a sensitivity at the level $\Delta\varphi_e \cong 1\times 10^{-4}$ rad. If W is the microwave power of the wave that returned to the emitting antenna after its retranslation by one of the satellites, then the standard quantum limit of fluctuations, $\Delta\varphi_{SQL}$, is known to equal /57/:

$$\Delta\varphi_{SQL} \cong \sqrt{\frac{\hbar\omega_0 f_{GW}}{W}}. \quad (66)$$

For the condition $\Delta\varphi_e \cong \Delta\varphi_{SQL} \cong 10^{-4}$ rad to be fulfilled at $L \cong 1.5\times 10^{13}$ cm and a transmitter power $10^8$ erg/s, antennas aboard the satellites should be of the order of $10^3$ cm in diameter and the power gain should be 200 dB. The latter requirement can be appreciably reduced, though in this case two high-stability auto oscillators would be needed for two satellites with $\Delta\omega_0/\omega_0 \cong <h_0^2>^{1/2} \cong 1\times 10^{-18}$. There are indications that this level of frequency stability is possible /58/. Note that the accumulation of data about phase fluctuation correlations during a long period of time in two branches of the radio interferometer will reduce the detection threshold.

Possible use of space radiointerferometry for gravitational wave detection is not restricted only to the cosmological background. Space interferometers can also be used to detect individual bursts generated by such astrophysical processes as supernova explosions, by two gravitating bodies passing each other, etc. It is essential that when a single burst, without memory ($h(-\infty) = h(+\infty) = 0$) (see /59–62/ for bursts with memory), is passing through an electromagnetic wave moving towards us from the source, the wave gains an additional phase shift:

$$\Delta\varphi_e \cong \frac{\omega_e}{2} \int h_{zz} dz \cong \frac{1}{2} h_{zz} \left(\frac{\lambda_g}{\lambda_e}\right) \tag{67}$$

If the emission and retranslation of an electromagnetic wave can be provided for a long time, the above mentioned phase shift will be memorized forever and the quantity $\Delta\varphi_e$ could be measured very accurately since that accuracy is determined by $N^{-1/2}$, where N is the number of photons used.

A similar system also operates in a laser-interferometer with mirrors multiply reflecting a laser beam /14,63,64/.

In the case of a space radiointerferometer the range of gravitational waves detected with its help shifts to low frequencies. This property makes space radiointerferometry a unique way to detect gravitational wave bursts whose duration varies from minutes to days.

## ACKNOWLEDGMENTS

The authors express their most sincere gratitude to A. Illarionov, D. Kompaneetz, L. Grishchuk, and M. Sazhin for useful discussion during the preparation of this paper; the authors would like to thank B. Bertotti and K. S. Thorne for critical remarks.

## REFERENCES

1. Braginsky, V. B., Kardashev, N. S., Novikov, I. D., and Polnarev, A. G. (1991) *Nuovo Cimento*, in press.
2. Rees, M. J. (1971) *MNRAS* **154**, 187.
3. Dautcourt, G. (1974) in Confrontation of Cosmological Theories with Observational Data, M. S. Longair (ed.), Reidel, Dordrecht, p. 299.
4. Estabrook, F. B., and Wahlquest, H. D. (1975) *Gen. Rel. Grav.* **6**, 439.
5. Sazhin, M. B. (1978) *Astron. J.* **55**, 65.
6. Armstrong, J. W., Woo, R., and Estabrook, F. B. (1979) *Astrophys. J.* **230**, 570.
7. Bertotti, B., and Carr, B. J. (1980) *Astrophys. J.* **236**, 1000.
8. Mashhoon, B., and Grishchuck, L. P. (1980) *Astrophys. J.* **236**, 990.
9. Carr, B. J. (1980) *Astron. Astrophys.* **89**, 6.
10. Hellings, R. W. (1981) *Phys. Rev.* **D23**, 832.
11. Bertotti, B., Carr, B. J., and Rees, M. J. (1983) *MNRAS* **203**, 945.
12. Braginskii, V. B., Mitrofanov, V. P., and Yakimov, V. N. (1985), Moscow State University Preprint No. 1.
13. Anderson, J. D., and Mashhoon, B. (1985) *Astrophys. J.* **290**, 445.

14. Thorne, K. S. (1987) in Three Hundred Years of Gravitation, p. 330.
15. Grishchuk, L. P. (1987) Proc. GR. 11, Stockholm.
16. Grishchuk, L. P. (198) *UFM* **156**, 297.
17. Boulanger, J. L., Le Denmat, G., and Tourrene Ph. (1988) *Phys. Lett.* **A126**, 213.
18. Sazhin, M. V., Blair, D. G., and James, S. K. (1989) Proc. Fifth Marcel Grossman Meeting, D. G. Blair and M. S. Buckyngham (eds.), World Scientific, Singapore.
19. Sazhin, M. B. (1978) *Astron. Circular* **1002**, 1.
20. Detweiler, S. (1979) *Astrophys. J.* **234**, 1100.
21. Mashhon, B. (1982) *MNRAS* **199**, 659.
22. Romani, R. W., and Taylor, J. H. (1983) *Astrophys. J.* **265**, L35.
23. Hellings, R. W., and Downs, G. S. (1983) *Astrophys. J.* **265**, L39.
24. Blandford, R., Narayan, R., and Romani, R. W. (1984) *J. Astrophys. Astron. (India)* **5**, 369.
25. Davis, M. M., Taylor, J. H., Weisberg, J. M., and Backer, D. C. (1985) *Nature* **315**, 547.
26. Krauss, L. M. (1985). *Nature* **313**, 32.
27. Carr, B. (1985) *Nature* **315**, 540.
28. Backer, D. C., and Hellings, R. W. (1986) *Ann. Rev. Astron. Astrophys.* **24**, 557.
29. Polnarev, A. G. (1988) Space Research Institute Preprint No. 1355.
30. Taylor, J. H. (1989) Trudi soveshania v Tbilisi i Moskve.
31. Landau, L. D., and Lifshits, E. M. (1973) Theory of Fields, Nauka, Moscow.
32. Misner, G. W., Thorne, K. S., and Wheeler, J. A. (1973) Gravitation, Freeman and Company, San Francisco.
33. Grishchuk, L. P. (1974) *ZhETF* **67**, 825.
34. Gradshtein, I. S., and Rizhik, I. M. (1977) Tables of Integrals, Sums, Series, and Products, Nauka, Moscow, p. 1108.
35. Starobinsky, A. A. (1979) *Pis'ma Astron.* **3**, 719.
36. Starobinsky, A. A. (1983) *Pis'ma Astron.* **9**, 579.
37. Hogan, C. J. (1986) *MNRAS* **218**, 629.
38. Deryagin, D. V., Grigoriev, D. Yu., Rubakov, V. A., and Sazhin, M. V. (1986) *Mod. Phys. Lett.* **A1**, 593.
39. Rubakov, V. A., Sazhin, M. V., and Veryshin, A. V. (1987) *Phys. Lett.* **115B**, 189.
40. Matzner, R. (1988), University of Texas Preprint.
41. Hogan, C. J., and Rees, M. J. (1984) *Nature* **311**, 109.
42. Vachaspati, T., and Vilenkin, A. (1985) *Phys. Rev.* **D31**, 3052.
43. Albrecht, A., and Turok, N. (1985) *Phys. Rev. Lett.* **54**, 1868.
44. Brandenberger, R. H., Albrecht, A., and Turok, N. (1986), Preprint No. SF-ITP-15.
45. Vilenkin, A. (1987) in Three Hundred Years of Gravitation, p. 499.
46. Bennett, D. P., and Boucher, F. (1988) *Phys. Rev. Lett.* **60**, 257.
47. Romani, R. W. (1988) *Phys. Lett.* **B215**, 477.

48. Dautcourt, G. (1969) *MNRAS* **144**, 255.
49. Doroshkevich, A. G., Novikov, I. D., and Polnarev, A. G. (1977) in Experimental Gravitation, Accad. Nazionale dei Lincei, Roma, p. 91.
50. Fabbri, R., and Pollack, M. D. (1983) *Phys. Lett.* **125B**, 445.
51. Adams, P. J., Hellings, R. W., and Zimmerman, R. L. (1984) *Astrophys. J.* **280**, L39.
52. Polnarev, A. G. (1985) *Astron. Zh.* **62**, 1041.
53. Starobinsky, (1985) *Pis'ma Astron. Zh.* **11**, 323.
54. Polnarev, A. G. (1986), Proc. GRG-11, Sweden.
55. Kardashev, N. S., and Slysh, V. I. (1988) in The Impact of VLBI on Astrophysics and Geophysics, M. J. Reid and J. M. Moran (eds.), Kluwer, Dordrecht, p. 433.
56. Yakovlev, O. I. (1985) Propagation of Radio Waves in the Cosmos, Nauka, Moscow.
57. Braginsky, V. B. (1967) *ZhETF* **53**, 1434.
58. Vessot, R.F.C. (1988) Workshop on Relativistic Gravitational Experiments in Space, Annapolis, Maryland.
59. Zeldovich, J. B., and Polnarev, A. G. (1974) *Astron. Zh.* **51**, 30.
60. Braginsky, V. B., and Grishchuk, L. P. (1985) *ZhETF* **89**, 744.
61. Braginsky, V. B., and Thorne, K. S. (1987) *Nature* **327**, 123.
62. Grishchuk, L. P., and Polnarev, A. G. (1989) *ZhETF* (in press).
63. Rudenko, V. N., and Sazhin, M. B. (1980) *Kvantovaja Electronika* **7**, 2344.
64. Kulagin, V. V., Polnarev, A. G., and Rudenko, V. N. (1986) *ZhETF* **91**, 1553.
65. Shklovsky, J. S. (1984) Stars, Nauka, Moscow.

# 21

# Radio Astronomy of the Next Century

## Y. N. Pariiskii
*Special Astrophysical Observatory, Nizhniy Arkhyz*

## I. GENERAL TRENDS IN RADIO ASTRONOMY INSTRUMENTATION

### a) Evolution of the Flux Density Limit of the Best Radio Telescopes with Time (see Fig. 21.1)

The flux density limit evolution may be approximated by the formula

$$\Delta P = A_p \, e^{B_p(T-T_o)/T}$$

where $A_p = 10^6$ Jy, $B_p = 0.46$, T is time in years, and $T_o = 1932$, radio astronomy's birth date. This picture shows that the range of radio astronomy expands exponentially with time (just as the Universe in the period of inflation!).

The most powerful emitters (such as very strong radio galaxies) may be observed at the end of our century up to the epoch of their formation. Later, less powerful galaxies may be seen up to the horizon, etc. We expect the same story for the Galactic population which begins and ends our Fig. 21.1. In a Euclidian Universe the range of the radio telescopes increases by a factor 2 every 3 years. This means that for objects which are at a distance more than 3 light years from the Earth and which are visible by a radio telescope the signal-to-noise will increase with time even if this object is moving away from us with almost the speed of light! Another prediction may be done for the nearby objects. If we can see an object during three years after its launch, we can see it forever (with the same emitting power on board). If the object is lost quickly after launch, it must appear inside the exponentially expanding range of radio astronomy sooner or later. Indeed, if the signal is decreasing as $1/R^2$, $R = vT$, but the sensitivity of the receiving Radio Telescope is increasing as

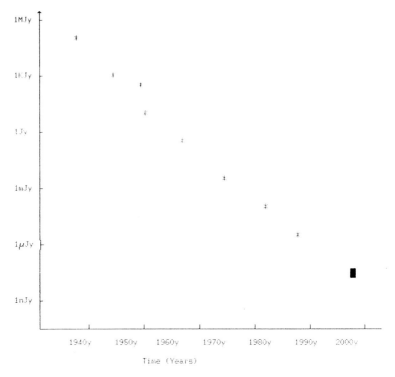

Figure 21.1. Range of radio astronomy versus time. The first point is the discovery of Galactic radio emission. The last point is VLA records. The black box is the expected level by year 2000. The Galaxy population will dominate again.

$$\Delta P = Ae^{-BT},$$

we always can find the T when the

$$S/N \gg 1.$$

## b. Evolution of the Resolving Power (see Fig. 21.2)

This may be approximated by the expression

$$\theta = A_{res} e^{-B_{res}(T-T_o)},$$

where $A_{res} = 1.3 \times 10^6$ ", $B_{res} = 0.44$, T is the current time in years, and $T_o = 1932$.

## c. Growth of the Collecting Area of the Radio Astronomy Community.

This is shown in Fig. 21.3 and it may again be approximated by the exponential law:

$$\Sigma = A_\sigma e^{-B\sigma(T-T_o)},$$

where $A_\sigma = 10$ m, $B_\sigma = -0.17$, T is the current time in years, and $T_o = 1932$.

We see that at the end of this century there will be about 1 million square meters of collecting surface on the Earth available for radio astronomy. We may hope that radio astronomers can use all this surface as a single giant radio telescope (if necessary).

We have mentioned radio telescopes which are built for radio astronomy.

In the next century a new kind of radio amateurs may appear, people having TV and

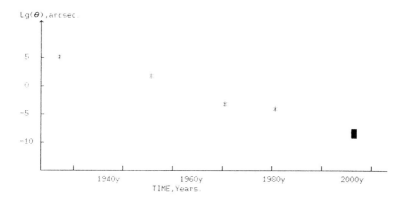

Figure 21.2. Evolution of the resolving power of radio telescopes. The expected level at year 2000 corresponds to A.U. array scales.

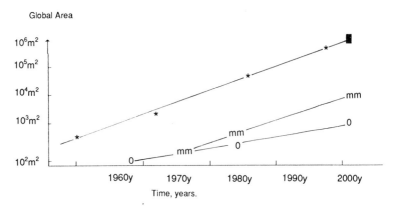

**Figure 21.3.** Growth of the collecting area of radio telescopes on the Earth. The last cross includes the projects under discussion. At year 2000 the expected area of all radio telescopes should be greater than 1,000,000 m².

PC can be organized through the VIDEO-NET which just appeared in the USA and Japan for the purpose of broad-band communication between users and a large computer. There are more than 100,000,000 TV-users on the Globe with $\lambda^2$ collecting area each.

The USA is now discussing the construction of new generation telescopes with a collecting surface about 200,000 m² at cm wavelengths, 20,000 m² at mm wavelengths, and 2000 m² at the wavelengths below 1 mm. This includes new ground and space radio telescopes and upgrading of the existing ones (see Appendix 1). A solution of the collecting area and resolution problems using nature can be found in Appendix 2.

### c. Evolution of the Brightness Temperature Sensitivity

This is shown in Fig. 21.4 and may be approximated by

$$\Delta T = A_t \, e^{-B_t(T-T_o)}$$

where $A_t = 390$ K, $B_t = 0.31$, T is the current time in years and, $T_o = 1932$.

In Fig. 21.4 we have used the best data on the anisotrophy of the cosmic background emission. The last point is the experiment COLD-90 which is under way at RATAN-600.

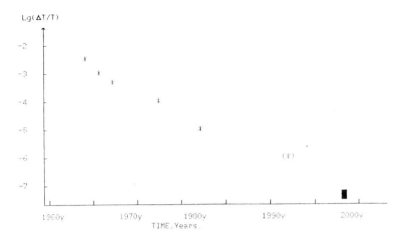

Figure 21.4. Evolution of the brightness temperature sensitivity. Best upper limits on the 3 K anisotropy measurements were used. The last point is the RATAN-600 COLD-90 project.

### d. Evolution of the Accuracy of Source Position Measurements

From Fig. 21.5 we have:

$$\Delta\alpha, \Delta\delta = A_{\alpha\delta}\, e^{-B\alpha\delta(T-T_o)},$$

where $A_{\alpha\delta} = 2 \times 10^5$ ", $B_{\alpha\delta} = 0.34$, and $T_o = 1932$ yr.

We have shown the position of the USSR "QUASAR" project on Fig. 21.5 in brackets. This project will be "just in time".

### e. Evolution of Image Quality (Dynamic Range of Maps)

The VLA may be used as the best example. We plot the progress in the dynamic range of this instrument (the ratio of weakest to strongest features on the final map) as a function of time in Fig. 21.6. About 30 db improvement was achieved in 10 years of VLA operation and another 10 db may be found in the next few years just through better algorithms and improvement of computer facilities.

Maps with better than 1: 1,000,000 brightness temperature resolution should appear at the beginning of the next century.

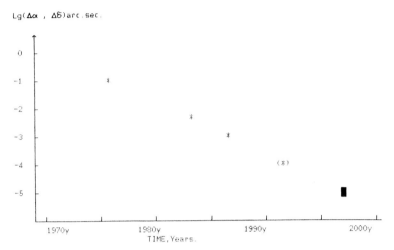

**Figure 21.5.** Evolution of the accuracy of source positions. The cross in brackets is the USSR QUASAR project. Few microarcsec absolute position measurements will be possible after the year 2000.

### f. Evolution of Receiver Noise

We have used the Arecibo 300-m radio telescope as an example in Fig. 21.7. Here receiver noise temperature is plotted as a function of time. We see that just now the curve crosses the 3 K cosmic background level. The rate of improvement may be expressed by

$$T_{rec} = A_{rec}\, e^{-B_{rec}(T-T_o)},$$

where $A_{rec} = 3 \times 10^5$ K, and $B_{rec} = 0.2$ ($\lambda = 21$ cm used). We may show that at 2000 yr the technology will be advanced enough to have receiver noise temperatures below the sky temperature at all wavelengths even above the atmosphere.

There is some problem with quantum noise. In the usual expression

$$\Delta T = T_{sys}/\sqrt{\Delta \nu \tau}$$

we have to use

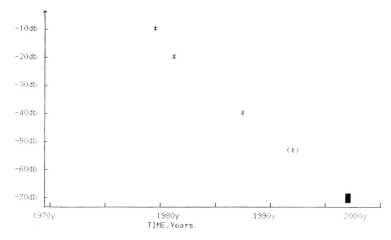

Figure 21.6. VLA progress in map quality (Dynamic range of the maps, the ratio of the strongest to weakest features on the maps).

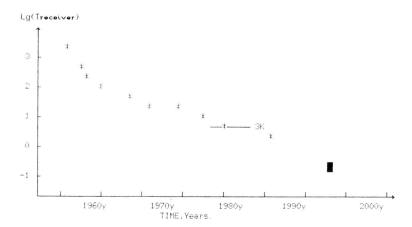

Figure 21.7. ARECIBO dish evolution of the receiver noise at 1.4 GHz. The 3 K level was crossed recently. There will be no technical problem to build receivers with less than the 3 K background noise.

$$T_{sys} = v/k$$

and this noise will dominate at wavelengths shorter than 3 mm. It means that to have an ideal receiver with $T_{rec} < 3$ K, the only way to increase the flux density sensitivity is the construction of a radio telescope with greater collecting surface.

### g. Progress of the VLBI Technique

Resolution progress is shown in Fig. 21.2. It is important that there are no technical limits on Space VLBI. Sensitivity of VLBI depends on bandwidth, integration time (coherence time), and on the collecting area of the telescopes. The later can be traced with the help of the Fig. 21.3. We hope that using a reference wave front (from a natural or artificial source), it will be possible to have as long an integration time as in the usual single dish observations at the end of this century. The single dish equivalent may be reached also as far as bandwidth is concerned. (MARK I, II, III, IV terminals reflects the change from a fraction of a MHz to GHz bandwidths).

It is very important to show how the VLBI technique crosses the radio domain and enters the optical domain at the beginning of the next century. We demonstrate this in Fig. 21.8.

In Table 21.1 we collected all evolution parameters of the radio astronomy from the very beginning to the year 2000.

Table 21.1. Summary of the evolution of radio astronomy

| Parameter | 1-year increment | Evolution from the beginning | Expected level at 2000 yr |
|---|---|---|---|
| Sensitivity | 1.5 | $3 \times 10^{13}$ | 0.03 $\mu$ Jy |
| Resolution | 1.5 | $1 \times 10^{13}$ | 0.03 $\mu$ arcsec |
| Brightness | 1.3 | $1 \times 10^{9}$ | 0.3 $\mu$ K |
| Position errors | 1.4 | $1 \times 10^{10}$ | 10 $\mu$ arcsec |
| Receiver noise | 1.2 | $1 \times 10^{5}$ | 0.3 K |
| Collecting area | 1.2 | $3 \times 10^{5}$ | 1 million m$^2$ |

## II. NATURAL LIMITS OF RADIO ASTRONOMY

### 1.

We believe that exponential growth of the resolution, sensitivity and other parameters of radio telescopes will hold up to the moment when the general exponential evolution of our civilization will stop. However, it is interesting to look closer at the limitation

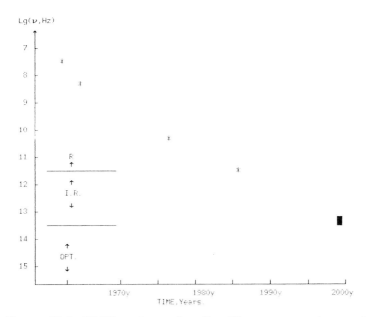

Figure 21.8. VLBI and wavelengths. We may expect operation of the VLBI technique not only in the infrared, but also in the optical after year 2000. (Black box). R, I.R., and OPT. indicates radio, infrared, and optical domains.

imposed by nature itself.
We shall discuss the following effects:

1. Confusion.
2. Scattering by the medium between the object and the observer.
3. Brightness irregularities of the Sky.
4. Physics of the objects under consideration.

The confusion limit was under discussion since 1950. News in this field appeared recently due to the invasion of flat-inverted-IR populations of radio sources. It results in the appearance of a well defined optimal frequency where the confusion effect is minimal (for a given resolution). As will be shown later, for all types of instruments there is an absolute sensitivity limit (in the flux density and brightness temperature as well) at a

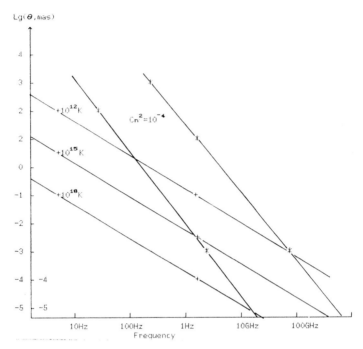

**Figure 21.9.** Scattering and brightness temperature limits (IAU Symp. #110 results were used.). $10^{12}$ K, $10^{15}$ K, $10^{18}$ K brightness limits correspond to the Compton limit, and records in the observed variability of pulsars to the other estimates. The strongest scattering was observed for Sgr A, a point source (black hole candidate). The upper right hand line at 1 mm wavelength may be used with the Space Array to remove scattering, see the text. The lower steep line corresponds to the Galaxy Pole region with $c_n^2 = 10^{-4}$ (see Symp. #110).

wavelength of about 1 cm. For RATAN-600 1 cm is also the best wavelength; the sensitivity at this point is limited by a small fraction of μJy (zenith mode). The second news was connected with the K. Kellerman statement that at the μJy level the distance between sources will be smaller than their size and a real saturation of the sensitivity may be reached (at any resolution).

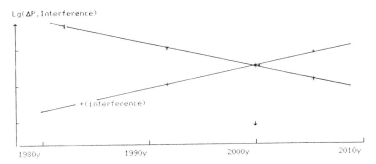

Figure 21.10. Interference and Sensitivity. Both evolve exponentially, but with different signs. When these two lines intersect, the range of radio astronomy will collapse exponentially.

Scattering effects are very strong at meter wavelengths. In very dense regions (e.g., Sgr-A, a point source) even at cm-mm-wavelengths the visible size may be limited by interstellar scattering effects. It gives a formal limit on the resolution of the radio telescope (e.g., a maximum base for space interferometry is now under discussion). News may come from the near field zone approach as happened with the seeing limit from our atmosphere and recently with interplanetary scattering effects.

One may show that with the increasing size of a radio telescope, the role of scattering effects may decrease and not increase as is usually discussed. If we are in the near field zone of the IS irregularities and our radio telescope is big enough to have the smallest dangerous scale of the ISM phase screen irregularity inside the Frenal zone, all scattering effects will be filtered out just as in the case of the troposphere. It may be shown that to escape from ISM scattering the size of the elements of the array should be close to the Earth's diameter and $\lambda < 1$ cm must be used.

Formal limits are shown in Fig. 21.11. We hope that for a high S/N ratio it is possible to remove the scattering effects. $T_b$ limits must be overcome by nature itself. Sky irregularities limitations are well known as far as atmospheric emission is concerned. This limit may be overcome by the beam switching mode observations, multi-frequency operation, water radiometry, or by interferometric methods.

For 1" resolution it was demonstrated by VLA groups; for 1' resolution there is suggestion to convert a single dish into a VSA (Very Small Array). RATAN-600 methods involved near-field zone effects (L-R-switching). Bad news appeared recently from the galaxy. It was found that even at high latitudes there are irregularities at the 1 mK level at 7.6 cm on a 1° scale with a nonthermal spectrum (RATAN-600 result).

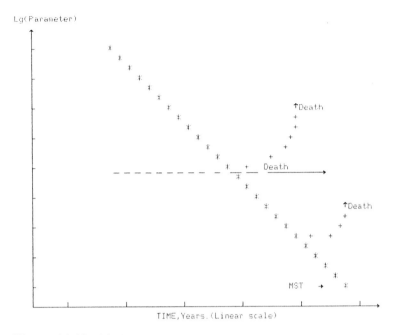

Figure 21.11. Birth and death of radio telescopes. The world level of radio astronomy is shown by the straight line marked by MST (Main Seq. Telescopes). The dashed line is the design period. The subsequent fate of the radio telescope depends on the funds available: no funds, funds for maintenance only, and funded enough to be as long on the MST line as possible.

Moreover, at short wavelengths Galactic cirrus emission has irregularities at 10"-10' scales with a strongly inverted spectrum. It means that again there is some optimal wavelength were these limitations are minimal.

At present it seems that this effect will act in the short mm region. At all wavelengths there must be an absolute limit on the brightness temperature sensitivity on all scales determined by the predicted 3 K-anisotropy at a level between 100–10 μK. New unpredicted large scale emitters appeared recently with a nonthermal spectrum identified with the walls of the Void structure.

Irregularity of the mm emission of the Galaxy may be caused by the irregularities of line emission even in continuum type observations.

Physical limitations were discussed many years ago in connection with the Compton threshold on the brightness temperature ($10^{12}$ K) of the synchrotron emission of RG and QSR. K. Kellermann in 1970 showed that for the 1-Jy population of radio sources it gives a formal limit on the resolution of the interferometers (just $\lambda/D$ where D is the diameter of the Earth). The main goal of space interferometry is to check the reality of this limit. Variable sources and pulsars can then definitely be good candidates for space VLBI Arrays with bases not very far from 1 A.U.

Quite different limitations may appear due to the very deep penetration of radio astronomy up to the epoch when radio sources were not yet born.

Strong emitters will be lost first. One can see from very deep VLA surveys that there are no QSRs at the submJy level. The Thompson scattering limitation ($\tau > 1$) was discussed in the 60 s (Kardashev). This will limit the depth of radio and optical surveys. However, new estimates of the density of the ionized gas and new deep (up to 29 mag.) optical surveys change the situation. We suggest that optical surveys may be limited by $z = 5.5$ due to dust absorption and only Radio Astronomy can go farther than 1" resolution for 21 cm (2.6 mm) redshifted lines that are very important for observations of these early galaxies.

## 2. Interference Problems (Electromagnetic Pollution)

Radio astronomy is developing exponentionally in parallel with the exponential development of the civilization on the Earth (energy and technology).

It means that electromagnetic pollution is also increasing exponentially. We demonstrate this in Fig. 21.11. From this figure we see that there is some crucial moment after which the real sensitivity will be less than before. It means that some experiments must be realized as quickly as possible- "Now or Never!". Only the back side of the Moon can help to escape from the "industrial limit". This problem may be also simplified by space radio astronomy, but even in this case a radio telescope has to be placed at distances from the Earth (noisiest planet possibly in the whole Galaxy) much greater than that of the Moon.

## Conclusion

Even a quick look at the situation brings us to the following conclusions.

1. Radio astronomy up to now has an extremely high exponential growth of potential without any sign of the deceleration of this process in the future 25 years. We may predict rather accurately the expected levels of sensitivity, resolution, quality of the images, etc. at the beginning of the next century.

2. Most of these parameters may be reached by the construction of the new generation of radio telescopes with collecting areas close to 1 million $m^2$ and with array sizes comparable with 1 A.U. Some parameters can be realized with presently existing

instruments by improving receivers, software, computer facilities, and image formation methods. VLA and ARECIBO telescopes showed how one can be on the Main Sequence of the best radio telescopes for many years inspite of the world exponential growth of radio astronomy potential. We reflect this situation using the analog with the stellar evolution process, see Fig. 21.11. To be on the main sequence as long as possible radio observatories must spend about 10% of the construction cost of the radio telescope every year on maintenance of the facility.

3. Very much can be done by improving computer facilities. It may be shown that not only image quality depends strongly on computer power, but also the flow of information as well. Better understanding of the image formation process shows that up to Gain factor losses may be recovered if all information available in the Wolf Coherent function is extracted. We shall come back to this problem later in the RATAN-600 section where this loss is very important.

4. Major uncertainties in the radio astronomy of the next century involve horizons connected with scattering effects and the unknown nature of objects.

5. IR and optics have to be ready to use radio methods (VLBI technique, etc.). Some suggestions in this direction were made 10 years ago in the POLYGAM project

## III. GIANT PROJECTS (SHORT SUMMARY)

The first suggestion to construct a 1 million $m^2$ and 1 A.U. size arrays was made in the USSR (Pulkovo group) in the 60s (Khaikin et. al. 1967). Below we list all main projects proposed during the last 25 years.

1. Decametric projects.
a) SPACE VLBI; 5 orbiting dipoles, 3-300 MHz (range (NASA).
b) SPACE ASTRO-ARRAY. The whole radio frequency range; 100 30 m dishes at many Earth diameter distances (NASA).
c) 1 million $m^2$ Ground Based Array with arcsec resolution (USSR, Kharkov-Pushino-Gorkij project); 10-100 m range.

2. Meter-wavelength projects.
a) Giant Equatorial Radio Telescope (GERT). 1 million $m^2$, arcsec resolution, $1\mu Jy$ sensitivity. A smaller version of this project is now under construction in India (35 25 m dishes, 60 km base, VLA type configuration with a central 1 x 1 km partially filled area; up to HI-frequencies.
b) See b) in item 1.
c) See c) in item 2
d) Arcsec Synthesis Array from 10 100 m telescopes, $\lambda = 1m$ (USSR, Pushino)

3. Decimeter wavelength projects.
a) Pulkovo 1964 RATAN-type project, 20 km size, 1 arcsec resolution, 2 million $m^2$ area, 100 m panels, 21 cm-1m range, 200 million dollars.

b) CYCLOPE project (USA). 1000 100 m dishes well packed. 20 cm CETI program.
b) USSR (IKI) 1 km-3 km dish in Space, low orbit, dm range.
c) Japanese project of the Global Dipole Array connected with a very large computer. All sky 15 years survey with 1mJy sensitivity and 1 milliarcsec resolution. We have mentioned that this 1970 project may appear again when VIDEO-NETS connecting TV and PC users will be realized for communication purposes. Now there are 100 million TV users.
d) A new (after Moon occultation method) Interstellar Interferometry method appeared using the recently discovered 1 A.U. lenses in the ISM. A pulsar light cylinder was resolved and 5µarcsec resolution achieved. ISS can be broadly used (first suggested by Kardashev).
e) A Lunar-Earth interferometry may be possible if sensitivity problems can be overcame. The signal from the first Frenal zone of the moon may interfere with the signal received by an Earth based radio telescope from the same source (first suggested by Ginzburg). Aperture synthesis gives good UV-coverage equivalent to a 400,000 x 700 km dish for low declination objects.

The Cornwell phase restoration method used for IPS can be used to convert the reflected emission into the original wave front structure.

4. Cm-wavelength projects.

a) Pulkovo cm RATAN-600 like Ring Radio Telescope; 0.5 million $m^2$ area, $\lambda > 3$ cm, 1 arcsec resolution
b) "Hubble-meter"-Pulkovo 1968 project for a space array, 1 A.U. in size, three dishes for extension of the near field zone up to the Hubble radius. Direct distance measurements should be possible through measurements of the wavefront curvature. Kardashev has shown that all cosmological parameters may be found if the redshifts of the objects are also known.
c) "POLYGAM" project. (USSR, SAO and many others). 3 70 m dishes and 10 25 m dishes in conjunction with a 30 m space telescope. Three independent projects are in progress now instead of "POLYGAM": RADIOASTRON, QUASAR, and a 3 70 m dish array, incorporated in the VEGA mission project.
d) QUASAR project (quickly developing) 10 32 m VLBI dedicated array for geodesy and time service; $\lambda > 1.35$ cm.
c) "RADIOASTRON" project, space VLBI, 77,000 km maximum base, $\lambda > 1.35$ cm, 10 m dish in space, many on the Earth.
d) "VSOP" (Japan project), 10 m at a 1 Earth radius orbit, $\lambda > 1.35$ cm.
e) "IVS"-international project (after Radio Astron mission). 30 m dish in 200,000 km apogee orbit, $\lambda > 3$ mm.
f) 300 m ARECIBO-type dish in Brazil; $\lambda > 1.28$ cm.

5. Mm-wavelength projects.

a) Space ASTRO-ARRAY, see item 1, a).
b) Mm VLA-type project

c) 70 m dish for 1 mm (USSR, first version under construction in Central Asia by Kardashev's group).
d) 60 m dish for less than 1 mm (USA, under consideration).
e) 54 m. Arecibo-type dish for 1 mm with 2.6 m on-axis optical telescope. Almost ready in Armenia, USSR.
e) Many mm and submm dishes and arrays in USA program.

A very important program of upgrading existing facilities to be on the Main Sequence of the best radio telescopes at the beginning of the next century can be found in the USA Radio Astronomy program for 1991-2010 (2,000 million dollar list, see Appendix 1).

## IV. RATAN-600 AND THE NEXT CENTURY

Now we shall try to find the position of the RATAN-600 among the best radio telescopes at the beginning of the next century if we could do our best to realize the potential of this unusual instrument.

### Absolute Limits.

1. Confusion

The confusion limit was calculated for different modes of operation and is shown in Table 21.2. The confusion limit makes the use of high sensitive receivers inefficient. Here $T_{eq}$ is the equivalent system temperature of the radio telescope limited by the confusion effect. A 10% bandwidth and $\tau=\lambda/D$ were used in these calculations. Even in the transit mode the output noise is dominated by confusion and not by receiver noise at $\lambda > 10$ cm, 20cm, 75cm in observations with a flat mirror, single sector, or in the zenith mode respectively. The highest sensitivity may be achieved at a 1 cm wavelength (0.03µJy, 0.3µK). We can compare the RATAN-600 absolute limit with the 100 m dish one (2mJy) and Arecibo (0.6 mJy) at the 6 cm common wavelength; RATAN-600 is more sensitive by a factor 100.

2. Absolute sensitivity limit.

News appeared recently about the optimum wavelength for deepest observation in the whole radio-infrared domain. We used the Brascheshini et. al. (1989) calculation of the confusion effects where best VLBI surveys and flat spectrum populations of radio sources are taken into account. We show new calculations in Fig. 21.12. We see that RATAN-600 can be used at the wavelength of maximum sensitivity and can reach the expected year 2000-level as far as brightness temperature and flux density are concerned.

Table 21.2. RATAN-600 equivalent confusion $T_{sys}$ (K)

| λ (cm) | Flat mirror mode | Single mirror mode | Zenith mode |
|---|---|---|---|
| 135 | 200,000 | 40,000 | 600 |
| 75 | 60,000 | 12,300 | 170 |
| 31 | 5,000 | 1,000 | 12 |
| 13 | 500 | 100 | 1.2 |
| 7.6 | 110 | 22 | <1 |
| 6 | 60 | 11 | <1 |
| 3.9 | 22 | 5 | <1 |
| 2.7 | 7 | 1.5 | <1 |
| 1.38 | <1 | <1 | <1 |

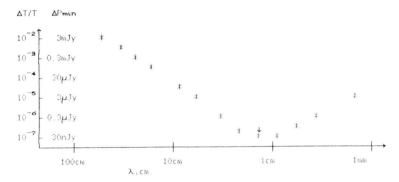

Figure 21.12. Absolute radio astronomy limit and RATAN-600 frequency range. Computed using Franceschini et al. (1989) results. The deepest observations are possible at about 1 cm within the RATAN-600 frequency range.

3. Frequency range limits.

Good accuracy of the panels was achieved recently (0.082 mm r.m.s.) and we hope that active optics methods will help us to maintain it for any weather conditions. It means that the maximum gain will be at

$$\lambda_{min} = 4\pi\varepsilon_{rms} = 1 \text{ mm}$$

if no other source of error exist. In the Single Sector mode we are limited by geometrical errors (0.2 mm), but in the most powerful Zenith mode there are no such limitations and we can have about 5000 m² at mm wavelengths,

RATAN-600 is limited in resolution not only by λ/D, but also by the seeng disk (about 1-2"). The new Ryle synthesis approach makes it possible to escape from this limit just as in phased array systems.

Extention of the panels to 11.4 m increases the long wavelength limit to 1 m (or 1.35 m if a special large secondary mirror is constructed). In this case we can realize 15,000 m² at decimeter and 10,000 m² at cm wavelengths (Fig. 21.13).

4. Dynamic range limits

Dynamic range (the ratio of the brightest to weakest features in the image) depends on the filling factor of the aperture which is better for RATAN-600 that for the VLA, which is limited to the 1:200,000 range.

5. Source Position accuracy measurements and Optical Identification Limit

Progress on the accuracy of source position measurements with RATAN-600 is shown in Fig. 21.14. We hope that with good receivers we shall have the error box small enough for very deep identifications.

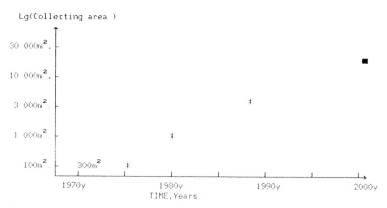

Figure 21.13. RATAN-600 collecting area: what we have in store. Three steps of increasing the area marked after the initial construction. The best one-sector result (1979), the first zenith mode construction (1988), and the absolute limit with new screens and with a special feed system (black box).

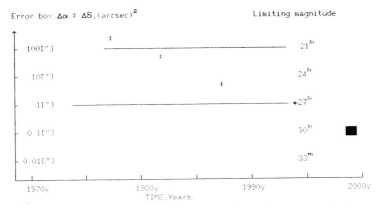

Figure 21.14. RATAN-600. Evolution of the "error box," black box - extrapolation to the year 2000. All sources with errors above the upper line can be identified on PSS prints but with errors above the lower line 6 m plates should be used; if the error box is below the lower line, optical identifications with Space Telescope Plates are possible.

## 6. Information Limits

The flow of information is limited now by the sensitivity, single beam mode of operation, bandwidth, resolution, and data reduction facilities. The absolute limit is fixed by $\lambda_{min}/D$, $\lambda_{max}/\lambda_{min}$, $\Delta\nu = c/\lambda_{min}$, and by the beam number.

Backend limitations of RATAN-600 are shown in Fig. 21.15. We are very far from an ideal system which can extract all information from the coherence analog-to-digital conversion, and on-line and off-line computer limitations.

The levels reached by MARK-systems are shown in the upper right corner.

We hope that at the beginning of the next century all these limitations will disappear.

Multi-beam operation using the phased focal plane array is now under discussion. An upper limit for the number of beams is about 10,000 (Pinchuk et. al. 1990)

We conclude with Table 21.3, where absolute limits on the RATAN-600 parameters are collected.

The last row shows the evolution of the one source of system temperature. The sky temperature is above 5 K at all RATAN-600 wavelengths.

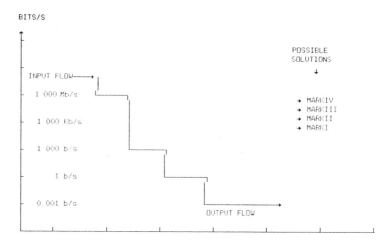

**Figure 21.15.** RATAN-600 and BACKEND limitations. The input flow is limited by the accuracy of the surface. This flow cannot be recorded due to u.h.f. limitations (first step), analog-to-digital conversion limitation (second step), on-line computer facilities (third step), and off line computer limitations. The Mark I, II, III, IV sequence shows the international activity in this field.

Table 21.3. Absolute limits on the RATAN-600 parameters

| Parameter | Began with | Improved to | Absolute limit | Future gain |
|---|---|---|---|---|
| Collecting area | 300 m² | 3500 m² | 15,000 m² | 4.5 |
| Resolution | 300 m² | 3500 m² | (0.2") | (25 millarcsec) |
| Sensitivity | 100mJy | 1mJy | < .1µJy | 10,000 |
| Brightness limit | 100mK . | 1mK | <1µK | >100 |
| Survey Source number | 100 | 10,000 | > 100,000,000 | >10,000 |
| Dynamical range | 1:100 | 1:10,000 | >1:200,000 | > 20 |
| Inform. flow | 1 bit/s | 1 Kb/s | 100 Gb/s | 100,000,000 Gb/s |
| $\lambda_{min}$ | 2 cm | 10-3 mm | 1mm | 3-10 |
| $\lambda_{max}$ | 6.5 cm | 31 cm | 1.35 m | 4 |
| Number of beams | 1-2 | 1-2 | > 10,000 | > 5,000 |
| Tracking time | 2 s | 2 min | > 1 day | > 700 |
| Ground Radiation | (70 K) | 5.5 K | < 1 K | 5.5 |

## Fundamental Problems to be Solved (Suggested by RATAN-600 Users).

1. SUN

a) Full disk, all day, all Stokes parameters, all frequency range with μsec time resolution (3-dimensional mapping + time)

b) "Stilsread"-type programs for Solar-Earth problems.

2. Planets

a) Moon: 3-dimensional mapping from the longest to the shortest wavelengths

b) Having good receivers (and tracking), planet satellites may be explored as our Moon.

c) Accurate temperature and position measurements of Mercury and Mars are also waiting for better times

3. GALAXY

a) Radio Stars (multi-freq. monitoring)

b) Pulsars: survey, ms-time keeping.

c) Continuum, spectral lines, polarization two-dimensional mapping of the Milky Way.

d) Small (and large) scale irregularities of the interstellar medium with small brightness temperature contrasts.

e) Variable stars and objects (as the Sgr-A point source)

4. Metagalaxy

a) All-sky surveys (later, shorter wavelengths).

b) Deep surveys of selected regions. Accurate position measurements. Identifications.

c) Active nuclei program (multi-freq.. patrol, spectrum-structure program, polarized remnants of activity, etc.)

d) Search for var. sources at different flux density levels.

e) Clusters of galaxies; S-,SZ-effects; steep spectrum sources

f) Radio source clustering; Void-structure.

g) Two-dimensional mapping of selected regions (COLD strip, Zenith strip, near pole region, etc.)

5. Cosmology

a) Lg N-Lg S, cosmological evolution of the radio activity in galaxies.

b) $\Delta T/T$: discovery, mapping and BIG COSMOLOGY.

c) HI-and CO-lines Forest program.

d) Spectral irregularities of the 3 K background

e) Search for the most distant objects (including the IR-radio program).

## Conclusion

We see that only a very small fraction of the potential of the Main USSR radio telescope, RATAN-600, has been realized. We estimate that even with 10% funds (from the capital investment in the construction of the RATAN-600 in the current rubles) every year, that is with minimal international standard funds, it is possible to be on the Main Sequence of the best world radio telescopes in the next 15-25 years. Some very important upgradings are needed in extra grants: mm-wavelength full operation, a Ryle synthesis mode of image formation, a full day tracking mode, and broad band connections with computer networks and users. Below we list the directions of activity where competition with world radio astronomy at the beginning of 21 century may be minimal and where RATAN-600 can help to fill the gap between dishes and arrays.

## RATAN-600 and High Priority Directions of the Modern Astronomy

1. As a reflector type of radio telescope all kinds of observations where a very broad frequency range (multi-color radio photometry) is needed are of high priority. Whole frequency range problems such as very deep redshift surveys are out of the competition.

2. Again, as a reflector, low brightness searches are possible. We demonstrated this in some RG studies and in the 3 K emission observations. It is better than usual reflectors (dishes) because of much stronger confusion limitations in the later case.

3. The large "daily field of view" in the transit mode of operation of RATAN-600 makes this instrument very efficient for survey type observations. The real efficiency depends strongly on the sensitivity of the radio telescope in this (small integration time) mode of operation. The best result may be reached if the single daily scans are dominated by confusion noise and not by receiver noise. Confusion noise is very small due to the unusual geometry of the RATAN-600. It is also important to note that the RATAN-600 frequency range includes the point of the absolute minimum of the combined radio-infrared confusion noise computed recently using IRAS and very deep VLA data at about 1 cm (see below).

4. As an aperture synthesis instrument RATAN-600 is without competition in the mapping of radio scales which are too big for VLA-type systems, but too small for parabolic dishes. In principal, the dynamical range ( the quality of the images) of such maps may be much better than in the VLA case due to a much greater filling factor of the aperture. Up to now we have not realized this advantage of the RATAN-600 (no software, no adequate computer facilities) except for the one-dimensional case.

## 21. RADIO ASTRONOMY OF THE NEXT CENTURY

## Radio Astronomy at the Special Astrophysical Observatory

In the northern Caucases, the 6 m optical telescope and the RATAN-600 are the main ground based facilities of the USSR. More than 600 people are on staff, and a large intellectual base was accumulated during the past 15 years. More than 80% of cm-and dm-results in radio astronomy (USSR) and almost all results in extra galactic optical astronomy (USSR) occurred with the help of these instruments and staff.

In 1979 we suggested to add to RATAN-600 new, complimentary facilities that included:

1. 128 m-dish for VLBI, spectroscopy, etc.
2. 20 m-dish for mm- wavelength molecular spectroscopy
3. 1 million $m^2$ pulsar field array
4. Aperture synthesis array connected with RATAN-600.

Some suggestions made have large interest, first of all the very big dish project, which would increase greatly the efficiency of the USSR national dedicated array (QUASAR) and the efficiency of the Space Mission as well. This largest dish in the world will be very popular in all international radio astronomy activity.

It is well known that the cost of a big observatory (in any country) is greater than the capital investments. It is also known that the stability and efficiency of the observatory is much higher if there are very different kinds of instruments which can react quickly to the changes of intellectual interests. That is why we believe that not only upgrading problems should be discussed, but also new facility perspectives as well.

## APPENDIX 1. USA LIST OF THE RADIO TELESCOPES UNDER PANEL DISCUSSION (PRICES IN MILLIONS)

| | | |
|---|---|---|
| 1. | VLBA, ready in 1992, 10 25 m dishes | $85 |
| 2. | Haystack upgrade, 37 m, up to 3 mm | $1.5 |
| 3. | Green Bank, 100 instead of 91 m up to 7 mm, 1995. | $75 |
| 4. | ARECIBO 305 m, upgrade, 1992, feed, screen, power | $23 |
| 5. | Berkeley-Illinois-Maryland array, 10 x 6 m dishes, 1991-1993, 300 m x 200 m (3 dishes now in operation) | $20? |
| 6. | OVRO mm-array upgrade, add 6 10 4 m for 2.6 mm, 180 m x 200 m (3 dishes now in operation). Large equipment changes | $20? |
| 7. | SAO submm. array, up to 0.35 mm., 0.1 arcsec. res; 6 x 6 m. 1996 | $40 |
| 8. | NRAO mm array, 3 km base, 40 x 8 m dishes, up to 0.07 arcsec. res, $\lambda < 1$ mm ($\nu > 350$ Ghz), 1996-2000. | $115 |
| 9. | VLA E-configuration, (upgrading), 200 m base, wide field, | |

| | |
|---|---|
| Galaxy mapping, 21 cm. | $9 |
| 10. VLA, VLBA upgrade. L, C, U, K band new receivers ($8), 50 cm-receivers ($3.5), 11 cm rec ($1.5), 7 mm rec ($3), VLBA 3.5 mm rec ($1.1), 1024 channel 1 GHz correlator ($11) | $37 |
| 11. VLA-VLBA -link and fiber optic link with Los Alamos. | $32 |
| 12. Large southern dish, Arecibo-type, 305 m, up to 1.28 cm | $100 |
| 13. Caltech submm. observatory, upgrade (Mauna Kea; 17 µ r.m.s.) (two engineers needed), 10 4 m dish at 14,000 ft. is ready | $2.5 |
| 14. 15/20 m mm radio-telescope $\lambda = 0.8$ mm, instead 12 m NRAO | $5.9 |
| 15. Submm-Tel 10 m dish, h = 3180 m 15µ-r.m.s. VLBY? | $15? |
| 16. Antarctica, Submm and Remote Observatory (ASTRO), 1.7 m dish at Pole (1992), 2.5 m IR, may be up to 30 m submm dish | $80? |
| 17. Large mm. tel., 20 m dish, 100µ rms, $\lambda = 2$ mm (MIT). | $60? |
| 18. OVRO, Solar Array Upgrade the present 2 x 27 m dishes by 8-14 2 m dishes, 30 cm–1.6 cm range (Caltech, NASA) | $3 |
| 19. SETI NASA "Microwave Observing Project", 34 m dishes of DSN for sky survey at 30 cm-10 cm range, 800 stars also | $117 |
| 20. SPACE VLBI. Earth for USSR, Japan; Intern. VLBI-Satellite | |
| a) USSR, Japan supports | $50 |
| b) IVS | $1000 |
| TOTAL: | $1910 |
| | |
| UPGRADE OF PRESENT RADIO TELESCOPES: | $ 228 |
| CONSTRUCTION OF THE NEW RT | |
| EARTH: | $495 |
| SPACE: | $1050 |
| SETI: | $117 |

The budget does not include 10% of the radio telescope construction cost every year in current dollars for each Radio Observatory.

(Feb. 15 1990, K. Kellermann information, J. Bahhcall Astr. Ap. Surv. Comm.)

## APPENDIX 2. NATURE AND RADIO TELESCOPES

Here we collect all suggestions to use nature itself as a radio telescope.

1. Moon and one-and two-dimensional mapping of sources.
2. Large craters as collecting surfaces. (Grote, Reber, F. Drake, L. Bahrah)
3. IPS, ISS as a tool for angular size measurements.
4. 1 A.U. refracting lenses in the ISM as very large radio telescopes.

5. Sea Interferometry.
6. Moon First Fresnel Zone as an Orbiting Space Radio Telescope (see text).
7. Venus Atmosphere as a Luneberg Lense. (Parijskii)
8. Earth's Ionosphere as a Giant Radio Telescope at long wavelengths. (Jennison)
9. Gravitational focusing.

## REFERENCES

Franceschini, A., Toffolatti, L., Danese, L., and De Zotti, G (1989) *Astrophys. J.* **344**, 1.

Khaikin, S. E., Kaidanovskij, N. L., Esepkina, N. A., et al. (1967) *Glovnoi Astronomicheskoi Observatorii USSR* **182**, 235.

Pinchuk, G. A., Parijskii, Y. N., Shannikov, D. V., and Mairova, E. K. (1990) S.A.O. Preprint No. 39.

# 22

# On Astronomy for the Twenty-first Century

## N. S. Kardashev
*Astro Space Center, Lebedev Physical Institute, Moscow*

Our review volume is dedicated to the memory of the remarkable scientists J. S. Shklovsky and S. B. Pikel'ner, who made significant contributions to the development of astronomy. I was lucky enough to work together with them for many years in the same group and even in the same room. This view and the perspectives for the development of astrophysics was formulated basically under the influence of these highly endowed people whose interests encompassed many regions of science.

## THE CONTRIBUTION OF THE TWENTIETH CENTURY TO ASTRONOMY AND PHYSICS

The rapid development of natural sciences in the twentieth century influenced the development of all aspects of life. The basic qualitative discoveries in the region of physics and astronomy are presented in Table 22.1 in chronological order. No pretense is made to completeness in the enumeration of the most important works indicated and their choice is connected with the subjective viewpoint of the author. Much of the data were taken from the books /1–4/. The basic result of the development of physics in the twentieth century was the possibility of generalizing all observed processes (from the microscopic to the macroscopic) using a logically connected set of a small number of fundamental laws. In particular, one can follow with the table the development of space science. Thus, the first ideas about extragalactic astronomy appeared only at the end of the nineteenth and beginning of the twentieth centuries. The well-known astronomer-observer Prof. J. Sheiner intuitively considered from the Potsdam Observatory that our star system is a spiral nebula in whose nucleus the solar system is located and the spiral forms the Milky Way. He also supposed that the star groups and nebular spots scattered in space are formations similar with our stellar system and that the latter would seem the same as a nebular spot from an unmeasurably large distance. Nebular spots with continuous spectra and dark absorption lines similar to the solar spectrum indicate the existence of independent stellar systems at large distances from us /5/. It is interesting that the Andromeda Nebula as an actual astronomical object was already known in the tenth century to the Persian astronomer Al-Sufi, but its extragalactic nature was established only in the twentieth century.

The range of radiation carrying basic information on objects of the universe was limited at the beginning of the century to the small visual range, but at the end of the century it encompasses the whole electromagnetic spectrum. During the same time a strategy of research of astronomical objects was formulated quite close to the strategy of physics in the laboratory with the exclusion of the principal impossibility of acting on the object of study. The search and detection of new types of astronomical objects was the first stage in the research. The second stage is the specially thought out posing of additional observations in different spectral ranges for clarifying the basic problem, what is the source of radiation that we detected, what are its basic physical parameters (material state, mass, size, distance, motions, and structure). Finally, the third stage of research is placing a given type of object in the Universe and its evolution, the mechanism of its formation, and the final stage of its development. At the end of the twentieth century all three stages of research have been carried out for the majority of classes of astronomical objects known at the beginning of the century. The greatest accomplishment is an understanding of the general pattern of the structure of stars, and the sources of their energy and evolution although a whole series of problems of physics of some types of stars still requires answers. At least the first two stages of research were carried out for a study of the solar system, the interstellar medium, and the structure of galaxies. In the twentieth century new astronomical objects were detected and a study of their structure was successfully performed. Pulsars, neutron stars, X-ray variable sources including supposed objects connected with black holes of stellar mass and sources of X-ray bursts, and galactic nuclei and quasars apparently connected with supermassive black holes are such objects.

The possible development of natural sciences and the future is tightly connected with changes in the social life of our civilization, and the development of experience and knowledge about spiritual and moral values. The positive changes occurring at the end of our century provide a basis to hope that many contemporary social and spiritual conflicts may be overcome and the civilization of the next century may direct a much larger fraction of its resources to the development of fundamental sciences than occurred in the twentieth century.

## POSSIBLE DEVELOPMENT OF TELESCOPES AND METHODS OF ASTRONOMICAL RESEARCH

There are many parameters describing the possible application of one or another type of telescope with a corresponding specification for each wavelength range. However, the primary parameters for all types of telescopes are the energy sensitivity (minimally detectable spectral flux density of radiation from the source at the observation point) and angular resolution. In the majority of ranges telescope sensitivity is presently limited only by its collecting surface area. Contemporary receivers of radiation are close to their theoretical limit of sensitivity although in the radio and infrared ranges a further increase of their sensitivity by 10–30 times is possible. For the optical and shorter wavelength ranges a quantum output of receivers close to unity is obtained. Here, besides increasing the telescope aperture, an important role will probably be played by methods of increasing the ratio of signal from the source studied to the noise due to radiation of the background by the creation of more refined optical systems approaching the diffraction limit. The application of radiation focusing telescopes for ranges harder than the optical is also possible. For this purpose it is proposed to use mirrors of oblique incidence for the reflection of hard radiation from crystal surfaces at small angles /6/ or zoned plates and even gravitational lenses. A radical increase of angular resolution is necessary for making images of compact sources, and high-accuracy measurements of their positions and characteristic motions. The development of interferometers is the most promising for this purpose. It is proposed to obtain an especially large return by the use of interferometers in space. The first successful radio interferometer between ground and space radio telescopes was already realized with the TDRSS satellite /7/. An angular resolution considerably exceeding the possibilities of radio interferometers on the Earth is proposed to be carried out using the space projects Radioastron /8/ and VSOP /9/ whose launches are expected in the '90s. The interferometer of Radioastron should work in the ranges 1.35, 6.2, 18, and 92 cm using a daily elliptical orbit with a maximum distance from the Earth of 80,000 km which determines the minimum width of the interference pattern as $3.5 \times 10^{-5}$ ". The interferometer of VSOP with ranges of 1.35, 6.2, and 18 cm is proposed to be launched into a satellite orbit with a period 6.06 hours and a maximum distance from the Earth of 26,000 km which corresponds to a minimum width of the pattern of $11 \times 10^{-5}$ ".

The development began in 1989 of the project IVS (International Very Long Base Interferometer Satellite), a space interferometer with a 20-m mirror for centimeter and millimeter ranges of the spectrum /10/. A space radio telescope with a mirror antenna of this size may already be considered as an element of a future space synthesis system because a pair of these instruments in space has sufficient sensitivity for carrying out observations without involving ground radio telescopes. Space interferometers in the infrared and optical ranges /11, 12/ are also discussed whose Earth prototypes already work successfully. There is no principal difficulty creating these systems at shorter wavelength ranges of the spectrum, ultraviolet, X-ray, and even γ-ray ranges. The main technical difficulty is apparently connected with the large requirements for telescope stabilization and the control of bases between them.

Possible parameters of future telescopes and interferometers are shown in Fig. 22.1 assuming that their collecting area A grows by 100 times in comparison with present dimensions and the base B reaches a billion kilometers for the radio, infrared, and optical ranges of the spectrum. It was assumed for wavelengths shorter than 1000 Å that the maximum base of interferometers is 1 km. The main impediment to realizing super-high

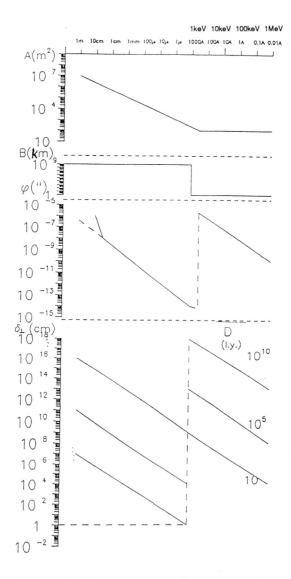

Figure 22.1. Probable parameters of telescopes in XXI century for different spectral ranges. A is the collecting area of telescopes, B is the base length of an interferometer, φ is the angular resolution, $\delta_\perp$ and $\delta_\parallel$ are linear resolutions

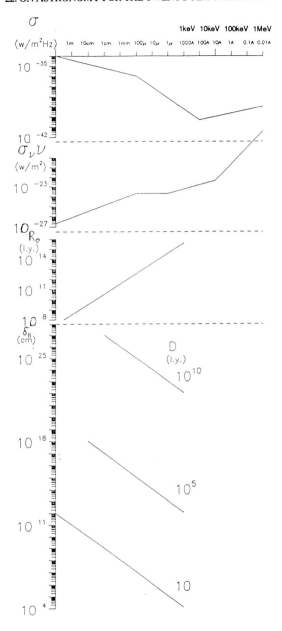

transverse to and along line of sight for different distances to the source (D), $\sigma$ and $\sigma_\nu$ are spectral and bolometrical sensitivities, and $R_0$ is the limiting distance which can be measured.

resolution in the radio range is the process of radio wave scattering on inhomogeneities of the interstellar and interplanetary plasmas. The diameter of the scattering circle of extragalactic sources in galactic latitude b for observations at frequency f is

$$\theta_{sc} = 0.2 (f\text{ GHz})^{-2} (\csc b)^{0.5} \text{ milli-angular second for } b > 20° \text{ and } f < 6 \text{ GHz}.$$

For shorter wavelengths and higher galactic latitudes an increasing and blurring of the image is not expected; however, random changes of the object coordinates are possible /13/.

Apparently, large mirrors of space telescopes will be created with the wide use of multielement automatically regulated constructions with a continuous control of the form of the reflecting surface. Similar constructions of adaptive telescopes have already been created on the Earth. The possibility of creating and exploiting optical and radio telescopes with a mirror diameter of several kilometers in space is described in /14,15/.

The limiting sensitivity in the spectral flux density of energy for ranges shorter than 1 mm may be estimated by supposing that it is determined by quantum fluctuations of the background cosmic radiation with the equation

$$\sigma_1 = \max \left( \left[ \frac{I_\nu \lambda^2 h}{4\pi A^2 \Delta t} \right]^{0.5}, \frac{h}{A \Delta t} \right),$$

where $I_\nu$ is the minimum intensity of the background, the integration time used $\Delta t = 10^5$ s ~ 1 day, and the receiver band $\Delta \nu \sim \nu$.

For the long wavelength part of the spectrum ($\lambda > 1$ mm) the sensitivity is determined by the classical fluctuations of the background

$$\sigma_2 = \frac{2kT_B}{A\sqrt{\Delta \nu \, \Delta t}},$$

where $T_B$ is the minimum brightness temperature of the background with the same assumptions on $\Delta \nu$ and $\Delta t$.

The angular resolution of interferometers is determined by the well-known relation $\varphi = \lambda/B$; the transverse linear resolution $\delta_\perp = \varphi D$, where for the distance to the source we took the values $10^{10}$, $10^5$, and 10 light years.

For interferometers with space bases it becomes possible to measure the curvature of the wave front or parallax shift and in the same way to determine the distance to even the furthest astronomical objects /16/. The accuracy of measurements along the line of sight will be

## 22. ON ASTRONOMY FOR THE TWENTY-FIRST CENTURY

$$\delta_{||} = \frac{\varphi\lambda D^2}{45B^2},$$

and the limiting distance which can be measured, $R_o = 4B^2/\psi\lambda$, where $\psi$ is the accuracy of an interferometric measurement of the difference in phases of signals from an astronomical object. The thus determined orientation values of sensitivity and angular resolution open completely new possibilities for carrying out research. In particular, observations may give information on all stars (with an absolute stellar magnitude $M_V < 17$) in our galaxy; for planetary systems the possibility of direct observation of objects similar to our Earth to distances ~1000 light years; for all galaxies of our type up to distances corresponding to their formation; on the surface of neutron stars and regions around the Swartzschild radius of black holes with their X-ray radiation to within distances to $10^5$ light years and for supermassive black holes to $10^{10}$ light years; on the basic properties of space-time, and global cosmological evolution, and to solve many of the fundamental problems of astronomy.

It is important to note still another direction of the development of the technology of astronomical research, the creation of multi-beam telescopes which accelerate the mapping of separate areas many times and the carrying out of complete sky scans. There are two directions of their development: one is the creation of multi-element mosaic receivers, these receivers (CCD matrices) have more than $10^6$ elements. The development of matrix receivers is also being carried out in all other ranges of the electromagnetic spectrum. An important immediate application of mosaic receivers is the creation of systems for detecting low contrast extended objects. For example, the problem exists of clarifying fluctuations of the relic cosmological background caused by an inhomogeneous distribution of matter in the universe at different stages of its evolution. Modern observations of radio brightness fluctuations give only an upper limit $\Delta T/T < 10^{-5}$ (according to measurements with the COBE satellite /17/ and surface observations /18/). With the use of limiting receivers with system noises $T \sim 2.7$ K a band width $\Delta f = 10$ GHz, and an integration time $\Delta t = 10^4$ s, a system of $N \sim 10^4$ receivers with horn antennas will be able to record a brightness contrast $\Delta T/T = \sqrt{\Delta f \Delta t N} \sim 10^{-9}$, i.e., $10^4$ times better than the existing level.

Another possibility is the creation of multielement interferometric systems of image synthesis. Here one supposes the development and launch into space of a large number of telescopes working as a single instrument and the creation of specialized processes for processing the information obtained from this space network. Ground radio astronomical synthesis systems work on the same principle, for example the VLA /19/ (27 mirrors of 25 m diameter with separation to 36 km), and the VLBA /20/ (10 mirrors of 25 m diameter with separation to the radius of the Earth). A further development of this approach is the construction of a ground system for millimeter range wavelengths MMA /21/ (40 mirrors of 8 m diameter with separation to 3 km). A system of interferometers consisting of N telescopes allows one to simultaneously obtain $N(N-1)/2$ Fourier harmonics of the image, i.e., to greatly accelerate the synthesis of an image area in comparison with the simple two-element interferometer. One can expect the creation of similar systems in combination with matrix receivers in all ranges of the electromagnetic spectra.

Thus the general promise of the development of astronomy in space can be represented in the form of three stages. In the first stage separate satellites with telescopes and orbits close to the Earth are created and search studies are carried out with two antenna interferometers (in particular Earth-Space interferometers). At this stage, together with the attainment of important scientific results, the technology is developed of creating large telescopes of different ranges and determining the most expedient sizes of interferometric systems.

At the second stage the creation of multielement interferometric systems is possible. The most optimal place for the positioning of a system is the region of the $L_2$ Lagrangian point, located in the antisolar direction at a distance of 1.5 million km from the Earth. This region (with a radius $10^6$ km) has important preferred properties (the Sun and the Earth are always on the same side and therefore more than half of the sphere of the sky is free from noise; here gravitational perturbations are also extremely small).

Finally, at the third stage it is possible to create a multielement interferometric interplanetary system.

For a synthesis of high quality images of a radiation source it is necessary to measure all harmonics of its spectrum of spatial frequencies. This can be satisfied by the simultaneous operation of N telescopes, and their shift in space during a time $\tau$ and retuning of frequency by $\Delta f$ (assuming that the image does not change during this time).

If the rotation time of the whole synthesis system P = 1 yr, N = 100, and $\Delta f/f = 0.1$, then the synthesis time $\tau$ will be:

$$\tau = \frac{\pi P}{(\Delta f/f)N(N-1)} = 1 \text{ day}$$

It is important to note that this relation does not depend on the radiation band, the maximum base of the interferometer, and sizes of the synthesized region. However, all these parameters are strongly connected with the detailed construction of telescopes and above all with the sensitivity and volume of information to be processed.

A possible configuration of astronomical synthesis systems at the end of the twenty-first century is shown in Fig. 22.2.

## CONCLUSION. THE MAIN PROBLEMS OF ASTRONOMY

We can only enumerate here those problems which now, at the end of the twentieth century, are the largest and determining in the development of many other particular directions of the science of the Universe. Practically all these problems are still very far

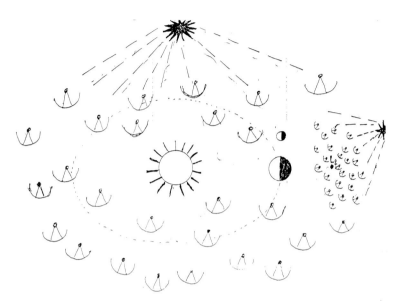

Figure 22.2. A possible configuration of astronomical synthesis systems at the end of the XXI century.

from solution. However, there is no doubt that their study may lead to results which will turn out to have a significant influence on the development of not only astronomy, but on the further development of our civilization. An enumeration of the main problems of astronomy has been discussed several times /22,23/. We condense these to a list of ten problems whose importance seems the greatest to the author.

1. Higher forms of intelligence in the Universe /23–25/.
2. The anthropic principle and the multiplicity of Universes /26–29/.
3. A cosmological model and the evolution of the present Universe.
4. Hidden mass /30, 31/.
5. A cosmological model and the evolution of the early Universe /32/.
6. Galactic nuclei.
7. Planetary systems and solid bodies in the Universe.
8. Gravitational wave astrophysics and relic gravitational waves.
9. Neutrino astrophysics and relic neutrinos.
10. The origin of cosmic rays.

It is difficult to foresee how much these problems will be advanced or even solved in the twenty-first century. However, the promise and refinement of methods of research described above provide a basis to hope for significant progress in their study. It is also necessary to note the possible appearance of fundamentally new means and methods. This relates first of all to problems 8, 9, and 10, whose development is difficult for the author to foresee (see /33–35/).

In conclusion we would like to note again the close interconnection of the accomplishments of astronomical science with the perspectives for the development of other sciences and society in general. Many of these questions were discussed in the books of I. S. Shklovsky "Universe, Life, and Intelligence" and "Problems of Modern Astrophysics" /24,2/. An understanding of the exceptional difficulties in solving conflicts on the Earth prompted him in the last years of his life to think about the large probability of the self-destruction of each civilization. This idea was first stated in the very first work on a search for extraterrestrial civilizations by S. Von Horner /36/. Similar thoughts were stated by S. B. Pikel'ner. In the opinion of the author there is a basis in our time to suppose that a general development may be managed without a fatal end at least in the perspective for the next century. Moreover, the hope appears on the threshold of the century for a new ordering of terrestrial civilization, characterized by the priority of the general good, and providing a radical bettering of the conditions of the existence of humanity as a whole and each separate personality on the basis of the gigantic economic, informational, and moral potential accumulated by our civilization during its whole history and especially in the course of the twentieth century. The possibility of detecting extraterrestrial civilizations in the twenty-first century simultaneously with the expected conquest of space and progress in biology and robotics even leads one to contemplate the reality of reaching immortality and the large probability of a radical change in the make-up of our civilization.

Table 22.1. Basic Discoveries of Physics and Astronomy in the Twentieth Century

| | |
|---|---|
| 1900 | Concepts of quantum theory (idea of the photon) and the fundamental constant of quantum action (M. Planck). |
| | The discovery of gamma rays (P. Villar). |
| 1901 | Spectral classification of stars (A. Caennon). |
| 1902 | Theory of gravitational instability (J. Jeans). |
| 1903–1912 | "Study of world spaces with jet devices" (K. Tsiolkovskii). |
| 1904 | Relativistic transformation of coordinates and time (H. Lorentz). |
| | Discovery of interstellar absorption lines (J. Hartmann). |
| 1905 | Postulation of quantization of electromagnetic radiation and the theory of the photo effect (A. Einstein). |
| | Special theory of relativity (A. Einstein). |
| –1913 | Detection of dwarf and giant stars (E. Hertzsprung, H. Russell). |

## 22. ON ASTRONOMY FOR THE TWENTY-FIRST CENTURY 367

|  |  |
|---|---|
|  | Law of energy distribution in the long-wavelength part of the radiation spectrum (J. Raleigh, J. Jeans). |
| 1906 | Discovery of the Lyman spectral series (T. Lyman). |
|  | Stellar atmosphere equilibrium theory (K. Schwarzschild). |
|  | First photoelectric observations of the moon and stars (J. Stebbins). |
| 1907 | Equivalence principle of gravitation and inertia (A. Einstein). |
| –1912 | Statistical study of stellar velocities (K. Swartzschild). |
| 1908 | Discovery of the magnetic field of sunspots (G. Hale). |
|  | Four-dimensional space-time of Minkovskii (G. Minkowskii). |
|  | Hypothesis of panspermi and a space origin of life on Earth (S. Arrhenius). |
| 1909 | Alpha particles—doubly ionized atoms of helium (E. Rutherford, J. Roids). |
| 1910–1914 | Discreteness of electric charge and the charge of the electron (S. Millikan). |
| 1911 | Discovery of superconductivity (H. Kamerling-Ones). |
| –1913 | Spectra-luminosity diagram of stars (E. Hertzsprung, H. Russell). |
|  | Discovery of the atomic nucleus (E. Rutherford). |
|  | Quantization of the magnetic moment (magneton) (P. Weiss, P. Langevin). |
| 1912 | Period-luminosity dependence of variable stars (Cepheids) (H. Leavitt). |
| –1915 | Experimental proof of the existence of the photon (R. Millikan). |
|  | Discovery of isotopes (J. Thomson, F. Soddy). |
|  | Measurement of galactic light velocities (V. Slipher). |
| –1914 | Detection of cosmic rays using piloted balloons (V. Hess, W. Kolhörster). |
| 1913 | Theory of hydrogen atom structure (N. Bohr). |
|  | Detection of reflecting nebulae (V. Slipher). |
|  | The existence of energy levels in atoms (J. Franck, G. Hertz). |
| –1915 | Detection of rotation of galaxies (V. Slipher). |
| 1914 | Theory of Cepheid pulsations (H. Shapley). |
|  | Figures of equilibrium of rotating stars and the possibility of fission or equatorial outflow (J. Jeans). |
| 1915–1916 | Theory of the hydrogen spectrum fine structure (A. Sommerfeld). |
|  | Discovery of white dwarfs (V. Adams). |
| 1916 | Theory of internal stellar structure (A. Eddington). |
|  | Theory of dissipation of planetary atmospheres (J. Jeans). |
|  | Theory of spontaneous and induced emission (A. Einstein). |
|  | Exact solution of the Einstein equations for a spherical mass, gravitational radius (K. Swartzschild). |
| 1917 | General theory of relativity and three effects of the general theory of relativity, gravitational waves, a model of a statistical Universe, and the cosmological constant (A. Einstein). |
|  | Detection of outbursts of new stars in galaxies and determination of distances to them (J. Ritchey, H. Curtis). |
| 1918 | Model of Galaxy and the distance to its center of 30,000 light years (H. Shapley). |
|  | Correspondence principle (N. Bohr). |
| –1919 | Model of stellar pulsations (A. Eddington). |

| | |
|---|---|
| 1919 | Nuclear reactions and the source of stellar energy (Z. Perrin). |
| | Discovery of nuclear reactions (the transformation of nitrogen into oxygen) and the discovery of the proton (E. Rutherford). |
| | Detection of the deviation of light in the gravitational field of the sun and the measurement of stellar diameters by an interferometer (A. Michelson, A. Eddington). |
| 1920 | Prediction of the existence of the neutron (E. Rutherford, W. Harkins). |
| | Verification of the General Theory of Relativity (A. Eddington). |
| −1921 | Theory of atomic ionization (M. Saha). |
| 1921−1922 | Explanation of the periodic system of chemical elements (N. Bohr). |
| | Crab Nebula—remnant of supernova explosion in 1054 (K. Lundmark). |
| −1929 | Theory of radiative transfer in stellar atmospheres (E. Milne). |
| 1922 | Scattering of light by electrons as the elastic collision of particles (A. Compton). |
| | Model of the expanding Universe (A. A. Friedman). |
| 1923−1924 | Mass-luminosity relation of stars (E. Hertzsprung, H. Russell, A. Eddington). |
| −1924 | Theory of particle-wave dualism (De Broglie waves) (L. De Broglie). |
| −1924 | Stellar temperature scale (E. Milne, R. Fowler). |
| 1924 | Nuclear spin and the theory of super-fine structure of spectral lines (W. Pauli, J. Uhlenbek, S. Goudsmith). |
| | Bose-Einstein statistics of particles with zero and integral spins (S. Bose, A. Einstein) |
| | Resolution of the galaxies M31 and M33 into stars and their extragalactic nature (E. Hubble). |
| | Pauli's exclusion principle (W. Pauli). |
| | Prediction of a bright ring around a gravitational lens (O. Hvolson). |
| | Idea of the origin of life as the result of chemical evolution (A. I. Oparin). |
| 1925 | Matrix quantum mechanics (W. Heisenberg). |
| | Classification of galaxies (E. Hubble). |
| 1926 | Equations of quantum mechanics (E. Shroedinger). |
| | Theory of luminosity of gaseous nebulae (H. Zanstra). |
| −1927 | Study of rotation of Galaxy (B. Lindeblad, Y. Oort). |
| | Determination of the temperature of planetary nebula nuclei (H. Zanstra, D. Menzel). |
| | Quantum mechanics of the helium atom (W. Heisenberg). |
| | Fermi-Dirac statistics of particles with half-integer spin (E. Fermi, P. Dirac). |
| 1927 | Uncertainty principle in quantum mechanics (W. Heisenberg). |
| | First observations of cosmic rays in the laboratory (D. V. Skobeltsyn). |
| | Exchange interaction (W. Heisenberg, P. Dirac). |
| | Discovery of electron diffraction and their wave properties (C. Davisson, L. Germer). |
| | Theory of alpha-decay as a tunnelling process (G. Gamov, E. Condon, R. Gerney). |
| | Quantum mechanics of the hydrogen molecule (F. London). |

| | |
|---|---|
| 1928 | Quantum mechanical equation of the electron in relativistic quantum mechanics (P. Dirac). |
| 1929 | The discovery of expanding Universe (running of galaxies) (E. Hubble). Theory of gravitational instability (J. Jeans). |
| −1931 | Invention of the coronograph for studying the Sun (B. Lyot). Determination of the chemical element content in the solar spectrum (S. Russell). |
| 1930 | Study of interstellar extinction of light (R. Trumpler). Discovery of Pluto (C. Tombaugh). |
| −1931 | Hypothesis of the neutrino (W. Pauli). |
| 1931 | Discovery of cosmic radio emission ($\lambda = 14$ m) (K. Jansky). Prediction of antiparticles (the production and annihilation of pairs) and monopoles (P. Dirac, C. von Weizsacker). |
| 1932 | Discovery of the neutron (J. Chadwick). Model of the neutron-proton structure of nuclei and strong interactions (D. D. Ivanenko, W. Heisenberg). Determination of the density of interstellar material (J. Oort). Discovery of deuterium (H. Urey). Discovery of the positron (C. Anderson). Infrared photographs of stellar spectra (P. Merrill). Nuclear transformation caused by neutrons (N. Feather, L. Meitner, W. Harkins). Invention of a wide-angle telescope (B. Schmidt). |
| 1933 | Calculation of the neutron mass and proof of its decay and transformation into a proton (F. and I. Curie). Model of polarization of vacuum (P. Dirac). Hypothesis of antimatter (P. Dirac). Observation of the annihilation of an electron and positron in cosmic rays (B. Blackett, G. Occhialini). |
| 1934 | Determination of the mass and size of the Galaxy (J. Plasket and D. Pears). The idea of nuclear chain reactions (L. Szilard, F. Joliot-Curie, L. Meitner). Prediction of the existence of neutron stars, their formation at the time of supernova explosions, and the generation of cosmic rays with it (L. D. Landau, F. Zwicky, B. Baade). |
| 1935 | Prediction of the existence of mesons and quanta of the nuclear field (H. Yukawa). |
| 1936 | Theory of beta-decay (G. Gamov). |
| 1937 | Detection of molecules in the interstellar medium (L. Rosenfeld, P. Swings). Magic large numbers of dimensionless combinations of universal constants (P. Dirac). Prediction of neutral currents (G. Gamov, E. Geller, N. Kemmering, G. Wentzel). The theory of thermonuclear reactions in stars (G. Behe). Prediction of interstellar magnetic fields and the mechanism of the acceleration of cosmic rays (H. Alfven). The discovery of the mu meson (K. Anderson, S. Neddermeier). |

| | |
|---|---|
| 1938 | The discovery of liquid helium superfluidity (P. L. Kapitsa). |
| | The interstellar gas participates in the rotation of the Galaxy (D. Pears, J. Plaskett). |
| −1944 | Invention of mirror radio telescope and radio mapping of the Galaxy (G. Reber). |
| 1939 | Theory of nuclear fission of uranium by slow neutrons (Ya. I. Frenkel, N. Bohr, J. Wheeler). |
| | The idea of frozen magnetic fields in plasma (H. Alfven). |
| | Calculation of the critical mass of uranium (F. Perrin et al.) |
| | Theory of hydrogen ionization in near-stellar nebulae (B. Stromgren). |
| | Theory of the chain reaction in uranium (L. Szilard et al.). |
| | Prediction of the possibility of black holes (R. Oppenheimer, H. Snyder). |
| 1940 | The synthesis of a transuranic element, neptunium (E. McMillan, F. Abelson). |
| | Model of the solar atmosphere (B. Stromgren). |
| 1941 | First model of solar structure taking into account thermonuclear reactions (M. Swarzschild). |
| 1942 | Discovery of solar radio emission (J. Hey, G. Southworth). |
| | Construction of the first nuclear reactor (E. Fermi). |
| | Magnetohydrodynamic waves in plasma (H. Alfven). |
| 1943 | The discovery of galaxies with a bright nucleus and broad lines (C. Seyfert). |
| 1944–1951 | Prediction and detection of radio lines of the basic states of interstellar hydrogen (H. Van de Hulst, J. S. Shklovsky, H. Even and E. Purcell, C. Müller and J. Oort, W. Christiansen and J. Hinman). |
| 1945 | First atomic explosions |
| | Tables of spectral lines for astrophysics (C. Moore Sitterley). |
| 1946 | Detection of the first far radio source (Cygnus A) (J. Hey, S. Parsons, J. Phillips). |
| | Detection of the magnetic field in stars (H. Babcock). |
| | Radio location of the Moon (J. de Witt, E. Stodola, Z. Bay, J. Mofenson). |
| | Theory of the variable solar radio emission as plasma oscillations (J. S. Shklovsky). |
| −1948 | Model of a hot Universe and the formation of chemical elements (G. Gamov, F. Hoyle). |
| | Detection of the frequency drift of solar radio bursts and the motion of their sources (J. Bolton, R. Peine-Scott, D. Yabsley). |
| 1947 | The discovery of globules (B. Bok, E. Reilly). |
| | The discovery of charged pi mesons (G. Lattes et al.). |
| | The theory of gravitational instability in an expanding Universe (E. M. Lifshitz). |
| −1948 | Discovery of stellar associations (V. A. Ambartsumyan). |
| 1948 | Detection of solar ultra-violet emission from a rocket (R. Tousey). |
| | The discovery of interstellar polarization (W. Hiltner, J. Hall). |
| | Principles of holography (D. Gabor). |
| | Model of the steady state Universe (F. Hoyle, H. Bondi, T. Gold). |
| | Discovery of the most powerful radio source, Cassiopeia A (M. Ryle, G. Smith). |

22. ON ASTRONOMY FOR THE TWENTY-FIRST CENTURY 371

| | |
|---|---|
| 1949 | Prediction of the probability of observing radio lines of interstellar molecules (J. S. Shklovsky). |
| | Identification of three brightest radio sources with supernova remnants and two with near peculiar Galaxies (J. Bolton, G. Stanley, O. Slee). |
| 1950 | The idea of the thermal isolation of plasma by magnetic field (I. E. Tamm, L. Spitzer). |
| | Dynamo model of the formation of the solar system (H. Alfven). |
| | Detection of radio emission of normal galaxies (M31) (R. Handbury Brown, C. Hazard). |
| −1951 | Theory of synchrotron radio emission of relativistic electrons in interstellar magnetic fields (H. Alfven and N. Herlofson, K. Kiepenheuer, V. L. Ginzburg, A. A. Korchak). |
| 1951 | Two-photon emission mechanism (transitions from the second level of hydrogen) of the continuous spectrum of nebulae (J. Greenstein, L. Spitzer, A. Ya. Kipper). |
| | Proof of the spiral structure of Galaxy (W. Morgan, D. Osterbrock, B. Sharpless). |
| | Detection of young stars surrounded by dust nebulae (G. Haro, G. Herbig). |
| | Theory of the thermal emission of the solar corona (J. S. Shklovsky, S. B. Pikel'ner). |
| | Testing of the first hydrogen bomb. |
| 1952 | Interferometer of intensities (R. Hanbury Brown, R. J. Twiss). |
| | Invention of a solar magnetograph (H. D. and H. W. Babcock). |
| | Proof of the statistical predominance of extragalactic radio sources (B. Mills). |
| | Idea of an adaptive telescope for exclusion of the influence of the atmosphere and optics errors (H. W. Babcock). |
| | First model of stellar evolution (A. Sandage, M. Swarzschild). |
| | Prediction of the thermal component of the radio emission of the Galaxy by interstellar plasma (I. S. Shklovsky). |
| 1953 | Prediction of a gaseous corona of the Galaxy (S. B. Pikel'ner). |
| | Accurate system of stellar photometry (H. Johnson, W. Morgan, D. Harris). |
| | Discovery of the period-luminosity relation for Cepheids and its application to the determination of extragalactic distances (A. Blaauw, M. Savedov). |
| | Theory of the continuous spectrum of the Crab Nebula from the optical to the radio as synchrotron radiation (J. S. Shklovsky). |
| | Experimental proof of the existence of the electron neutrino (F. Reines, C. Cowen). |
| 1954 | Quantum generators or masers (N. G. Basov, A. M. Prokhorov, C. Townes). |
| | Measurement of the optical polarization of the radiation of the Crab Nebula and verification of its synchrotron nature (V. A. Dombrovskii, M. A. Vashakidze). |
| | Molecular model of the genetic code (G. Gamov). |
| | Theory of the radiation spectrum behind a shock front in interstellar gas (S. B. Pikel'ner). |
| 1955 | Beginning of subnuclear studies on accelerators (R. Hofshtadter). |

|  |  |
|---|---|
|  | Atomic frequency standard (L. Essen). |
|  | Discovery of the antiproton (O. Chamberlain, E. Segre, S. Wiegand, T. Ypsilantis). |
|  | CPT symmetry of elementary particles (W. Pauli). |
|  | Thermonuclear formation of chemical elements in stars (J. and M. Burbidge, W. Fowler, F. Hoyle). |
| 1956 | Discovery of the antineutron (B. Cork, G. Lambertson, O. Piccioni, W. Wenzel). |
|  | Theory of a Fermi liquid (L. D. Landau). |
|  | Scheme of evolution of planetary nebulae (J. S. Shklovsky). |
| 1957 | Spectral classification of galaxies (N. K. Mayall, W. Morgan). |
|  | Launch of the first artificial satellite (S. P. Korolev). |
|  | Discovery of the nonconservation of parity in weak interaction (T. Wu). |
|  | Theory of superconductivity with the formation of Cooper pairs (J. Bardeen, L. Cooper, J. Shrieffer). |
|  | Theory of tunneling phenomena in semiconductors (L. V. Keldysh). |
| –1958 | Unified theory of weak and electromagnetic interactions (J. Schwinger, S. Glashow, A. Salam, J. Ward). |
| 1958 | Discovery of the nuclear gamma resonance or Mossbauer effect (R. Mossbauer). |
|  | Theory of neutrino oscillations (B. M. Pontkorvo). |
|  | The principle of laser operation (C. Townes, A. Schavlov). |
|  | The discovery of interacting galaxies (B. A. Vorontsov-Velyaminov). |
|  | The discovery of the Earth's radiation belts (J. Van Allen, S. N. Vernov, A. E. Chudakov). |
|  | Detection of radio bursts of UV Ceti type stars (A. Lovell). |
| 1959 | Supposition of communication with an extraterrestrial civilization at 21 cm (G. Kokkoni, P. Morrison). |
|  | Radar observations of the Sun (V. Eshleman, R. Barthle, P. Gellagher). |
|  | Luna-1 is the first interplanetary station, Luna 2 is the first station making contact with the Moon. |
|  | Flight of an automated station Luna-3 and first pictures of other side of the Moon |
| 1960 | Creation of the hydrogen maser, the most accurate frequency standard (N. Ramsey). |
|  | First search of radio signals from extraterrestrial civilizations (F. Drake). |
|  | Creation of the first laser (T. Maiman). |
| 1961–1962 | Hypothesis of massless particles (Goldstone boson) and the theory of symmetry breaking (J. Goldstone, A. Salam, S. Weinberg). |
| –1973 | Anthropic principle in cosmology (R. Dicke, B. Carter). |
|  | First manned space flight (Yu. A. Gagarin). |
|  | Hypothesis of the explosion of one star as a supernova in a binary system and the ejection of the second one (A. Blaauw). |
|  | Theory of the laser in nuclear gamma transitions (L. A. Rivlin). |
|  | Theory of self-focusing of electromagnetic radiation (G. A. Askaryan). |

|  | Discovery of cosmic gamma radiation (Explorer-11) (J. Arnold, A. Metzger, E. Anderson, M. Van Dilla). |
|---|---|

Discovery of cosmic gamma radiation (Explorer-11) (J. Arnold, A. Metzger, E. Anderson, M. Van Dilla).

Proof of the evolution of the Universe with radio source counts (M. Rail, R. Clark).

The discovery of the muon neutrino (G. Danby et al.).

1962 Detection of the strongest X-ray source Scorpio X-I and the background X-ray cosmic radiation (B. Rossi, R. Giacconi, H. Gursky, F. Paolini).

–1963 First holograms (Yu. N. Denisyuk, E. Leith, Y. Upatnieks).

1963 Discovery of quasars (M. Schmidt).

Model of supernova explosions dominated by energy release in the form of neutrinos (W. Fowler, F. Hoyle).

Detection of the radio line of the OH molecule from the interstellar medium (C. Townes).

Detection of X-ray radiation from the Crab Nebula (H. Friedman).

Model of a rotating black hole (R. Kerr).

1964 Hypothesis of a field describing the physics of particles (P. Higgs et al.).

Hypothesis of quarks (M. Gell-Mann, G. Zweig).

Discovery of CP invariance breaking (J. Kronin, V. Fitch, J. Christenson, et al.).

Discovery of the interstellar scintillation of cosmic radio sources (E. Hewish).

Detection of strong infrared radiation from quasars and galaxies (H. Johnson, G. Neugebauer, R. Leighton).

1965 First accelerators with colliding beams (colliders in Novosibirsk and Stanford).

Craters of Mars with photos from the station Mariner-4 (W. Pickering).

Detection of the relic radio emission predicted by the model of a hot Universe (A. Penzias, R. Wilson).

Detection of space masers (H. Weaver, D. Williams, N. Dieter, W. Lum, E. Gundermann).

Detection of protoplanetary dust clouds near some stars.

Detection of clusters of galaxies (H. Shapley)

Hypothesis of "the color" of quarks and gluons (N. N. Bogolyubov et al., M. Hahn, Y. Nambu).

1966 Landing of an automated station on the surface of the Moon (Luna-9).

1967 Model of the origin of breaking of symmetry of the Universe (A. D. Sakharov).

Discovery of pulsars (J. Bell, A. Hewish).

1968 Map of the sky in the 100 MeV range with the satellite OS O-3 (G. Clark, G. Garmire, W. Kraushaar).

Model of a pulsar as a rotating neutron star with a strong magnetic field (T. Gold).

Piloted flight to the moon on Apollo-10 (T. Stafford, J. Young, E. Cernan).

1969 Manned expedition on the Moon (N. Armstrong, E. Aldrin).

Two-phase model of the interstellar medium (S. B. Pikel'ner).

1970 First satellite X-ray observatory (Uhuru) (R. Giacconi et al.).

| | |
|---|---|
| 1971 | Discovery of the annihilation line of positrons from the region of the Galactic Center (W. Johnson, F. Harden, R. Haymes). |
| 1972–1973 | Cosmic gamma background radiation, from the satellite SAS-2 (D. Kniffen et al.). |
| –1974 | Concept of supersymmetry of elementary particles (D. V. Volkov, B. Zumnio et al.). |
| | Tunneling of the scalar field in the spontaneous breaking of symmetry (D. A. Kirzhnits, A. D. Linde). |
| –1976 | Prediction of the birth of monopoles, strings, and walls in the early Universe (Ya. B. Zeldovich, D. A. Kirzhnits, I. Kobzarev, L. B. Okun, T. Kibble). |
| 1973 | Hypothesis of the existence of gluons (M. Gell-Mann, S. Weinberg, A. Salam et al.). |
| | Discovery of neutral currents and currents in weak interaction (F. Hazard et al.). |
| | Discovery of cosmic gamma X-ray pulses with the satellite Vela (R. Klebesadel, I. Strong, R. Olson). |
| 1974 | Discovery of particles of the quark-antiquark connecting state (charmonium) (S. Ting, B. Richter). |
| | Theory of black hole radiation (S. Hawking). |
| | Transformation of white holes into black ones (Ya. B. Zeldovich, I. Novikov, A. Starobinskii). |
| | Map of the surface of Mercury from the station Mariner-10 (W. Pickering et al.). |
| | Theory of the spectrum of solar flare radiation (S. B. Pikel'ner). |
| 1975 | Discovery of the tau-lepton and the tau-neutrino hypothesis (M. L. Perl et al.). |
| –1982 | Map of the sky and catalog of gamma-ray sources from the satellite COS-B (K. Bennett, W. Hermsen, K. Pinkau, et al.). |
| | Detection of X-ray bursters from neutron stars (W. Lewin, Y. Doty, G. Clarketal). |
| 1976 | Discovery of D-mesons, experimental verification of the quark model (I. Peruzzi et al., G. Goldhaber et al.). |
| 1977 | Discovery of ipsilon-mesons (S. Herb, L. Lederman, et al.). |
| 1978 | Einstein X-ray observatory (R. D. Giacconi et al.). |
| 1979 | Theory of distortion of pulsar rotation under the action of the background gravitational radiation (S. Detveiler). |
| | Discovery of a gravitational lens in space (D. Walsh, R. Carswell, R. Weymann). |
| | Discovery of the nonconservation of parity (K. Prescott et al.). |
| 1981 | Model of inflation in the early Universe (A. Guth). |
| 1982 | Discovery of the pulsar 1937+21 with the shortest period of 1.5 ms (D. Backer et al.). |
| 1983 | Detection of an intergalactic HI hydrogen cloud (G. Helou, E. Salpeter, Y. Terziah). |
| | Discovery of intermediate vector W and Z bosons (G. Arnison, M. Banner, P. Bagnaia, et al.). |

Detection of gamma radiation in the $10^{15}$ eV range from the source Cygnus X-3 (M. Samorski, W. Stamm, J. Lloyd-Evans, et al.).

Detection of an extragalactic pulsar (in the Large Magellanic Cloud) (P. McCullough et al.).

1984 Detection of the radiation of gravitational waves by a pulsar in a binary system (J. Weisberg and J. Taylor).

1985 Hypothesis of superconducting strings (E. Witten).

Strings may create density fluctuations for the formation of galaxies (A. Vilenkin).

1986 Discovery of gigantic optically luminous arches around some galaxies (R. Lynds, V. Petrosian).

1987 Explosion of a supernova star in the Large Magellanic Cloud detected by light, neutrino, and radio bursts.

1988 Detection of a super cluster of galaxies, the great attractor (L. Lynden Bell et al.).

The discovery of an extragalactic radio ring, the result of gravitational lensing (J. Hevitt et al.).

1989–1990 Measurement of the spectrum of cosmological microwave background radio emission and its isotropy from the satellite COBE (J. Mather et al.).

1990 Experimental proof that only three sorts of neutrinos exist (CERN).

Detection of a probably periodic distribution of galaxies with redshift? (T. Broadhurst et al.).

## REFERENCES

1. Struve, O., and Zebergs, V. (1962) Astronomy of the 20th Century, Macmillan, New York.
2. Shklovsky, J. S. (1988) Problems of Contemporary Astrophysics (in Russian), Nauka, Moscow.
3. Kaplan, S. A., and Pikel'ner, S. B. (1979) Physics of the Interstellar Medium (in Russian), Nauka, Moscow.
4. Khramov, Ju. A. (1983) Physics, Biographical Reference (in Russian), Nauka, Moscow.
5. Scheiner, J. (1899) *Astrophysical J.* **9**, 149.
6. Mitrophanov, I. G. (1984) *Adv. Space Res.* **3**, 533.
7. Levy, G. S., et al. (1989) *Astrophysical J.* **336**, 1098; Linfield, R., P., et al. (1990) *Astrophysical J.* **358**, 350.
8. Kardashev, N. S., and Slysh, V. I. (1988) In The Impact of VLBI on Astrophysics and Geophysics, M. J. Reid and J. M. Moran (eds.), p. 433, Kluwer, Dordrecht.
9. Hirabayashi, H. (1990) In Proc. VSOP International Symposium, ISAS.
10. Altunin, V. I., et al. (1991) Report on the Assessment Study, European Space Agency SCI (91)2.

11. Faucherre, M., et al. (1989) In Diffraction-limited Imaging with Very Large Telescopes, D. M. Aloin and J. M. Marriotty (eds.), NATO ASI series, p. 274.
12. (1987) Proc. ESA Workshop, ESA SP-273, Nordwijk, The Netherlands.
13. Goodman, J,. and Narayan, R. (1989) *MNRAS* **238**, 995.
14. Sheffield, Ch., and Kondo, Y. (1981) *Destinies* **3**, 88.
15. Buyakas, V. I., Shklovsky, J. S., et al. (1979) *Acta Astronomica* **6**, 175.
16. Kardashev, N. S., Parijsky, Yu. N., and Umarbaeva, N. D. (1973) *Izvestija SAO* **5**, 16.
17. Smoot, G. F., et al. (1991) *Astrophysical J.* **371**, L1.
18. Readhead, A.C.S., et al. (1989) *Astrophysical J.* **346**, 566.
19. Napier, P. J., Thompson, A. R., and Ekers, R. D. (1983) *Proc. of the IEEE* **71**, 1295.
20. Romney, J. D.. (1988) In Impact of VLBI on Astrophysics and Geophysics, M. J. Reid and J. M. Moran (eds.), p. 461, Kluwer, Dordrecht.
21. (1990). Proposal to the National Science Foundation submitted by Associated Universities, Inc.
22. Oda, M. (1990) In Observatories in Earth Orbit and Beyond, Y. Kondo (ed.), p. 427, Kluwer, Dordrecht.
23. Dyson, F. J. (1988) Infinite in All Directions, Harper and Row, New York.
24. Shklovsky, J. S. (1987) The Universe, Life, and Intelligence (in Russian), Nauka, Moscow.
25. Kardashev, N. S., and Strelnitsky, V. S. (1988) In Bioastronomy—The Next Steps, G. Marx (ed.), pl. 295, Kluwer, Dordrecht.
26. Barrow, J. D., and Tipler, F. J. (1988). The Anthropic Cosmological Principle, Oxford University Press, Oxford.
27. Tipler, F. J. (1991) Final Anthropic Cosmological Principle: The Omega Point Theory, in press.
28. Barrow, J. D. (1991) Theories of Everything (The Quest for the Ultimate Explanation), Clarendon Press, Oxford.
29. Sakharov, A. D. (1976) *Pis'ma Zh. Eksp. Teor. Fiz.* **5**, 32.
30. Kardashev, N. S. (1990) *MNRAS* **243**, 252.
31. Peebles, P.J.E., and Silk, J. (1990) *Nature* **346**, 233.
32. Linde, A. D. (1990) Elementary Particle Physics and Inflationary Cosmology (in Russian), Nauka, Moscow.
33. (1990) Workshop JPL, Annapolis, Maryland, April 19–20.
34. Braginsky, V. B., et al. (1991) Space Radio Interferometry and Gravitational Waves, this volume.
35. Berezinsky, V. S., et al. (1990) Astrophysics of Cosmic Rays (in Russian), V. L. Ginzburg (ed.), Nauka, Moscow.
36. Von Hoerner, S. (1961) *Science* **134**, 1839.

# Index

Active Galactic Nuclei (AGN), 253, 260
Angular resolution, 37, 38, 40, 42, 44, 123, 128, 133, 135, 148, 237, 239, 252, 262, 358–360, 362–363
Anthropic principle, 298, 365, 372

Barionic asymmetry in the Universe, 113, 284
Black holes, 9, 92, 251, 254, 358, 363, 370

Clusters of galaxies, 103–104, 111, 113, 117, 128, 160, 225, 229, 231–233, 254, 257–260, 262, 266, 280–281, 283–284, 286, 351, 373, 375
Cosmic background radiation, 38, 48, 93, 334, 336
Cosmic rays, 25, 110, 131, 148–150, 365, 367–369
Cosmological constant, 367
Cosmological models, 278, 282, 324
Cosmological redshift, 40–41, 47, 98, 205–206, 250, 343, 345, 352, 375
Cosmology, 12, 25, 37, 272, 277–280, 286, 291, 295–296, 300–301, 315, 351, 372
Crab Nebula, 9, 13, 93–96, 120, 190, 223, 235–236, 247, 252, 262, 368, 371, 373
Chemical evolution, 368

Dark matter, 265, 278, 280, 287–290, 293
Deceleration parameters in cosmology, 204, 290

Element synthesis in the Universe, 183, 365, 370

Friedman Universe, 295–296, 300, 368. *See also* Cosmological models
Future telescopes, 359, 360

$\gamma$-ray astronomy, 47, 92, 359
Galactic center, 115, 122, 141–142, 144, 148, 156, 247, 374
Galaxies, 4, 7–9, 20, 32, 39, 45, 48–49, 55, 81, 99, 103–106, 109–111, 113, 117, 123–124, 126–128, 148, 161, 163, 170, 185, 205, 225, 231, 253–272, 277, 279, 281, 284, 286, 296, 299, 331, 343, 351, 358, 363, 367–373, 375
Galaxy, the, 9–10, 13, 32, 49, 62, 81–82, 87–88, 91–93, 97, 105, 111, 117, 120, 122–123, 131, 139–149, 163, 185, 229, 232, 234–236, 247, 261–262, 332, 340–343, 351, 363, 367–371
Giant molecular clouds, 231
Gravitational collapse, 167–174, 177–179, 182, 184–186
Gravitational lenses, 260, 359, 368, 374–375
Gravitational radiation (waves), 316, 326, 374
Great Attractor, 278–279, 375

HI regions. *See* Interstellar neutral gas.
HII regions. *See* Interstellar ionized gas
Hidden mass, 365
Hubble constant, 47, 191, 284, 319

Immortality, 366
Inflation of the Universe, 278, 287, 290–291, 295–297, 300, 324–325, 331, 374
Infrared (IR) astronomy, 1, 7, 8, 55, 61, 74, 88, 152, 182, 306–310, 346, 359, 369, 373
Intelligent life in the Universe, 10, 19, 20, 303, 304, 311
Interferometry, 37–40, 42, 49, 87, 94–95, 99, 128, 152, 163, 307, 315, 323–326, 328, 341, 343, 345, 355

377

# INDEX

Intergalactic matter, 110–111, 113, 117, 225, 257, 284, 374
Interstellar dust, 4, 88, 147, 152, 341
Interstellar ionized gas, 9, 37, 90, 131, 134, 370, 373
Interstellar neutral gas, 26, 87, 104–107, 132, 147, 369–370, 373
Interstellar matter, 2, 4, 8–9, 13, 26, 34, 48, 54, 58–60, 62, 81, 87–88, 90, 92–94, 103, 117–118, 133, 137, 141, 145, 148, 151–152, 168, 223–240, 351, 358, 362, 366, 369, 371, 375
Interstellar magnetic field, 91–92, 106–107, 118, 369
Interstellar molecules, 87, 105, 132, 147, 369–370, 373

Magellanic Clouds, 126, 148, 167, 170, 185, 189, 196, 201, 215, 223, 229, 231–232, 236, 375
Masers, 38, 41, 48–49, 122, 135–136, 141, 151–163, 258, 371–373
Metagalaxy, 81, 258, 260, 295–296, 298–301, 351
Microwave Background Radiation (MBR), 37, 277, 282, 375
Molecular hydrogen. *See* Interstellar neutral gas

Neutral hydrogen. *See* Interstellar neutral gas
Neutrinos, 169–171, 173–175, 177–182, 184–185, 196, 211, 281, 284–285, 287, 289, 293, 300, 365, 369, 371–375
Neutron stars, 9, 38, 119, 170, 175, 177–178, 181, 190, 211, 233, 235, 247–252, 358, 363, 369, 373–374
Nonthermal emission, 26, 68, 71–73, 75, 78, 82, 91–94, 139, 148, 246, 341–342
Novas, 248
Nucleosynthesis. *See* Element synthesis in the Universe

Optical astronomy, 1, 353
Origins of life in the Universe, 19, 307, 367–368

Parabolic reflector telescope, 248, 261, 268, 307–309, 344, 346, 353, 362, 369
Planets, extrasolar, 303–311, 365, 367
Planets, in the solar system, 8, 10, 37, 63–79, 119, 343, 351, 363
Planetary nebulae, 9, 13, 156, 230–231, 236, 368, 372
Prominences, 119
Protostars, 152–153, 159–160, 163
Pulsars, 1, 7, 9, 38, 49, 81, 92, 95–96, 117, 119, 122, 190, 236, 248–250, 295, 340, 343, 351, 358, 373

Quarks, 373–374
Quasars, 1, 7, 9, 37, 38, 41–42, 45–49, 81, 99–101, 111, 148, 248, 253, 256–267, 284, 286, 295

Radio galaxies, 7, 9, 39, 45, 48–49, 81, 99–100, 117, 128–129, 253–263, 265–266, 270, 286, 331
Radio interferometry, 37, 39, 42, 45, 48, 82, 84, 315, 323–326, 328, 359
Radio recombination lines, 9, 11, 37, 46, 131–148, 151, 155, 159, 161, 204, 344, 370–371
Radio stars, 7, 39, 262–263, 351
Redshift, 40–41, 47, 98, 205–206, 250, 343, 345, 352, 375
Relativistic particles, 9, 38, 41, 45, 94, 118, 189, 203–204, 209–212, 219–220, 224, 233–234, 239–240, 262, 287, 371

Search for extraterrestrial intelligence (SETA), 303–304, 308, 310–311, 366, 372
Sensitivity of telescopes, 39, 48–49, 82–83, 123, 135, 140, 148, 248, 250, 252, 279, 326–327, 331, 338–352, 358–359, 362–363

Seyfert galaxies, 253–254, 260–261, 264, 266–267, 370
Solar activity, 34, 68, 73, 119, 246
Solar chromosphere, 1, 26, 52–54, 57–59, 61

Solar corona, 1, 2, 8–9, 11, 26, 53–55, 57–62, 69, 71, 81, 96–97, 119, 209, 246, 371
Solar flares, 26, 247, 374
Solar photosphere, 53–55, 59, 184
Solar system, 9, 37, 68, 75, 78, 118–119, 306, 315, 325, 357–358, 371
Space interferometry. *See* interferometry
Star evolution, 92
Star formation, 103–105, 110–111, 113, 141, 152–153, 155–156, 159–160, 225, 231–233, 255, 258, 265, 272
Stellar evolution, 92, 168–169, 181, 224, 226–229, 237, 344, 371
Stellar winds, 26, 62, 153, 160, 170, 189, 198, 202, 204, 210, 212, 214–216, 220, 223–231, 236, 240
Sun, 8, 11, 54–58, 60–62, 69, 78, 81, 96, 97, 119, 142, 246–247, 251, 306–307, 351, 364, 368–369, 372
Sunspots, 118–119, 367
Superluminal motion, 45–47
Supermassive objects, 254, 358, 363
Supernova remnants (SNR), 9, 38, 81, 92, 104, 117, 120–121, 224–225, 231, 233–240, 247–248, 262, 368, 371
Supernovae, 9, 62, 92, 94, 103–105, 110, 120, 167–171, 177–186, 189–220, 223–240, 248, 262, 327, 369, 372–373, 375
Synchrotron radiation, 13, 38, 41, 120, 224, 233–234, 236, 239, 247, 371

Synthesis telescopes, 128

21-cm neutral hydrogen line, 2, 9, 13, 147, 336, 343–344, 372
Time-scale of evolution of the Universe. *See* Universe, evolution

Ultraviolet (UV) astronomy, 7, 8, 54–59, 77, 88, 147, 155, 182, 216, 246, 261, 359
Universe, barion dominated, 113, 284
Universe, evolution, 13, 63, 77–78, 88, 98, 103–104, 110–111, 170, 183, 253–272, 277–278, 295, 301, 358, 363, 365, 373
Universe, inflationary. *See* Inflation of the Universe
Unresolved problems in astronomy, 42, 109

Very Long Baseline Interferometry (VLBI), 37–49, 152, 155, 156, 163, 190–191, 193, 196–197, 201, 206–207, 209, 315, 338–339, 343–346, 353–354
Voids, 271–272, 277, 294, 342, 351

White dwarfs, 13, 119, 235, 251, 367
Wolf-Rayet stars, 229

X-ray astronomy, 1, 7, 55, 103–105, 110, 113, 119–120, 224–225, 239, 245–252, 373–374